Defence in Animals
A survey of anti-predator defences

To Janet

Defence in Animals

A Survey of anti-predator defences

M Edmunds MA DPhil

Longman

Longman
1724-1974

Longman Group Limited
Burnt Mill
Harlow
Essex CM20 2JE

Distributed in the United States of America by Longman Inc.,
New York

Associated companies, branches and representatives
throughout the world

First published 1974

ISBN 0 582 44132 3
Library of Congress Catalog Card Number: 73—92246

Printed in Great Britain by
Whitstable Litho Ltd

Preface

In this book I attempt to describe the various means by which animals escape being killed and eaten by predators. Two of the most obvious anti-predator defences used are physical weapons, such as teeth and horns, and colour, for example camouflage. It is easy to see how physical weapons could be used in defence, but there are very few detailed studies demonstrating how they are actually used and how effective they are. The significance of colour, on the other hand, has attracted the attention of biologists and naturalists for more than a century, and there is a wealth of literature on cryptic, warning and mimetic coloration. Much of this is summarized in two classic books, *The Colours of Animals: Their Meaning and Use Especially Considered in the Case of Insects*, by E. B. Poulton (1890), and *Adaptive Coloration in Animals*, by H. B. Cott (1940). Cott's monumental work is still essential reading for the student of animal defence today, but in 1940 the sciences of ecology and ethology were in their infancy. In the past twenty years behavioural and ecological studies of animals have shed a great deal more light on the evolution and effectiveness of colour and other anti-predator defences, and on predator—prey relationships in general. I have attempted to summarize as much as possible of this literature, but obviously a book of this length cannot include reference to every relevant paper. I have probably given undue weight to some examples and not enough to others for the simple reason that I feel that I can write with more confidence and authority about animals with which I am personally familiar. It is for this reason that many of the examples I use come from tropical Africa and comparatively few from tropical America, Asia and Australia.

Over 800 animals are referred to in this book, and it is unlikely that the reader will be familiar with all of them. In the index to English and scientific names I have therefore indicated to which class or phylum each animal belongs.

The bibliography includes reference to many original research papers, but also to such secondary sources as books and review articles. These latter often provide the best introduction to the literature on any particular topic.

Finally, the reader with no statistical training should not be put off by the frequent reference in both text and figures to n, p and χ^2. n simply refers to the number of observations or of animals in a sample. p (probability) is an estimate of the chances of an event occurring such that $p = 1$ means it will occur every time; $p = 0.01$ means that it will be expected to occur in about 1 in 100 trials; and $p = 0.05$ means that it will be expected to occur in about 1 in 20 trials. Any figure of p which is less than ($<$) 0.05 indicates that the figures are unlikely to have been obtained by chance alone, whilst if p is greater than ($>$) 0.05, this indicates that the figures may have been due to chance. χ^2 is simply a mathematical test which gives an estimate of probability.

I have much enjoyed delving into the literature, writing and illustrating this book, but even more I enjoy watching living organisms in the wild. If this book succeeds in arousing in others a wonder and fascination for living things so that they attempt to ask and answer questions as to how, and why, then it will have served its purpose.

It is a pleasure to acknowledge the help and encouragement I have received from Professor D. W. Ewer and the staff of the Department of Zoology, University of Ghana, over the past ten years: staff and students have regularly produced curious beasties for me to examine and have asked awkward questions for me to ponder over. This book would certainly not have been written if I had not had the opportunity of living in the tropics, and I am especially grateful to my wife Janet for reading and criticizing several drafts of the book, and to Dr R. F. Ewer for critically reading earlier versions of Sections 1 and 2. I am also very grateful to Dr N. Smythe for permission to publish his beautiful colour photograph of the New World hawkmoth caterpillar *Leucorampha*; to Dr M. W. F. Tweedie for his photograph of the incredibly cryptic gecko *Ptychozoon*; and to Dr I. Sazima for his photograph of the dramatic display of the toad *Physalaemus* which has been used in the text and also, with modifications, on the cover.

Contents

List of plates

Plate 5   facing page 206

(*a*) Aposematic reduviid bug *Phonoctonus lutescens*
(*b*) Aposematic reduviid bug *Platymeris rufipes*
(*c*) Aposematic arctiid moth *Arctia caja*
(*d*) Aposematic six-spot burnet moth, *Zygaena filipendulae*, possibly cryptic from a distance

Plate 6   facing page 207

(*a*) Deimatic display of grasshopper *Phymateus cinctus*
(*b*) Deimatic display of praying mantis *Statilia apicalis*
(*c*) Brush-tailed porcupine (*Atherurus africanus*): spines on body and quills on tail which can be rattled as a deimatic display
(*d*) Spider *Caerostris albescens*: cryptic on a branch

Plate 7   facing page 238

(*a*) Deimatic display of young white-faced owl *Otus leucotis*
(*b*) Deimatic display of hawkmoth caterpillar *Leucorampha*
(*c*) Peacock butterfly (*Nymphalis io*): eyespots used for deimatic display on upper surface of wings
(*d*) Deimatic eyespot display of hawkmoth *Platysphinx constrigilis*

Plate 8   facing page 239

(*a*) and (*b*) Caterpillar of oleander hawkmoth (*Deilephila nerii*): (*a*) normal posture, cryptic; (*b*) deimatic display with conspicuous eyespots
(*c*) and (*d*) Caterpillars of hawkmoth *Hippotion eson*: (*c*) brown form in deimatic posture with pronounced eyespots; (*d*) green form in cryptic position

Introduction

'No man is an Iland, intire of it selfe; every man is a peece of the Continent, a part of the maine;' (Donne, 1624).

Just as no man is an island, so also no animal is an island; every animal interacts with a variety of other animals during the course of its lifetime. One important type of interaction is competition, both between individuals of the same species and between individuals of different species. Competition may be for food or for some other resource such as shelter, a mate, or a place in which to reproduce. It has been studied intensively by ecologists both in the field and in the laboratory, and is of tremendous importance in evolution: those animals which compete most successfully are the ones which are most likely to survive long enough to reproduce, hence their genes are likely to be represented in the gene pool of succeeding generations.

A second type of interaction is partnership in which an association between two (or more) animals improves the ability of one or both of the animals to survive. Partnerships may benefit only one of the animals (commensalism), or both of them (mutualism), and the association may or may not be obligatory (Gotto, 1969; Burton, 1969).

A third type of interaction between animals occurs when one animal is eaten by another animal — these are predator–prey interactions and parasite–host interactions. A succession of interactions of this type leads ecologists to the concept of the food chain in which animal A at the top of the food chain eats animal B, animal B eats animal C, and animal C in turn eats a plant which is at the base of the food chain. In practice simple food chains of this sort are very rare and probably only occur on a few isolated islands or in the tundra where there are comparatively few species in the community. In most communities we find interactions in the form of a food web with each animal species in the community feeding on several different species of animal or plant, and in turn being eaten by several different predators and parasites. In such a community there will be competition between the various predators for each prey species, and if there is a shortage of available food then the individual predators that survive are likely to be those which are best able to find and catch the various species of prey. Hence many predators have evolved very specialized and efficient methods of capturing particular species of prey, for example the cheetah which runs down the fastest gazelle by sheer speed, or the spitting spider *Scytodes* which ejects a stream of viscid silk from its jaws (*chelicerae*) a distance of 2–3 cm to pin down flies. Many gazelle, however, manage to evade the cheetah, and some flies are so alert that they fly off before *Scytodes* gets within spitting distance. Thus there is also competition between the various prey organisms to avoid being eaten. Those prey individuals that survive to reproduce are likely to be the ones which are best able to avoid being eaten by predators. Any adaptation which reduces the chances of success of an attack by another animal is a defensive adaptation or a defensive mechanism. Animals also have adaptations which protect them from physical and chemical factors in the environment (such as desiccation for a terrestrial animal). It is convenient to define *protective adaptations* as adaptations which protect an animal from hostile physical, chemical and biological factors in the environment, and to restrict *defensive adaptations* to those protective adaptations which protect it against attacks by other animals (Edmunds, 1966a).

Just as many animals have evolved elaborate defensive mechanisms so many predators have evolved mechanisms of countering the prey's defences. There is in effect an arms race with the prey evolving more elaborate or more diverse defensive adaptations and the predators developing methods of coping with these defences. Such predator–prey 'coevolution' is likely to be simple only in those few communities where the food web is simple and resembles a classical food chain. Most prey species have a variety of predators and most predators can eat a variety of prey species, so the defensive adaptations of animals are usually directed at several different predators. The entire defensive repertoire of an animal can be called its defensive system, and this is just as important as are the more conventional digestive, reproductive and excretory systems. A defensive system may comprise sensory components concerned with detection of predators, motor components concerned with some form of active escape behaviour, and structural features such as armour plating or spines.

Defensive adaptations may be directed against parasites as well as against predators. Internal parasites may be countered by a variety of foreign body reactions including the production of antibodies (see, for example, Caullery, 1952, and Smyth, 1962). External parasites may be removed by special behavioural patterns such as grooming one another as in monkeys, or the cleaning behaviour of cleaner shrimps and cleaner fish which remove parasites and fungi from other fish (Limbaugh, 1961). Sometimes there is no clear distinction between anti-predator defences and anti-parasite defences: thus the praying mantis *Tarachodes afzelii* guards her newly emerged young from ants in exactly the same way that she guards the egg case from attacks by ovipositing parasitic Hymenoptera and Diptera (Ene, 1962; Edmunds, 1972).

Defensive adaptations can also be directed against the predators of the eggs or young of an animal. This is parental care, which at its simplest involves producing fewer but larger yolked eggs, and at its most elaborate, viviparity and protection of the young after birth. Defensive adaptations can also be directed against other individuals of the same species — this is intraspecific defence. Intraspecific fighting is common in mammals and in many other animals, but usually there are ritualized forms of combat and characteristic submissive postures which prevent the occurrence of serious injury (Eibl-Eibesfeldt, 1961). These behavioural and morphological adaptations which protect animals from damage during intraspecific aggressive contests are of course defensive mechanisms, but unless they are also used in defence against predators they are not discussed further in this book. Similarly, anti-parasite defences and parental care adaptations are not considered further apart from a few special cases.

The concept of defence can be extended to the plants which are eaten by herbivorous animals. Any plant that is protected from being eaten will be at a selective advantage relative to other plants, and hence a wide variety of anti-herbivore defences have evolved. Many plants have spines on the leaves and stems which render them unpalatable to herbivorous mammals. Other plants have glandular hairs which are defensive against insects. For example the wild potato *Solanum polyadenium* has hairs which discharge a sticky substance when they are ruptured by an aphid, and this completely immobilizes the insect so that it starves (Gibson, 1971). *Passiflora adenopoda* has hooked hairs that hold and injure young caterpillars of *Heliconius erato* so that they die from starvation and from loss of blood (Gilbert, 1971). Many plants have chemical defences, such as terpenoids, steroids, cyanides and rotenoids, which are stored in the leaves or other tissues, and which make them toxic or unpalatable to animals. Some milkweeds contain chemicals known as cardenolides which cause vertebrates to vomit so that a vertebrate which eats milkweeds actually gets little or no nourishment from the plant. The neem tree *Azadirachta indica* contains a chemical that protects the leaves from attack by grasshoppers (Gill and Lewis, 1971), and many other plants contain chemicals that protect them against insect or vertebrate herbivores (Whittaker and Feeny, 1971, Levin, 1971, and Krieger *et al.*, 1971, all give further references). Indeed it has even been suggested that the success of the flowering plants (angiosperms) may be attributed, at least in part, to their numerous chemical defences which do not occur in conifers (gymnosperms) (Ehrlich and Raven, 1967). Many plants also have ecological defences against herbivores, such as intertree fruiting synchrony, so that although many young seeds or seedlings may be eaten, there will be so many that the herbivores cannot possibly eat all of them. For example *Sterculia apetala* trees all fruit at the same season, and though there is very heavy mortality of seeds which fall close to the parent trees a few usually survive. Furthermore, a few seedpods are often carried away by squirrels which then open them, find the pods have internal irritant hairs, and drop them. These pods have been carried so far from a *Sterculia* tree that the usual seed-eating herbivores do not find them, and so they may germinate successfully (Janzen, 1972).

Other plants have even more complex defences involving the use of certain animals to keep away other herbivorous animals. Extrafloral nectaries are a good example. These are supplies of sugary fluid, usually on the leaves, which probably evolved to attract ants to the plant, the ants in turn being so aggressive that they drive away foliage eating (and other) insects (Leston, 1972).

Some animals have overcome these defences of plants. It is well known that goats will eat thistles and nettles (*Cirsium* and *Urtica* spp.) without apparent harm, and many species of caterpillars feed on toxic plants either by having a high tolerance level to the plant toxins or by possessing detoxifying enzymes (see Krieger *et al.*, 1971; Levin, 1971). Natural selection has favoured, on the one hand, those plants with the most effective defences, and on the other, those herbivores which are best able to overcome these defences. As a result there has been coevolution of plants and their herbivores, just as we shall see in later chapters that there has been coevolution of prey and their predators in animals.

A more complex example of plant defence is that of the swollen thorn acacia (*Acacia cornigera*) in Central America (Janzen, 1966). The acacia tree provides shelter for the ant *Pseudomyrmex ferruginea* which lives in hollows excavated in the thorns, and it also supplies food, for all stages of the life cycle of the ant, from nectaries on the leaves and from curious glandular structures called beltian bodies. The ants drive off any insects that attempt to land and feed on the acacia, and both the spines and the ants are likely to prevent herbivorous mammals from feeding on the plant. No doubt the original plant defence was the thorns which protected it from attack by mammals, though not from phytophagous insects. Hence the association with ants evolved to the benefit of both tree and ant. Another well documented example of plant–animal coevolution is that of conifer trees and pine squirrels (*Tamiasciurus* spp.). It is probable that the pine cone evolved as a defence of the seeds against seed-eating mammals (Smith, 1970). As the squirrels evolved more efficient methods of extracting the seeds from the cones, so the cones evolved into more elaborate protective structures.

Thus the evolution of plant–herbivore systems is a striking example of the reciprocal effects of evolutionary change on two or more species of organism, and it is clear that there is a close parallel between the evolution of herbivore–plant systems and the predator–prey systems discussed in Section 4.

It is evident from this brief survey of defences that it would be possible to compile several large volumes on the subject — indeed Cott

(1940) has already given a long and excellent account of just a small part of the subject of animal defence, namely protective coloration, and much further work has been done in the thirty years since he wrote. In this book therefore I shall restrict discussion to defences of animals against predators. I shall consider anti-predator defences in two categories: primary defences and secondary defences. Primary, or indirect, defences operate regardless of whether or not there is a predator in the vicinity and they function to decrease the chances of an encounter with a predator from taking place (Robinson, 1969a; Kruuk, 1972). Secondary, or direct, defences operate when a prey animal encounters a predator, and they have the function of increasing the prey individual's chances of survival during the encounter.

In Sections 1 and 2, I deal with the various different forms of primary and secondary defence. Defences of animals that live in groups or in close associations with other animals are considered separately in Section 3. In Section 4, I consider the defensive systems of animals as a whole including such evolutionary aspects as the selection of predators with more efficient prey—capture adaptations and of prey with more efficient defence systems. Finally, I attempt to draw some conclusions relating to the ecology and evolution of anti-predator defensive systems.

Section 1
Primary defence

Robinson (1969*a*) defines primary defensive mechanisms as those defences which operate before a predator initiates any prey-catching behaviour. I have followed Kruuk (1972) and slightly modified this definition so that I consider primary defences to be defences which operate regardless of whether or not there is a predator in the vicinity. The predator may not detect the prey at all (anachoresis and crypsis), or it may be capable of detecting the prey but fail to recognize it as something edible (aposematism and batesian mimicry). Usually the prey is not aware of the presence of the predator, but in some cases if the prey is aware of the predator, the primary defence mechanism may be exaggerated. For example, stick-mimicking mantids such as *Danuria* and *Hoplocorypha* often rest with the forelegs apposed and partly flexed, but when disturbed (as by a predator) they promptly protract them fully so that the resemblance to a stick is intensified (Fig. S1.1). Similarly, grazing herbivorous mammals such as rabbits commonly freeze when alerted by a possible predator. The reason why these animals do not normally rest in the best concealed position is because it conflicts with some other essential activity — in both of these examples with feeding.

The function of primary defence is to decrease the chance that an encounter will take place between an animal and a potential predator. Six categories of primary defence can be recognized: anachoresis, crypsis, aposematism, batesian mimicry, group defence and defensive associations. In group defence and defensive associations it is not always easy to distinguish primary from secondary defences. Therefore I shall discuss the primary defences anachoresis, crypsis, aposematism and batesian mimicry in this section, and consider both primary and secondary aspects of group defence and defensive associations in Section 3.

There is a certain amount of apparent overlap between the first four categories of primary defence which I shall consider, but the categories are nevertheless distinct since they are based on different principles. Thus it may be very difficult to say whether a particular insect is merely cryptic on a leaf or whether it is a leaf-mimic. But if it is cryptic then it is protected because the predator fails to distinguish it from its

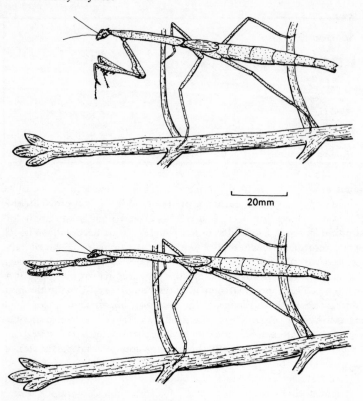

20mm

Fig. S1.1   Female praying mantis *Danuria buchholzi* in typical prey-capture posture (above) and in best concealed position (below).

background, whereas if it is a leaf-mimic then the predator may distinguish it from its background but mistake it for a leaf which is not associated with food. Another animal may be aposematic with respect to one predator, but cryptic with respect to another which is immune to its distasteful attributes. This may be possible because one predator hunts and recognizes its prey by sight whilst the other does so by scent. Most of the examples given in this section are of defence against visually hunting predators, because we know most about this type of predator–prey interaction. But a few cases are known of defence against predators hunting with the aid of other senses.

Chapter 1
Anachoresis

Animals which live in crevices or holes can be called *anachoretes*, and the phenomenon can be called anachoresis (from the Greek ἀναχωρέω — I retire, withdraw from the world, become a recluse). Anachoretes avoid stimulating the sensory receptors of predators simply because the predator does not come across them. Some anachoretes spend almost their entire life hidden from predators, for example earthworms, moles, burrowing lizards and snakes, and burrowing legless amphibians (Gymnophiona). Such animals rarely come above ground unless flooded out or unless the earth is baked hard during a drought. Some species of earthworm do come above ground at night-time to copulate or to feed, but others apparently live their entire life below the surface. There are also many insects which live in the soil but typically it is the larva that lives underground whilst the adult can fly — the wireworm and leather-jacket are well known examples, the former metamorphosing into a click beetle, the latter into a crane-fly. Ant-lions (*Myrmeleon* spp.) dig a pit in sandy soil and capture ants which fall into it and cannot climb the friable walls of the pit. Ant-lions remain below ground until they emerge as adult neuropteran insects.

All of these animals can be described as more or less permanent anachoretes, but other animals are anachoretic for just part of the day or night. Many small arthropods living amongst leaf litter remain hidden during the day but emerge to forage for food at night-time, for example scorpions, millipedes, centipedes and some species of carabid ground-beetle. Such species have poorly developed eyes for detecting predators by day, but they have well developed senses for detecting nocturnal predators. Many small mammals such as voles, rabbits, bush-babies and badgers also lie hidden in holes during the day when visually hunting predators could easily detect them, and they forage for food at night when they are less easy to see. Actually these generalizations over-simplify the situation. Living in a hole or burrow may also be an adaptation for reducing water loss in terrestrial animals, particularly in those living in deserts, and it also provides protection during inactive periods of rest or sleep when the animal is not so alert and capable of taking evasive action as it is at other times. On the other hand, many small birds roost in holes at night where they are protected from nocturnal carnivores, but they are active during the day. Although they may be conspicuous in the day-time, they have good eyesight for detecting predators, and well developed secondary defensive mechanisms (such as flight).

In the sea many crustacea, polychaetes and molluscs burrow in the substrate of the sea bed. Bivalves such as *Tellina* and *Mya* live their entire lives (apart from the planktonic larval stage) buried in sand or mud with just the siphons protruding to enable them to feed and obtain oxygenated water. The polychaetes *Chaetopterus* and *Arenicola* also feed entirely within their burrows. Other anachoretic polychaetes and crustacea have to emerge either partially or completely in order to feed. Such animals live in a retreat which may be a naturally occurring crevice, a hole or tube made by the animal using it, or a hole made by a different animal. Some polychaetes secrete mucoid tubes to which sand grains are attached (Terebellidae, Sabellidae, Sabellariidae), and others secrete calcareous tubes (Serpulidae). A few species actually excavate a tube by boring into calcareous rocks and shells (e.g. *Polydora* and *Potamilla*). Presumably boring is aided by chemical secretions. Amongst the bivalves *Lithophaga* bores into calcareous rocks by chemical means, whilst *Pholas* bores, by the mechanical rasping of the shells, into a variety of different types of rock (Nicol, 1960). Since predators can burrow in sand or mud and so capture burrowing polychaetes and bivalves it is of obvious protective value to be able to tunnel into rocks where such predators cannot follow.

Some anachoretes are commensal, relying for primary defence on the tube (or other form of retreat) made by another animal. The tubes of many burrowing invertebrates contain commensals. Thus the burrow of the echiuroid *Urechis* may contain a scale worm, a crab, a goby fish and a bivalve (Fig. 1.1) (MacGinitie and MacGinitie, 1949). Most of these animals derive protection from the association, and they may also obtain food from the currents produced by the host. Pinnotherid crabs sometimes occur in the tubes of *Urechis* or *Chaetopterus*, and also in the mantle cavity of bivalves. *Pinnixa*, *Scleroplax* and young *Pinnotheres* are probably all filter feeders which derive protection from the association but do not harm the host. Adult *Pinnotheres*, however, feed on the food strings on the gills of the host (*Ostrea*, *Mytilus*, or some other bivalve) and they may damage and eat the gills as well (Stauber, 1945; Christensen and MacDermott, 1958; Dales, 1966). So some associations which were originally protective have evolved towards parasitism. In all of these examples the animal is utilizing the tube or shell of an anachorete for its own defence. Cases where some other attribute of the host animal (such as spines or stings) are utilized in defence are discussed in Section 3.

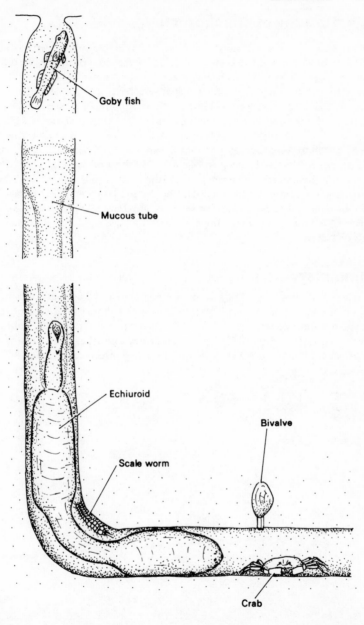

Fig. 1.1   The echiuroid worm *Urechis caupo* in its burrow with associated commensals, all of which derive protection against predators on the sea bed by living in the burrow. The commensals are: the crab *Scleroplax granulata;* the bivalve *Cryptomya californica;* the scale worm *Hesperonoë adventor;* and the goby fish *Clevelandia ios*. (*Redrawn from* MacGinitie and MacGinitie, 1949, Fig. 58.)

Limitations of anachoresis

One important limitation of anachoresis is that it conflicts with other activities of the animal such as feeding or reproduction. Anachoretes must either emerge to feed and mate, or they must have specialized feeding habits (such as having a suspension feeding current) and either be gregarious or have external fertilization. But a feeding current gives a clue to the presence of the animal, and may attract predators to search more closely. A further disadvantage is that once a predator has found one anachorete, either by persistent search or by chance, then it may learn the habits of the prey and search out others in the same way (this 'hunting by searching image' is further discussed on p. 41). So anachoretes cannot be too abundant or else predators may learn how to find them, and they often have secondary defensive mechanisms (such as escape holes) to protect them from those predators that do manage to find them.

Summary

Anachoresis is the habit of evading predators by living concealed in a hole or some other retreat. Some animals spend almost their entire lives hidden from predators, and these have very specialized feeding habits. Other animals emerge to feed, mate, and carry out other essential activities, so they must also have other more active means of escaping from predators.

Chapter 2
Crypsis

Animals which are camouflaged to resemble part of the environment are said to be *cryptic*, and the phenomenon is called *crypsis*. In this book I am further restricting crypsis to animals which are not normally distinguished from the environment by a predator although they are visible. If the animal resembles an inanimate object or a plant such that a predator may distinguish it from its background but fails to recognize it as being edible, I have considered the defence to be batesian mimicry, not crypsis.

A cryptic animal must harmonize with its background. Good examples are the green colour of many grasshoppers and caterpillars, and the transparency of the majority of animals in the plankton of the surface layers of water. At greater depths in the sea (500−1 000 m) there is little light since most light at the red end of the spectrum has been absorbed. Most planktonic animals at this depth are dark red, appearing black in the dim bluish light when viewed from above or from the side, and so these too are camouflaged (see Hardy, 1956, for examples).

Perfection of crypsis by predator selection

Although cryptic animals have a general resemblance in colour to their background, it is probable that predators recognize prey by certain characteristic features such as the rounded body, legs or eyes (Robinson, 1973; see also p. 260). Hence simple colour resemblance may not be sufficient to prevent a predator from detecting an animal. Being three-dimensional, animals are often rounded in cross-section, and with light normally coming from above there is a ventral shadow on the body which could be conspicuous to a predator. Many cryptic animals are counter shaded (darker dorsally, paler ventrally) so that this ventral shadow is obscured (Fig. 2.1). Animals which normally rest upside-down have reversed counter shading with the dorsal surface pale and the ventral (upper) surface dark (e.g. the pelagic gastropod *Ianthina* (Fig. 2.2), tree sloths, and some species of the catfish genus *Synodontis*).

Pelagic fish such as herring and mackerel may be approached by a

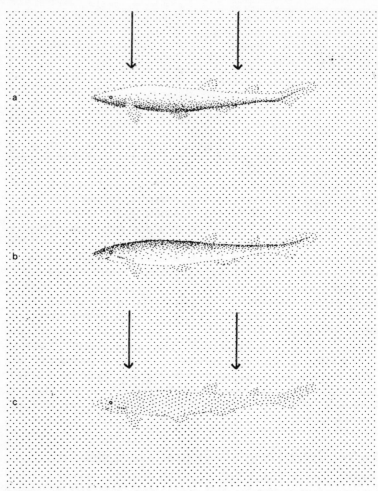

Fig. 2.1   The appearance of uniformly and counter shaded fish in different light regimes: (*a*) uniformly coloured fish with light coming from above (arrows). Notice the ventral shadow, and bright dorsal surface due to reflected light; (*b*) counter shaded fish with diffuse lighting; (*c*) counter shaded fish with light from above. Notice how the ventral shadow is counterbalanced by the fish's colour so that it is inconspicuous. (*After* Cott, 1940, Fig. 1.)

predator from any direction, and their coloration renders then inconspicuous from almost any direction too. There is a layer of black pigment dorsally whilst laterally and ventrolaterally there are silvery reflecting plates (reflectors), and below this is a further silvery reflecting layer, the argenteum (Fig. 2.3). Silvery fish are actually better camouflaged than are normally counter shaded fish. Sharks have typical counter shading (Fig. 2.1): when viewed from above or from any angle

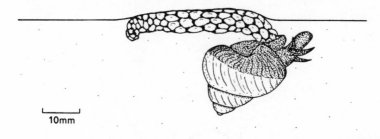

10mm

Fig. 2.2 The gastropod *Ianthina ianthina* clinging to its bubble float at the surface of the sea. The animal has been drawn with side lighting to show the reversed counter shading. The body and the stippled part of the shell are violet in colour.

above the horizontal, the dark dorsal surface reflects about the same amount of light as is reflected up from the depths of the sea, so the fish is camouflaged. From any point below the fish, however, a predator looking up will see the outline of the shark in dark silhouette against the bright downwelling light. Hence countershading only conceals a shark from predators above or on the same level as itself (Fig. 2.4).

A silvery fish such as a herring is also camouflaged when viewed from above — the dark pigment blends with the dark background of the depths of the sea and reflects little light. But it is important that no flashes of silver from the sides should be visible from above to betray the presence of the fish to a predator. This is ensured by the reflectors being aligned vertically so that when viewed from above they cannot reflect light (Fig. 2.3). When viewed from any point lateral to the fish the near-vertical reflectors automatically reflect light of similar intensity to that passing on either side of the fish. Only from directly below is the herring easily seen by a predator (Denton and Nicol, 1965, 1966). The silhouette in ventral view can be minimized by the fish being laterally flattened with a tapering ventral keel, but it cannot be completely eliminated even with another silvery reflecting layer (the argenteum) below the lateral reflectors (Fig. 2.4), and in any case such a shape conflicts with the shape best adapted to fast swimming.

Some bathypelagic fish, crustacea and cephalopods camouflage the silhouette in ventral view with a battery of uniformly spaced light emitting organs (photophores) which direct light ventrally (Fig. 2.5) (Fraser, 1962; Clarke, 1963). Denton (1970, 1971) has shown that the quality of light emitted is almost identical to that found at the depths where these fish live (500–1 000 m). In the hatchet fish *Argyropelecus*, beneath the photophores there are reflecting tubes which taper ventrally and are half silvered on the outer surface but fully silvered on the inner surface. The result is that the most intense light is emitted ventrally and less intense reflected light is emitted ventrolaterally and laterally (Fig. 2.6). The angular distribution of light emitted matches

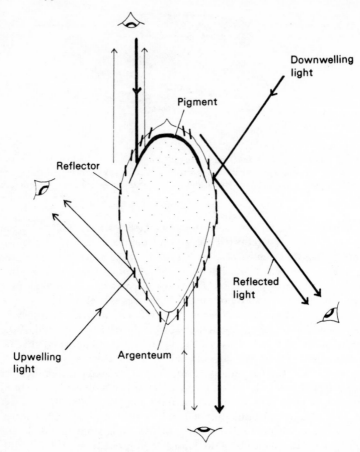

Downwelling
light

Pigment

Reflector

Reflected
light

Upwelling
light

Argenteum

Fig. 2.3 Diagram to show the light received by an observer looking at a herring (*Clupea harengus*) from different directions. Thick arrows indicate high intensity downwelling light, or such light reflected from the fish; thin arrows indicate low intensity upwelling light or low intensity reflected light. The fish is shown in transverse section with the orientation of the reflectors indicated, but not drawn to the same scale as the fish. Note that light reflected from the fish is of similar intensity to light coming past the fish when viewed from above or from the sides, but not when viewed from below. For further explanation see text. (*After* Denton and Nicol, 1965, and Denton, 1970.)

Fig. 2.4 Diagrams of counter shaded fish, silvery fish, and silvery fish with photophores to show the positions from which they would be visible as a silhouette against a brighter background. The fish are seen in cross-section with dorsal pigment, lateral reflectors, and ventral photophores indicated. The length of each arrow indicates the relative intensity of light from that particular direction. The areas stippled are the areas from which an observer would see the fish as a conspicuous silhouette. (Counter shaded and silvery fish redrawn from Denton and Nicol, 1965, Fig. 12.) ▶

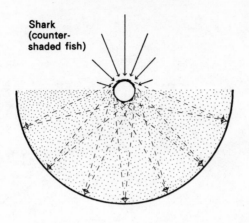

Shark
(counter-
shaded fish)

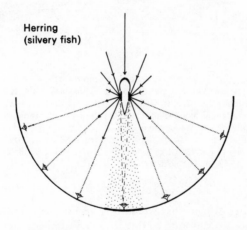

Herring
(silvery fish)

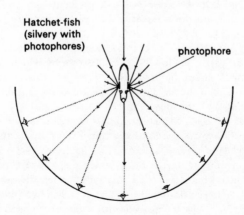

Hatchet-fish
(silvery with
photophores)

photophore

Fig. 2.5   Diagram to show how ventral photophores conceal an animal when it is viewed from below. On the left: the squid *Histioteuthis*, the fish *Argyropelecus*, and the shrimp *Euphausia* viewed from below against a background of downwelling light appear as black silhouettes. On the right: the same three animals with their photophores operating. When they are observed with the eyes out of focus, the photophores and the black body merge into the background so that there is no conspicuous silhouette (*Argyropelecus*: after Clarke, 1963, Fig. 3; *Euphausia*: after Hardy, 1967, p. 168; *Histioteuthis*: based on a plate in Herring and Clarke, 1971.)

the intensities of light that would be seen by a predator looking vertically or diagonally upwards, so this gives very efficient camouflage. Since *Argyropelecus* also has dorsal black pigment and vertically aligned silvery reflectors on the sides, it is well camouflaged to predators from all directions (Figs. 2.4, 2.7).

In addition to causing a ventral shadow on an object, unidirectional light will also cast a shadow on the substrate on which the object may be resting. With lateral lighting (as at dawn and dusk) the shadow may be very large and hence be a conspicuous recognition mark for a predator (Fig. 2.8). Shadow caused by side lighting is concealed by flattening of the body in many animals living on tree trunks. The tree living gecko *Ptychozoon kuhli* has a dorso-ventrally flattened tail (Figs. 2.9a and b). This species is known as the flying gecko because there is a lateral membrane on the body which helps it to glide from tree to tree. This membrane is curled ventrally under the body when the animal is at rest. The flattened tail, however, has two functions: it gives increased surface area (compared with a tail of rounded cross-section)

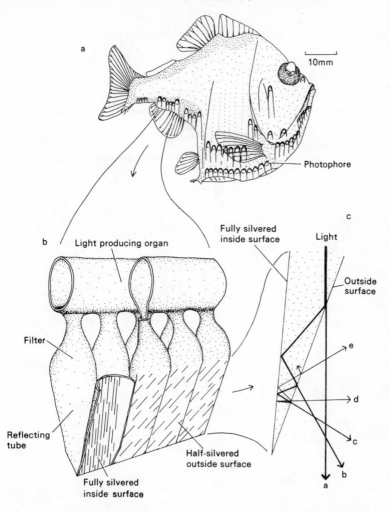

Fig. 2.6   Light producing organs (photophores) of the hatchet fish *Argyropel-ecus aculeatus*: (*a*) shows the position of the photophores; (*b*) diagram of a battery of photophores; (*c*) section through the reflecting tube of a photophore showing how the light emitted downwards is more intense than that emitted laterally. The outer surface of the tube is half-silvered whilst the inner surface is fully silvered and gives total reflection. Each double reflection of light results in progressive diminution in the intensity of light emitted (a–e). (a *original* b *and* c *redrawn from* Denton, 1970.)

when gliding, and it also prevents the formation of lateral shadow when resting on a tree trunk (Annandale, 1905; Tweedie, 1960). The mantid *Theopompella* is another flattened animal found on tree trunks (Plate 3*d*). In this insect the wings extend laterally at the sides of the body

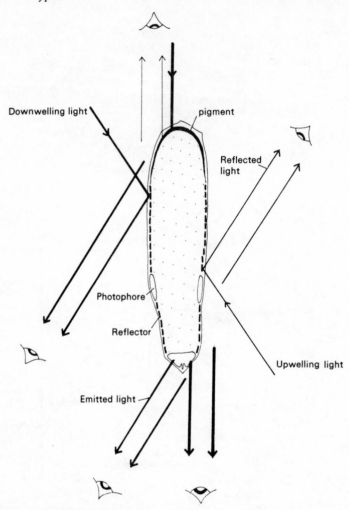

Fig. 2.7   Diagram to show the light received by an observer looking at a hatchet fish (*Argyropelecus*). Thick arrows indicate high intensity light (downwelling light, reflected downwelling light, or emitted light), thin arrows indicate low intensity light (upwelling light or low intensity reflected light). Note that the light reflected from the fish when viewed from any direction is of similar intensity to that passing by the fish. Compare with Fig. 2.3. (*Based on data in* Denton, 1970, 1971.)

and obliterate any lateral shadow. Some caterpillars are also flattened and have laterally directed hairs which conceal any lateral shadow (Plate 1*a*).

The body contour or outline of an animal may be another important means of recognizing prey for visually hunting predators. The character-

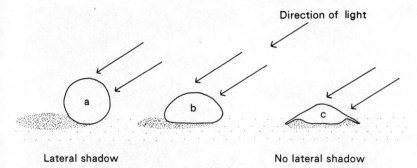

Fig. 2.8 Diagram to show how lateral shadow from oblique lighting can be eliminated: (*a*) rounded object with long shadow; (*b*) flattened object with much smaller shadow; (*c*) flattened object with angled flanges so that shadow is completely eliminated.

istic body contour can be concealed by the development of projections and processes which break up the outline, for example the frogfish *Histrio histrio* which lives amongst *Sargassum* weed (Fig. 2.10). Alternatively the contour may be obscured by disruptive colour marks so that these meaningless marks are more conspicuous than is the true outline. This is well seen in the pattern of certain wading birds which nest amongst stones such as the ringed plover (*Charadrius hiaticula*). The eggs, chicks and adult of this species all camouflaged by disruptive patterns (Cott, 1940). Some of the colour patterns adopted by the cuttlefish are also disruptive (Figs. 2.20*c* and *d* and see p. 29). Butterfly fish (*Chaetodon* and *Heniochus* spp.) often have disruptive patterns of black and white vertical bands which break up the outline of the fish (Fig. 2.10), but these colours may also have intraspecific signalling functions as well.

The eye of vertebrates is another conspicuous feature which may be a recognition mark by which predators detect an otherwise cryptic prey. The eye is commonly concealed by a disruptive line, as in the ringed and kentish plovers (Fig. 9.1), the oryx, and in various species of frogs (Cott, 1940). Although this is the most widely accepted explanation of the function of eye stripes, it has been pointed out that there is a strong correlation between animals which possess an eye stripe and the habit of feeding on swift moving prey (Ficken *et al.*, 1971). Since the eye stripe in birds normally passes from the eye to the beak, it is possible that it acts as a sighting line which increases the chances of successfully capturing active prey. It is possible that eye stripes perform both of these functions in some birds and are thus important in prey capture as well as in predator avoidance; but eye stripes in many other animals (e.g. grasshoppers and some fish) do not pass from the eye to the mouth (see Fig. 2.10), and in these they must be camouflaging in function.

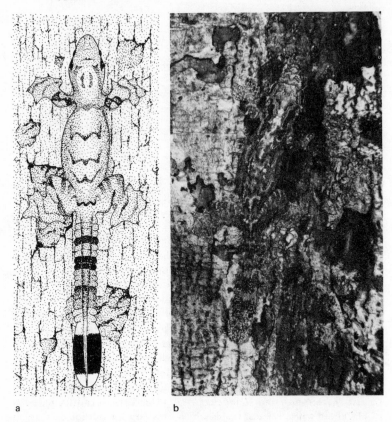

a   b

Figs. 2.9a and *b*   The highly cryptic flying gecko *Ptychozoon kuhli* resting on a
tree trunk. Note the disruptive coloured markings on the body and tail, and the
tapering flanges on the tail which eliminate shadow. ((*a*) *After* Tweedie, 1960,
Fig. 2 and plate VI. (*b*) photograph supplied by M. W. F. Tweedie, Malaya.)

Thus crypsis may involve very elaborate adaptations. The fact that
these adaptations occur suggests that crypsis is of protective value and
that predation on the least well concealed individuals has been the
selective factor responsible for perfecting the crypsis.

Survival value of crypsis

A number of experiments have been performed which demonstrate the
advantage to a species of being cryptic. One of the earliest was carried
out by Di Cesnola (1904) using green and brown morphs of the
European mantid *Mantis religiosa* with wild birds as predators. Di

Fig. 2.10   The sargassum fish *Histrio histrio* (=*Antennarius*) (left) and the butter-fly fish *Heniochus macrolepidotus* (right). The body outline is obscured by projections in the sargassum fish and by disruptive marks in the butterfly fish. (*Redrawn from* Cott, 1940, Figs. 23 and 70.)

Cesnola tethered twenty green and forty-five brown mantids to green vegetation, and twenty-five green and twenty brown mantids to brown vegetation. After eighteen days he found that all of the conspicuous insects had been eaten whilst all those which harmonized in colour with their background survived. Beljajeff (1927) repeated the experiment and obtained similar results (Fig. 2.11), but when he repeated it yet again crows came and took almost all of his mantids of all three colours used (Cott, 1940). This result was not easy to explain at the time, but following the work of Croze (1970) on the hunting behaviour of carrion crows, it is now clear that the birds developed a searching image for all three morphs of the mantid, and so preyed on yellow, green and brown ones irrespective of the background. Perhaps the density of experimental insects was very high in this particular series of experiments so that the predators were quickly able to form searching images for the prey. But apart from this result, the experiments on mantids show that predators take more of conspicuous prey than they do of cryptic prey.

   Similar results have been obtained in selection experiments using different coloured grasshoppers as prey with chickens, turkeys, wild birds or chameleons as predators (Eisentraut, 1927; Isely, 1938; Ergene, 1951, 1953). In one series of experiments Ergene presented a chameleon with ten yellow and ten green *Acrida turrita* on either a green or a yellow background. In every one of sixty-three such experiments the chameleon took more of the conspicuous insects and fewer of the matching ones (summarized in Fig. 2.12). In another series of experiments Ergene showed that the waldrapp ibis (*Geronticus eremita*) also

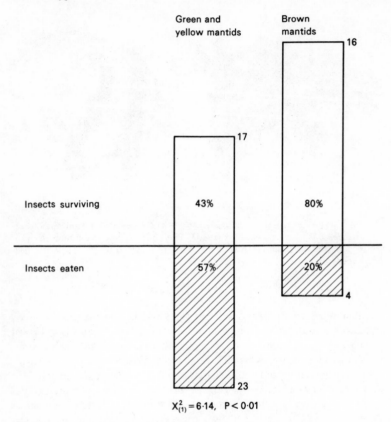

$$X^2_{(1)} = 6.14, \quad P < 0.01$$

Fig. 2.11   Predation by wild birds on green, yellow, and brown *Mantis religiosa* presented on a brown background. (*Data from* Beljajeff, 1927.)

took more conspicuous insects than of those which matched their background.

Sumner (1934) conducted a similar series of experiments using dark and pale mosquito fish (*Gambusia patruelis*) in black and white tanks as prey with penguins (*Spheniscus mendiculus*) as predators. His results (summarized in Fig. 2.13) show that the birds took more dark fish from the white tanks and more pale fish from the black tanks. Similar results were obtained using a night heron and a sunfish as predators (Sumner, 1935). Hence in any population of prey animals those individuals that are the most conspicuous are likely to be preyed on first, and the most cryptic individuals have the best chance of surviving. Proof of the selective advantage of crypsis in natural populations is described for *Biston* and *Cepaea* on pp. 47 and 50.

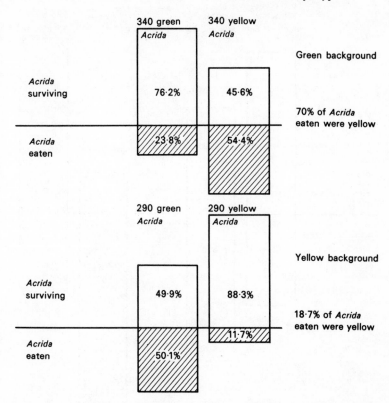

Fig. 2.12 Predation by a chameleon on green and yellow grasshoppers (*Acrida turrita*) presented on green and on yellow backgrounds. (*Data from* Ergene, 1950*a*.)

Limitations of crypis

1. To remain undetected a cryptic animal must not move

Movement of an otherwise cryptic animal may enable a predator to detect it. Hence locomotion is either very slow, with rocking from side to side so that the lateral movements of the animal resemble the gentle movements of plants by the wind, or the animal may remain motionless until disturbed and then move very rapidly (secondary defence) before freezing and once again resting motionless.

2. Crypsis conflicts with other essential activities

As is the case with anachoresis, crypsis may conflict with other essential activities of the animal. For efficient protection a cryptic animal must

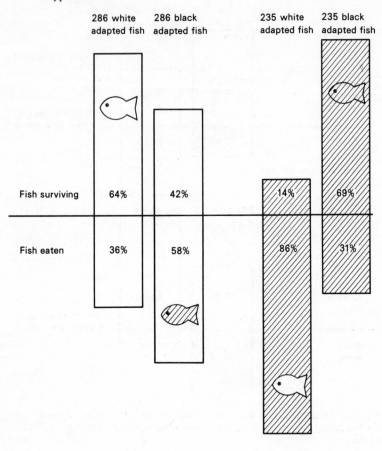

Fig. 2.13 Predation by penguins on black and white adapted mosquito fish (*Gambusia patruelis*) in black and white tanks. (*Data from* Sumner, 1934.)

remain motionless, but in such a position it cannot collect food. Many cryptic animals are motionless during the day but feed actively at night-time or at dawn and dusk. Mantids wait motionless for prey to come near, or they may slowly stalk prey with much 'teetering' backwards and forwards before they pounce suddenly to make a capture. But the most efficient prey-capture posture is with the forelegs flexed, and this may conflict with the optimal cryptic posture for some species (Fig. S1.1).

Many cryptic species are active at night and restrict their reproductive activities to this time also. But reproduction often involves the use of visual signals which can only be received in daylight. An efficient visual signal directed at a conspecific can also attract a predator, so many animals which are cryptic have visual recognition signals which are normally hidden. Some stimulus from another conspecific then releases the behaviour pattern which renders the sex recognition signals visible (for example the speculum of ducks is only clearly visible during courtship displays). In many passerine birds there is a clear conflict between selection for crypsis and selection for conspicuous sexual signals. The result is very often that the female is cryptic with few recognition marks since she often spends more time brooding eggs where crypsis is of paramount importance, whilst the male may be conspicuous with several signal marks. In the reed warbler (*Acrocephalus scirpaceus*) both sexes are cryptic with no obvious recognition marks, and both sexes brood the eggs in an open nest. In the chaffinch (*Fringilla coelebs*) the female is cryptic with white shoulder patches which are normally partly hidden when at rest (they may even be disruptive), whilst the male has conspicuous colours and does not brood. Titmice are usually brightly coloured in both sexes with conspicuous recognition marks, but since brooding is in a hole or crevice there is no advantage in being cryptic at this time. The balance between the selective advantages of crypsis in the two sexes when brooding, singing and feeding, and the advantages of conspicuous signal marks for use in territorial and courtship situations varies in different species of birds. It is not easy to formulate any generalizations, but these examples do illustrate the point that there is conflict between selection for crypsis and selection for conspicuousness.

3.   An animal will be very conspicuous if it rests in the wrong place

(a)   *Selection of appropriate background*

An animal may be cryptic in one habitat but very conspicuous in another. A green grasshopper is well concealed so long as it remains on green vegetation, but it becomes conspicuous if it rests on earth or sand. Moths, which are cryptic when resting on tree trunks, are very conspicuous when they are attracted to light and rest on the white walls of buildings. Clearly selection by predators is likely to eliminate any animal which settles in the wrong place, so there must be some mechanism that ensures that an animal normally rests on the appropriate background. Such a mechanism could be an innate and genetically fixed background preference for each individual animal, or it could be based on some form of colour comparison between part of the body of the animal and its background followed by movement until the two colours are almost the same. Sargent (1968) found that when given a choice of a black or a white background, the dark moth *Catocala antinympha* rests more often on the black surface whilst the pale *Campaea perlata* rests more often on the white surface. Sargent repeated the experiments after painting the scales round the eyes and on the thorax of *Catocala* white, and those of *Campaea* black. In a third choice experiment he cut off the wings of *Campaea* (Sargent, 1969*a*). None of these treatments affected the choice of background, so he concluded that the difference in choice of background between these two species is a genetically determined preference for a particular colour and is not due to a behavioural matching response.

In the moth *Cosymbia pendulinaria* Sargent (1968) found that both the typical form and a dark coloured form produced by subjecting the larvae to cold temperatures prefer to rest on a white background rather than on a dark one, so in this species the preference is also genetically determined and is not affected by the phenotypic appearance of the insect. In the peppered moth *Biston betularia* Kettlewell (1955) showed that the melanic forms rest more often on black whilst the typicals rest more often on white backgrounds, but it is not known if this preference is innate and determined by the same gene that determines the colour of the insect (or by a closely linked gene), or if there is some form of colour comparison of part of the insect with its background.

Schinia florida is a yellow moth that normally rests on the yellow flowers of evening primrose (*Oenothera biennis*) where it is cryptic. Sargent (1969*b*) showed that if given a choice it prefers to rest on *Oenothera* to other yellow flowers, and that even when the colour of the flower is hidden by a muslin bag it still comes to rest on the muslin round the *Oenothera* rather than on other yellow flowers. In the absence of *Oenothera*, *Schinia* shows no preference for yellow flowers over flowers of other colours. Evidently in this moth crypsis is brought about by choice of background based on the scent characteristics of the preferred plant, not by its colour.

Rhododipsa masoni is another moth which rests on flowers in North America, in this case on *Gaillardia aristata*. The flower heads have yellow central disc florets, red outer disc florets, and yellow ray florets. The moth has red forewings and yellow head and thorax. Thus if it rests radially on the flower with the head either inwards or outwards, it is cryptic, but if it rests tangentially it is not (Fig. 2.14). Brower and Brower (1956) found that seventeen out of twenty-one moths were resting in the best concealed position (with the head and thorax over yellow florets and the wings over red florets). They point out that the moth is too heavy to rest on the outer ray florets (where it would be conspicuous), and is unlikely to rest in the centre of the yellow disc florets since then it could not feed. Only the red disc florets are open with nectar exposed, so the best position for feeding is with the body crossing these red florets. It is probable that as *Rhododipsa* evolved its close association with *Gaillardia*, including its characteristic feeding posture, so natural selection perfected its colour pattern by eliminating those insects with inappropriate distributions of red and yellow.

Grasshoppers and mantids can also select a particular coloured back-

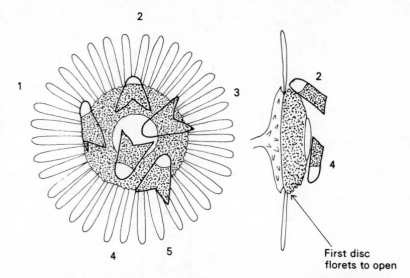

Fig. 2.14   Diagram to show resting positions of *Rhododipsa masoni* on *Gaillardia aristata* flowers. Central disc florets, ray florets, and head and thorax of moth are yellow (shown white); outer disc florets and wings of moth are red (shown stippled). Moths 1, 2 and 3 are cryptic whilst moths 4 and 5 are conspicuous. (*After* Brower and Brower, 1956, Fig. 2. © by the University of Chicago.)

ground. Ergene (1950*a*) found that in green habitats green morphs of the grasshopper *Acrida turrita* were much more frequent than yellow morphs, but in yellow or brown habitats the yellow morphs were more frequent than the green. She gave green and yellow insects a choice of green or yellow backgrounds on which to rest and found that significantly more insects chose to rest on a background of similar colour to their body than on a background of a different colour (Fig. 2.15). Ergene (1950*b*) performed similar experiments with the grasshopper *Oedipoda miniata* which has four colour morphs in Turkey, each associated with a different type of soil. Again, each morph showed a strong preference for resting on soil of similar type and colour to that of its normal environment, and each morph was most cryptic to the human eye when on its preferred soil.

In Ghana two of the commonest mantids in the savanna regions are *Sphodromantis lineola* and *Miomantis paykullii*. Both species can be either green or brown in colour. Green *Miomantis* show a definite preference for resting on a green rather than a brown surface, and brown insects show a significant, though less pronounced, preference for resting on brown rather than on green (Fig. 2.16) (Barnor, 1972). Since the colour of *Miomantis* is environmentally determined and not genetically fixed (see p. 35), it is clear that individual insects do not have an innate preference for a particular colour. When the back half of the eyes of *Miomantis* are painted over, green insects no longer show a

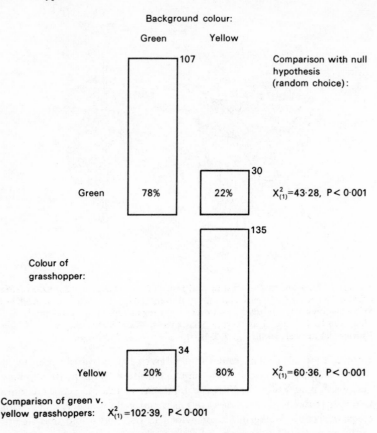

Fig. 2.15　Choice of green or brown backgrounds by green and yellow grass-hoppers (*Acrida turrita*). (*Data from* Ergene, 1950*a*.)

preference for green but tend to prefer brown instead (Fig. 2.17). This could be interpreted as suggesting that the insects normally compare their own colour with that of their background; when the back of the eyes is covered they can no longer see their own colour so the preference is lost. Control insects, however, in which the front half of the eyes was painted, behaved just like the experimentals (Fig. 2.17), so it is unlikely that choice of background is made by matching colours. Hence the mechanism by which this species chooses a matching background is not known, but it could be an endogenous (innate) preference for a particular colour which can be altered by factors which also cause a change in the mantid's own colour.

Green *Sphodromantis lineola* also show a significant preference for resting on green leaves rather than on brown leaves, but the behaviour of brown insects does not differ significantly from random, nor does it

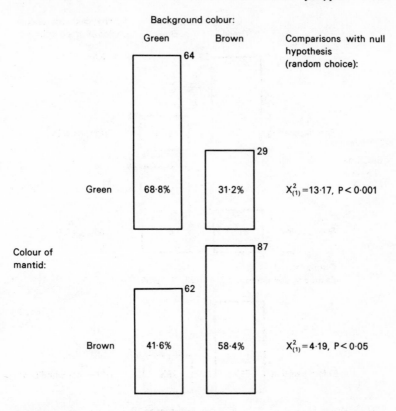

Fig. 2.16 Choice of green or brown backgrounds by green and brown mantids *Miomantis paykullii*. (*Data from* Barnor, 1972.)

differ significantly from that of the green insects (Fig. 2.18). Hence the green insects clearly select a green background, but it is not certain if the brown ones also prefer green or if they settle on green and brown at random (Barnor, 1972). A possible explanation for this apparently maladaptive behaviour is given on p. 36, but it is likely that background selection in *S. lineola* involves an endogenous preference for green rather than brown, though further work is required to confirm this.

Cryptic animals not only rest on an appropriate background, but they also adopt characteristic resting attitudes in which their crypsis is most effective, and they remain motionless in this position. Moths which normally rest on tree trunks with the head upwards have disruptive marks running parallel to the body axis, whilst moths which normally rest sideways have the disruptive marks running transversely. In both cases the disruptive marks are parallel with the lines found on

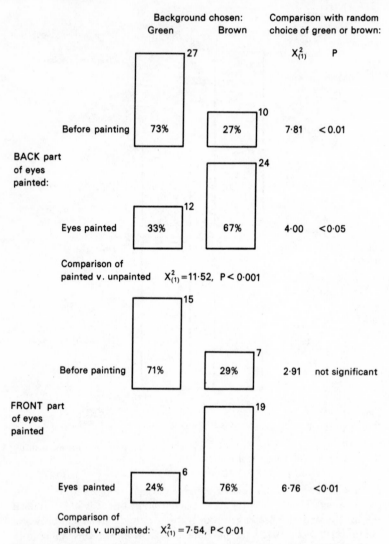

Fig. 2.17 Choice of background by green mantid *Miomantis paykullii*, with normal eyes, with back half of eyes painted black, and with front half of eyes painted black. (*Data from* Barnor, 1972.)

the bark. Sargent (1969c) studied the resting behaviour of two bark-like species of moths when placed in a cylindrical chamber with a choice of either horizontal or vertical black and white stripes (black tapes on a white surface). *Melanolophia canadaria* normally rests sideways on tree trunks and most moths in the cylinder rested in the correct position to give good camouflage with relation to the stripes (i.e. sideways on the

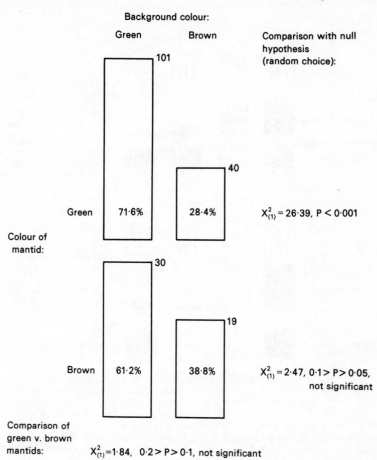

Fig. 2.18   Choice of green or brown backgrounds by green and brown mantids *Sphodromantis lineola*. (*Data from* Barnor, 1972.)

vertical stripes and head upwards on the horizontal stripes) (Fig. 2.19). When the stripes were covered with a film of cellulose acetate so that there were no vertical or horizontal irregularities on the surface, the moths' resting positions were at random with relation to the stripes (Fig. 2.19). Thus the choice of the most appropriate resting posture by this moth is attained by the use of tactile, not visual, stimuli. The moth *Catocala ultronia* also normally rests correctly aligned to stripes, but it normally rests with the body axis vertical, and it has a strong preference for vertical over horizontal stripes (Fig. 2.19). When the surface irregularities were hidden by an acetate film, all insects rested vertically, but there was no preference for resting on vertical rather than on horizontal stripes. Hence in this species the resting position is innate and

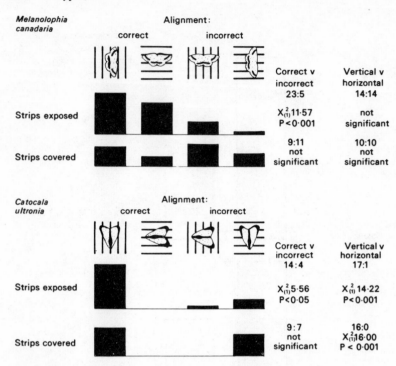

Fig. 2.19   Resting positions of *Melanolophia canadaria* and *Catocala ultronia* on vertical or horizontal black and white stripes. 'Correct' alignment is in a position such that the markings on the moth are parallel with those on the tape strips; 'incorrect' alignment is in a position with the moth's markings not coinciding with those on the background. The histograms give the numbers of moths resting correctly and incorrectly on vertical and horizontal stripes, with and without the tape strips covered with a cellulose acetate film. (*Data from* Sargent, 1969*c*, Fig. 1.)

independent of any feature of the substrate, but the place where it chooses to rest is determined by substrate features detected by tactile sense organs.

In nudibranch molluscs there is also evidence for selection of background, but in this case choice of background is really choice of food. *Rostanga pulchra* is a red dorid which shows a distinct preference for feeding on the red sponge *Oplitaspongia pennata*, and the mottled *Archidoris montereyensis* prefers to feed on the brownish yellow sponge *Halichondria panicea*. Both of these dorids find their preferred food by chemical means, not by colour (Cook, 1962). Similarly the red eolid nudibranch *Catriona aurantia* shows a distinct preference for feeding on *Tubularia indivisa* when given a choice of hydroids, and it is very well camouflaged when on this species. It is very rare to find *C. aurantia* on any other species of hydroid (Braams and Geelen, 1953; Edmunds,

1966*a*). A great many other nudibranchs are associated with one or a few species of prey, and many are cryptic when on their normal food (see Thompson, 1964, and Edmunds, 1966*a*, for references to the older literature).

Thus many cryptic animals have complex behavioural mechanisms which ensure that they rest in the place where they are best concealed. Species with even more specialized resting postures are further discussed on p. 122 since they can be regarded as being mimetic — as explained on p. 1, it is convenient to consider animals with highly specific similarities to part of the environment as being mimetic rather than cryptic, although the distinction between the two is somewhat arbitrary.

(b)   Colour change to match a particular background

As a consequence of cryptic animals being camouflaged in some habitats but not in others, such animals can only occur commonly in those places where they are well concealed. The more perfect is the resemblance between an animal and its resting place, the more restricted in occurrence is such a species likely to be if it is to remain cryptic. One way out of this dilemma is for an animal to change its colour so that it can harmonize with two or more different backgrounds. The chameleon is a well known example, though the range of patterns and colours available to most species of chameleon is not large, and no systematic study has been conducted to see what circumstances elicit each colour pattern.

Flatfish can also change colour to match a variety of different backgrounds. The flounder, for example, can be a variety of shades of grey or brown to match sandy or muddy substrates, and there may be large or small disruptive spots or blotches if the background is at all pebbly. Such colour change takes place slowly over a period of several days. More rapid colour change occurs in cephalopods, some of which can change colour in less than a second. The best known species is the cuttlefish *Sepia officinalis* studied by Holmes (1940) (see also Boycott, 1958). When swimming it usually adopts a zebra-striped pattern with the dark and pale bands breaking up the body outline (Fig. 2.20*a*). When resting on a sandy bottom it adopts a sandy coloured, mottled pattern (Fig. 2.20*b*), but on pebbles it may take on one of several disruptive patterns of dark and pale blotches and marks (Fig. 2.20*c* and *d*).

A number of arctic birds and mammals show seasonal changes in colour with a brown summer plumage (or pelage) cryptic on earth and sparse vegetation, and a white winter plumage cryptic on snow. Examples are the snowshoe hare and the ptarmigan. Of course if the background changes from white to brown before the animal gets its summer plumage it will be very conspicuous for a time. It has, however, been suggested that it is of advantage for the male ptarmigan to remain

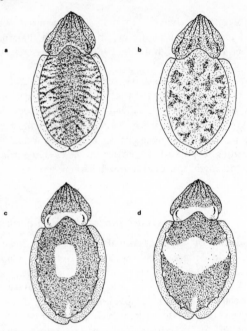

Fig. 2.20   Colour patterns adopted by the cuttlefish *Sepia officinalis* on different backgrounds: (*a*) zebra-patterned when swimming; (*b*) mottled pattern when on a sandy substrate; (*c*) and (*d*) two disruptive patterns when on stones or pebbles. (*Redrawn from* Holmes, 1940.)

white whilst the brown female is incubating eggs so that attacks by predators are directed at him rather than at her (see Cott, 1940, for details). If a few males are killed at this time it is of little significance since they have already mated, but the female, who has to rear the eggs and young, still requires protection.

A number of insects are also able to change colour, but this can usually occur only at moulting. Poulton and others have shown that the pupae of butterflies such as *Pieris brassicae, Danaus chrysippus* and *Papilio* spp. can be either green or brown depending on the immediate surroundings of the caterpillar just prior to pupation — hence this is an environmentally induced polymorphism. Clarke and Sheppard (1972*a*) have studied pupal polymorphism in the swallowtails *Battus philenor* and *Papilio polytes*. In both species there is a significant association between pupal colour and background colour, but with quite a lot of errors — mostly of green pupae on brown twigs (Table 2.1). Since cater-pillars in dark boxes or on red or transparent objects develop into green pupae, it appears that green pupae should be regarded as 'normal'. If there is much brown colour or bright light in the environment just prior to pupation, then the pupae produced are brown (Fryer, 1913). But

Table 2.1   Colour of pupal sites and pupae of *Battus philenor* in California. (*Data from* Clarke and Sheppard, 1972*a*.)

Pupal colour	Pupal sites		
	trunks and branches over 50 mm *diameter*	*brown twigs* <12 mm *diameter*	*green twigs* <12 mm *diameter*
Brown	23	16	0
Green	1	11	4

thin brown twigs do not always produce sufficient stimulus to exceed the threshold required to switch from green to brown, so there are more 'errors' in pupae on thin than on thick brown twigs.

Although pupal colour is an environmentally induced polymorphism it must be genetically controlled and should therefore be modifiable by selection. Clarke and Sheppard reared one population of *Papilio polytes* in which they destroyed all pupae which did not match their background — hence they selected for those that did match their background. In a second population they selected for non-matching pupae by destroying all those that matched their background. The pupal colours produced in these two populations after five generations of selection are given in Table 2.2. These figures can be analysed in a

Table 2.2   Pupal colour of *Papilio polytes* after five generations of selection. (*Data from* Clarke and Sheppard, 1972*a*.)

Population		Background colour			
		Green		Brown	
	Pupal colour	*green*	*brown*	*green*	*brown*
Matching pupae selected		569	1	64	33
Non-matching pupae selected		316	3	74	31

number of different ways. By comparing the number of pupae which match their background with those that do not, it is clear that the population selected for matching pupae has significantly more matching pupae than has the other population (Fig. 2.21). Comparison of the backgrounds on which the insects pupated shows that in the population selected for matching pupae significantly more pupae were on green backgrounds than occurred in the other population (Fig. 2.22). Hence selection for matching pupae (such as might occur with heavy predation) can increase the proportion of pupae matching their background in colour, but it does so largely by increasing the number of caterpillars which choose to rest on green backgrounds. There is no evidence that selection alters the reaction of the caterpillars to a brown background.

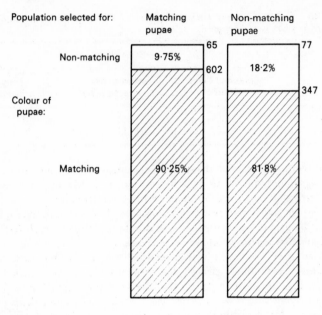

Population selected for: Matching pupae   Non-matching pupae

Non-matching   9·75%   65 / 602   18·2%   77 / 347

Colour of pupae:

Matching   90·25%   81·8%

Comparison of matching v. non-matching pupae in the 2 populations: $X^2_{(1)}=16·21$, $P < 0·001$

Fig. 2.21   Pupal colour of *Papilio polytes* after five generations of selection showing the numbers of pupae which match their background. For further explanation see text. (*Data from* Clarke and Sheppard, 1972*a*.)

Presumably different populations of such insects will experience different spectra of predators and hence they will evolve different proportions of green and brown pupae.

In the North American butterfly *Papilio polyxenes* the factors determining pupal colour are more complex. This insect has two or three generations during the summer months, then overwinters as a chrysalis. In the summer caterpillars may pupate either on green or brown objects and so it will be of advantage if pupal colour is determined by conditions seen by the caterpillar just prior to pupation. As with *P. polytes* and *Battus philenor*, thin green twigs induced caterpillars to turn into green pupae, thick brown branches induced them to turn into brown pupae, whilst brown twigs resulted in some 'errors' with both green and brown pupae being formed (West *et al.*, 1972). Probably it is the thinness of the object rather than its colour that is important in determining the colour of the chrysalis. This experiment was performed with sixteen hours of daylight, similar to conditions in a temperate summer, and all of the pupae hatched within about two weeks. When the experiment was repeated with eight hours of daylight, to simulate conditions in late autumn, almost all the pupae produced were brown and entered

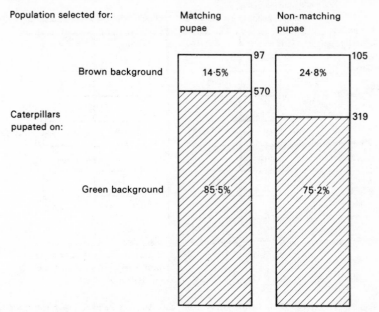

Population selected for: Matching   Non-matching
                            pupae   pupae

Fig. 2.22   Pupal colour of *Papilio polytes* after five generations of selection showing the numbers of larvae which pupated on green and on brown backgrounds. (*Data from* Clarke and Sheppard, 1972*a*.)

a state of diapause, not hatching out for several months. In this butterfly the control of pupal colour is adapted to its ecology: short day length induces all caterpillars to produce brown diapausing pupae which are likely to be cryptic on leafless trees in winter and which do not hatch out until warmer weather returns; long day length causes caterpillars to produce green pupae if they are on thin twigs but brown pupae if they are on thicker branches, and none will enter diapause.

Environmentally induced polymorphisms also occur in many orthopteroid insects which have green and brown forms. In the grasshopper *Acrida turrita* Ergene (1950*a*) found that green larvae placed on a yellow background change to yellow at the next moult. However, judging from Rowell's work on other species of grasshopper, it appears that the critical factor determining whether a grasshopper changes colour is not the quality of light reflected from the background, but the quality of incident radiation received by the insect. The factors which determine whether a grasshopper becomes green or brown are not the same for all species, and their interaction is extremely complex (summarized by Rowell, 1971).

In the American grasshopper *Syrbula admirabilis*, green nymphs can

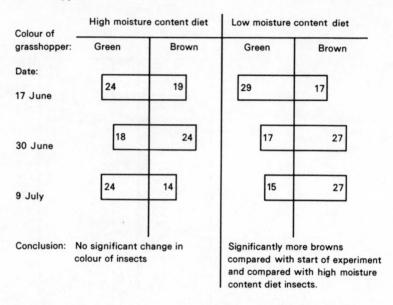

	High moisture content diet		Low moisture content diet	
Colour of grasshopper:	Green	Brown	Green	Brown
Date:				
17 June	24	19	29	17
30 June	18	24	17	27
9 July	24	14	15	27
Conclusion:	No significant change in colour of insects		Significantly more browns compared with start of experiment and compared with high moisture content diet insects.	

Fig. 2.23   Effect of moisture content in the diet on coloration of the grasshopper *Syrbula admirabilis*. Both groups were fed on Bermuda grass which had about 66 per cent water in the high moisture content group and about 38 per cent water in the low moisture content group. (*Data from* Otte and Williams, 1972.)

Relative humidity:	> 90%		30–40%	
Colour of grasshopper:	Green	Brown	Green	Brown
Date:				
17 June	29	32	35	22
30 June	25	35	10	38
9 July	25	20	11	23
Conclusion:	No significant change in colour of insects.		Significantly more browns by 9 July compared with start of experiment and compared with high humidity insects.	

Fig. 2.24   Effect of relative humidity on coloration of the grasshopper *Syrbula admirabilis*. (*Data from* Otte and Williams, 1972.)

be induced to change to brown either if fed on a diet with low moisture content (Fig. 2.23), or if kept in a low humidity environment (Fig. 2.24) (Otte and Williams, 1972). Since in nature grass with high moisture content and high relative humidity are correlated with green vegetation whilst grass with low moisture content and low relative humidity are correlated with dry, brown vegetation, it is clear that use of moisture content of food or air as a determinant of body colour is adaptive. Unfortunately it is not known if *Syrbula* also changes colour in response to background colour (the experiments described by Otte and Williams to test this lasted only three to eight days and cannot be regarded as conclusive).

In mantids the two most important factors determining whether or not a nymph changes colour are relative humidity and incident light — particularly ultra-violet light. The relative importance of these two factors varies in different species: in *Miomantis paykullii* relative humidity is of more importance than light in determining the colour an insect becomes (Barnor, 1972), whereas in *Sphodromantis lineola* and *Mantis religiosa* light is of equal or greater importance than relative humidity (Barnor, 1972; Jovancic, 1960). These differences can be related to the ecology of the species concerned. Thus *Miomantis* lives in grass where there may be very considerable changes in colour over a period of a few days (e.g. the green growth following a storm after a prolonged drought). Under these conditions a change in the environment is likely to be closely correlated with change in relative humidity, hence there is selective advantage accruing to insects in which relative humidity determines their colour at the next moult. Changes in the relative frequencies of green and brown morphs throughout the year show that this species is indeed very sensitive to changes in the weather: the frequency of greens varies from 80 or 100 per cent in June (the month with the most wet days) to as low as 0 or 20 per cent in parts of the dry season (November to March) when there may be only one or two wet days in a month (Fig. 2.25, and see also Edmunds, 1972).

Sphodromantis lineola, however, rests in shrubs and trees, and brown morphs are never common. In more than four years of observation at Legon, Ghana, they have only been found in the dry season, and even then they occur at low frequency in the population (Fig. 2.26, and see also Edmunds, 1972). Different species of savanna trees may flush at different seasons of the year and a green *Sphodromantis* nymph, which cannot quickly move from one tree to another, may be in a tree which, over a period of a few days, loses all of its green leaves. It may be several days or weeks (especially in the dry season) before new green leaves appear, and such an insect will be very conspicuous. Flushing is not necessarily correlated with rain, nor with humidity, hence it would not be advantageous for the green/brown coloration to be determined by humidity. But there is a very striking difference between the amount of incident light and heat reaching a mantis on a bare tree and the

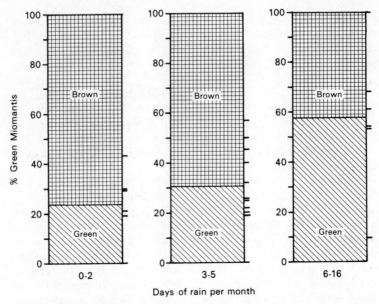

Fig. 2.25   Relationship between the days of rainfall per month at Legon, Ghana, and the percentage of *Miomantis paykullii* caught which were green. The actual percentages of green insects in each month are indicated on the right side of each block.

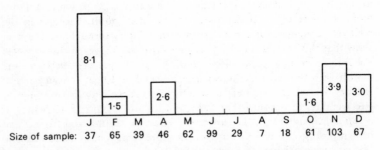

Size of sample: 37   65   39   46   62   99   29   7   18   61   103   67

Fig. 2.26   Percentage frequency of the brown morph of *Sphodromantis lineola* in different months of the year at Legon, Ghana, from November 1967 till April 1973.

amount reaching a mantis on a tree with leaves. Jovancic (1960) claims that strong incident radiation causes *Mantis* to turn brown, and if this is also true for *Sphodromantis lineola* it would cause insects on bare trees to turn brown whilst insects on green ones would remain green. In this species it is therefore advantageous to use light rather than humidity to determine coloration, but much further confirmatory work is required. The fact that the brown mantids apparently have a preference for resting on green rather than on brown is of no great importance since there will be little green vegetation anyway on a bare tree, and since green morphs will be more conspicuous and at a strong selective disadvantage to the brown insects under these conditions.

Environmentally induced polymorphisms also occur in savanna insects such as some grasshoppers and mantids (e.g. *Galepsus toganus*) which produce blackish forms when the grass has been burnt (fire melanism), but not at other times of the year (Hocking, 1964). Although most of these changes occur only at moulting, Burtt (1951) found that adult *Aulacobothrus wernerianus* (Acrididae) turned black two to four days after being placed on burnt ground. Further study on the mechanism of colour change in fire melanics is wanted, but the survival value of the phenomenon in terms of camouflage from predators is obvious.

A much simpler form of colour change occurs in nudibranch molluscs such as *Aeolidia papillosa* in which the colour of the animal is determined by the pigments present in the food. Hence after feeding for a few days on *Actinia equina* (a red sea anemone) the digestive gland in the dorsal papillae becomes red, and the slug is then cryptic when amongst a colony of *Actinia*. Similar results have been described for other eolids by Haefelfinger (1969).

In all of these cases the ability of the animal to change colour means that it is not restricted to just one or a few types of habitat as in the case with cryptic animals which cannot change colour.

(c)   Modification of background to improve crypsis

Another way in which an animal can improve its chances of survival is by altering its environment so that it is itself less easy to detect. Web-building spiders normally rest in the centre of the web, where they are conspicuous to predators, or in a retreat to one side of the web, where they are hidden (anachoresis) but where they cannot so quickly reach and capture prey that has been only temporarily entangled in the web. A third solution is to sit in the middle of the web but concealed under a leaf or twig that the spider has itself placed there. This occurs in some species of *Theridion* (Fig. 2.27) and of *Tetragnatha* (Bristowe, 1958; Hingston, 1927a), whilst *Araneus alsine* hides in a leaf above the centre of the web (Wiehle, 1927). Species of *Cyclosa* and *Uloborus* add to the web one or more blobs or bands of silk, some of which may contain oval egg masses or remains of prey but which are of similar size and shape to the spider itself. It is likely that a predator may attack one of these dummies instead of the spider (Fig. 2.28). Finally species of *Argiope* often build a zigzag white webbing device in the form of a diagonal line or cross: / or X (Plate 3a). Marson (1947a) found that in *Argiope pulchella* the device was most elaborate and complete when the web was built in brightly lit places, but that in dull places it was rarely a complete X, and in very dull places there might even be no device at all.

Fig. 2.27   Web of *Theridion* sp. from Ghana. Note the platform with scattered threads above it and the leaf retreat in the centre of the web. The spider can be seen inside the rolled leaf with its legs in contact with a vertical thread to the platform. The details of the web are somewhat simplified for clarity.

He interpreted the devices as being camouflage devices since the spider often rests with her legs aligned to the arms of the device, and since both her body and the devices are white (or rather, her body contains white and dark marks whilst the device is white with gaps through which the dark background is visible). Robinson and Robinson (1970) argue that the diagonal and cross devices of *A. argentata* are not defensive since they are more likely to draw attention to the object at the centre of the cross than to conceal it. They placed crickets in the centre of artificially constructed cross devices and found that birds took these in preference to crickets without devices. However, they did report a

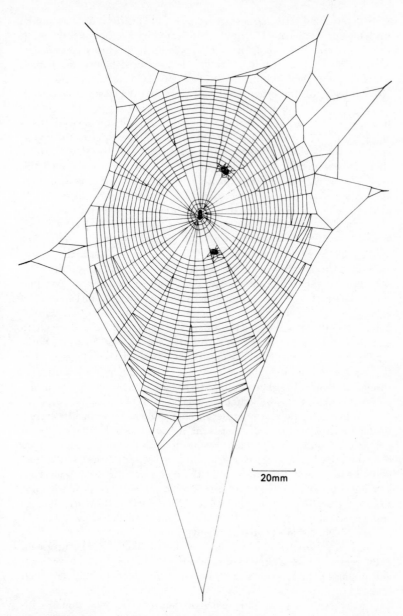

Fig. 2.28   Web of *Cyclosa* sp. from Ghana. The spider is resting in the centre of the web, but there are two dummy spiders consisting of remains of prey also in the web. After a further week this particular individual had nine dummy spiders in a row vertically above the hub, all attached to the same radius, and spaced more or less evenly from hub to frame.

statistically significant aversive response to the artificial devices on first presentation — any novel stimulus associated with the cricket would be of some protective value for a few trials until the birds had become used to it. They concluded that the devices must have a mechanical function, but they could not produce any evidence to support this. Ewer (1972), on the other hand, considers that the very similar devices of *A. flavipalpis* are probably defensive because they provide conceal-ment, and the variety of different arrangements of the device reduces the chances of a predator being able to learn to associate the appea-rance of the device with presence of prey. The Robinsons' observation that birds were initially repelled by the artificial device supports this view. If correct, one would expect devices to be most successful if the spiders are relatively rare since then no predator is likely to find suffi-cient to learn to associate pattern with prey, but if the spiders are very common there should be no, or very few, webs with devices. There is some indication that this is what occurs. All of the *A. flavipalpis* studied by Ewer built devices and this spider is not normally a very abundant species. But only 35 per cent of the webs of *A. argentata* observed by the Robinsons had devices, and judging by the fact that they examined 2 500 webs the density of this species must have been very high. In Jamaica the same species builds devices more often (85 per cent with devices, $n = 244$) (Marples, 1969), which is difficult to explain if the device is supposed to have a mechanical function, but is readily explicable if it is related to density and predation pressure. Thus the evidence indicates that the device is of defensive importance, though it is possible that the proximal factor determining when and where it is placed may be something mechanical.

Another animal which modifies its environment to perfect its crypsis is the sandwich tern which nests in dense groups in the centre of colonies of gulls or of other sea birds. An unusual feature of its behaviour is that it scatters faeces all over its nesting area — the host gulls do not normally defaecate in the colony and it is normally con-sidered to be advantageous for birds to keep the nest site clean so as not to give a clue to predators that there is food nearby. However, the eggs and chicks of the sandwich tern are coloured so that they are better concealed on the faeces spattered environment than they are on clean pebbles. Since predators of the gull colony are likely to concentrate on gulls' eggs and young they may overlook the differently coloured eggs and young of the sandwich tern. This situation is more fully described on p. 248.

Finally some animals carry parts of their environment around with them and so become cryptic. Larvae of the reduviid bug *Acanthaspis* feed on ants or other insects and stick the empty skins of their prey or pieces of bark or stones on their backs. Caddis larvae build cases of stones or vegetation which are cryptic on the bed of a pond or stream. In the sea the gastropod *Xenophora* sticks pieces of shell onto its own shell and thus becomes cryptic on shell gravel. The crab *Dromia* carries a sponge over its carapace so that it looks like a sponge rather than a crab, and many spider crabs (masking crabs) place algae, hydroids, bryozoans or other debris on their backs. The debris is held in place by setae or legs and it may actually settle and grow on the exoskeleton. In all cases the crab is concealed by the mask of material so the habit is of obvious protective value.

4. Predators which hunt by searching image may quickly learn to find cryptic prey

The principal disadvantage of crypsis is that predators may either accidentally or through systematic search find a prey animal, develop a searching image for that particular species, and then search out the rest of the individuals in the environment. This is unlikely to happen if the prey species is rarely encountered as the predator will soon forget the searching image. So cryptic species must be spaced widely if they are to gain full protection from their primary defence.

There are many examples known of cryptic animals being spaced out: hawkmoths lay one or two eggs, then fly some distance before laying any more; young spiders and young peppered moth caterpillars scatter in the wind; camberwell beauty caterpillars (*Vanessa antiopa*) are gregarious and probably aposematic, but the pupae are cryptic and just before pupation they crawl off in different directions to pupate in isolation. Larvae of the mantids *Tarachodes afzellii*, *Oxypilus hamatus* and *Pseudocreobotra wahlbergi* have intraspecific displays which serve to space out individuals. The displays in the three species involve different signal marks so have presumably evolved independently: in *Pseudocreobotra* the most important feature is a 'target' mark dorsally on the abdomen, whilst in *Tarachodes* and *Oxypilus* it is colours on the forelegs but these are displayed in different ways in the two species (MacKinnon, 1970; Edmunds, unpublished). That predator selection is responsible for this spacing out is suggested by the fact that the fish *Vimba vimba* deposits spawn in compact masses if there are no predators, but scatters it when predators are present (Tinbergen *et al.*, 1967; Croze, 1970).

Proof that spacing out is of survival value was obtained by Tinbergen *et al.* (1967) and by Croze (1970). Tinbergen and his colleagues laid out painted chicken's eggs in 3 x 3 grids with eight of the nine eggs in each series partly hidden under soil and vegetation. In one series of three experiments they found that with an inter-egg distance of 50 cm, twenty-four out of twenty-seven eggs were taken by carrion crows (89 per cent) but with an inter-egg distance of 800 cm only five out of twenty-seven eggs were taken (19 per cent). Hence there is more intense predation on the more crowded population. Croze (1970) repeated the experiments using coloured flour and lard prey on 4 x 4 grids. He

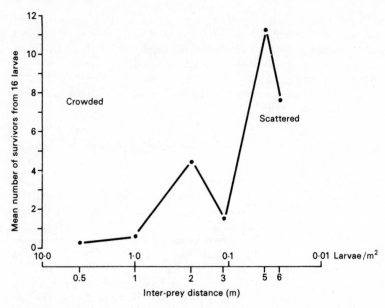

Fig. 2.29   Mean survival of flour and lard larvae in crowded and scattered populations when exposed to predation by crows. (*Redrawn from* Croze, 1970, Fig. 4.)

confirmed that the number of prey surviving increases as the density of the prey decreases (Fig. 2.29). He further showed that it takes less time to find prey in the crowded than in the scattered populations, so that in terms of reward to the predator it may not be profitable to persist in searching for prey which are at very low density. The predators in fact search over a very specific area close to where they found the last prey (area-associated search) rather than at random over a very wide area.

Further experiments, using painted mussel shells baited with meat, showed that crows can form a searching image very quickly and that it is specific for a particular shaped and coloured object. When the mussel shells are presented without a reward the crows continue to inspect them for several days, but the behaviour gradually wanes unless reinforced with an occasional reward.

If hunting by searching image occurs commonly it is likely that predator selection will normally set a limit to the density of a cryptic prey species. If the prey is too common, hunting by searching image will occur and predation will be very intense so that the proportion of prey surviving will be low. If the prey is less common, predators will not hunt by searching image (or if they do they will soon give up and the searching image will wane), predation will be very much less, and the proportion of prey surviving will be high.

One way in which a scattered cryptic species can become more common but without suffering increased predation is by occurring in

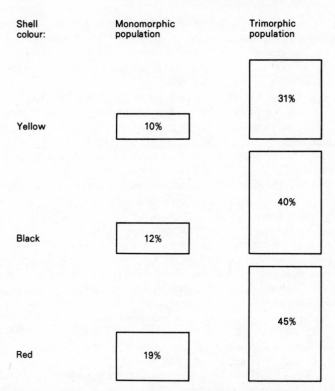

Fig. 2.30   Percentage survival of painted mussel shells in monomorphic and in trimorphic populations when exposed to predation by carrion crows. (*Data from* Croze, 1970.)

several different colour forms. This is polymorphism. Croze (1970) exposed sets of twenty-seven painted baited mussel shells to predation by carrion crows. Sometimes the shells were monomorphic (all the same colour), sometimes they were trimorphic (nine red, nine yellow and nine black). He found that crows took significantly more monomorphic prey than of polymorphic prey (Fig. 2.30). It was clear that the predators formed searching images for all three morphs, yet 39 per cent of all trimorphic prey survived whilst only 13 per cent of all monomorphic prey survived. Hence a polymorphic population suffers less predation than a monomorphic population of the same density, and one can regard polymorphism as a means of reducing the hunting success of the predators. Alternatively polymorphism can be regarded as a means of increasing the density of a population in a particular habitat without increasing the numbers actually killed by predators. If the two morphs each occupy different habitats in the same environment (see p. 23), then the overall density of the species can be higher for the same level of predation. It is hardly surprising then that many cryptic animals are polymorphic, and in species which occur at very high density, such as the sandy beach bivalve *Donax* (Moment, 1962) and the African moth *Achaea lienardi*, it is difficult to find two individuals of the same pattern. Other examples of polymorphism are discussed elsewhere in this chapter.

When one morph in a polymorphic population is much commoner than another morph it is likely that predators may build up a searching image for the commoner morph but not for the rarer morph. Hence more of the common morph should be preyed on relative to its frequency in the population than of the rarer morph. This is 'apostatic selection' (Clarke, 1962a) or 'reflexive selection' (Moment, 1962), and it is a form of frequency-dependent selection (Murray, 1972). Allen and Clarke (1968) exposed green and brown flour prey to predation by wild birds. In the first series of experiments the ratio of green to brown bait offered was 9:1, but of the bait eaten by birds significantly more than 90 per cent were green. In the second series of experiments the ratio of green to brown was 1:9 and of the bait taken by birds more than 90 per cent were brown. However, there was much variation in behaviour between individual birds as well as between different species. For example, on one day when the ratio of prey offered was nine green: one brown, on eight consecutive visits individual blackbirds took a total of thirty-eight green and no brown prey. On the ninth visit a blackbird took twelve brown and no green; on the next four visits birds took a total of thirty-three green and no brown; and finally a bird took twelve brown. Thus individual birds were concentrating on prey of a specific colour and ignoring other prey of different appearance. Most birds concentrated on the green prey because it was the most frequently encountered morph, but one or a few birds had a searching image for brown prey and completely ignored the more numerous greens.

These experiments show that, assuming two morphs are equally cryptic, a rare morph will have a selective advantage relative to a common morph when the population is exposed to predators which hunt by searching image. The African banded snail *Limicolaria martensiana* appears to be a case in point.

The shell of *Limicolaria* is typically streaked, but three pallid forms also occur in some populations. Owen (1965) showed that the degree of polymorphism in Uganda is correlated with density (Table 2.3). Owen suggested that in a dense population predaceous birds build up a searching image for the commonest form present in the population so that a rarer morph is less heavily preyed on even if it is slightly more conspicuous than the common streaked form. In very dense populations birds may build up searching images for two or three morphs, so selection will favour a greater degree of polymorphism than in sparse populations. Conversely in sparse populations the proportion of snails preyed on is less, predators will not develop a searching image, and the most cryptic morph will be at a selective advantage to the less cryptic

Table 2.3 Percentage frequencies of different forms of *Limicolaria martensiana* in four populations (A–D) in Uganda. (*Data from Owen, 1965.*)

Population	streaked	% Occurrence in population pallid A	pallid B	pallid C	Density of population (snails/m²)
A	55	21	20	4	>100
B	78	11	11	0	8
C	93	7	0	0	5
D	100	0	0	0	<1

morphs. Support for this hypothesis came when one habitat under study was changed (Owen, 1969). The shrubs and tall grasses were cleared and the habitat became more open. Before clearance, in 1963, the four morphs had frequencies of 68, 15, 13 and 4 per cent, and the density was $26/m^2$. But after clearance in 1968 the density had fallen to $2/m^2$, and the morph frequencies were 79, 6, 14 and 1 per cent. In this case the principal predators were open-billed storks (*Anastomus lamelligerus*) which appear to have been acting as apostatic (frequency-dependent) predators.

Crypsis is normally considered to imply that an animal harmonizes in colour with its background so that it is not easily found by predators which hunt by sight. Some predators, however, detect prey with the aid of other senses. The prey of predators which hunt by smell can presumably be cryptic and well protected if they are odourless or have the same smell as their immediate environment such as the plant on which they live. Just as visually hunting predators may hunt common prey by searching image, so predators that hunt by smell may also form searching 'images' for a particular smell and ignore prey with a different smell. Soane and Clarke (1973) trained house mice for three days to eat bait scented with either peppermint or vanilla, and then offered them a choice between peppermint and vanilla bait of identical colour. Fifteen mice conditioned on peppermint took significantly more peppermint than vanilla bait (eighty-eight peppermint to fifty-seven vanilla, $\chi^2_{(1)} = 6.63$, $p < 0.02$) whereas fifteen mice conditioned on vanilla took significantly more vanilla (thirty-five peppermint to ninety-seven vanilla, $\chi^2_{(1)} = 29.12$, $p < 0.001$). The mice were therefore taking more prey with a familiar smell than prey with a novel smell. If prey whose predators hunt by smell are common, then selection should favour the evolution of scent polymorphisms, just as it favours colour polymorphisms in species whose predators hunt by sight.

In the deep sea there are some fish with highly developed lateral-line systems (Marshall, 1954). Lateral-line systems can detect water displaced by other objects near the fish (Dijkgraaf, 1963), and some deep-sea fish probably detect prey in this way. It is possible that the elongated body shape of many deep-sea fish is an adaptation which renders them less easily detected by the lateral-line organs of nearby predators. This must remain speculation until we know more about the functioning of lateral-line organs, but it seems reasonable to suppose that deep-sea fish may hunt by searching image of particular patterns of water movement.

Some cryptic animals that have been intensively studied

We have seen that a great many animals are camouflaged, and many of these are polymorphic for colour. In a few cases the selective advantage of camouflage has been demonstrated. Ecological and evolutionary studies have been carried out on natural populations of a number of cryptic animals and it has been found that the factors affecting each population and each species are very complex: predator selection for the most efficiently camouflaged individuals is only one of many factors involved in the evolution of a population. In this section I shall describe a few cases where detailed ecological studies have been carried out on polymorphic, cryptic animals.

The peppered moth *Biston betularia*

There are two common colour morphs of the peppered moth in Europe: the typical form which is mottled grey and white and is cryptic when resting on lichen covered tree trunks, and the black melanic form which is cryptic when resting on soot blackened tree trunks such as occur in industrial regions. Before 1850 less than 1 per cent of peppered moths in Britain were of the melanic form (*carbonaria*). During the nineteenth century, however, there was a tremendous growth of industry, particularly in the north and midland regions of Britain, and the soot and fumes from the factories killed off many of the lichens on trees and replaced them with a black sooty deposit. Under these conditions the melanic moths were the better camouflaged form and they increased in frequency compared with the typicals. By 1895 95 per cent of the moths at Manchester in North-West England were melanic, and today the figure is about the same. In South-West England, however, the typical insects still form 90–100 per cent of the population. The melanic form, *carbonaria*, differs from the typical in a single gene which is dominant. There is also another partially melanic form, *insularia*, which is controlled by another allele at the same locus, dominant to the typical but recessive to the *carbonaria* allele (information about the peppered moth is summarized by Ford, 1964). Kettlewell (1956) was able to demonstrate that the high proportion of melanics in industrial areas is largely the result of selective predation of the more conspicuous typical morph by birds. For his study he chose an industrial wood near Birmingham in which the frequencies of the three forms were 85 per cent *carbonaria*, 4.8 per cent *insularia* and 10.1 per cent typical, and an unpolluted wood in Dorset where there were no *carbonaria*, 5.4 per cent *insularia* and 94.6 per cent typicals. He released large numbers of marked typicals and *carbonaria* in the two areas and attempted to recapture them a few days later at mercury vapour traps. The results, given in Fig. 2.31, demonstrate that the typicals survived better in Dorset whilst the melanics survived better in Birmingham. Kettlewell also released equal numbers of insects of both morphs onto trees and watched them during the course of the day. In Dorset he observed twenty-six typical and 164 *carbonaria* eaten by a variety of insectivorous birds whilst in Birmingham forty-three typical

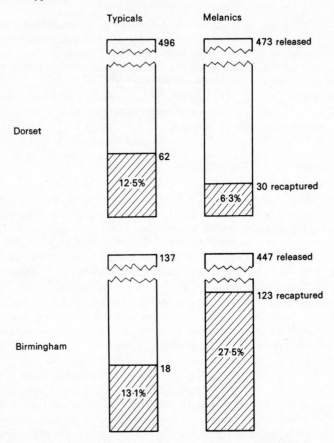

Fig. 2.31   Percentages of typical and melanic peppered moths (*Biston betularia*) recaptured in Dorset and in Birmingham. (*Data from* Kettlewell, 1956.)

and only fifteen *carbonaria* were eaten by birds. Clearly the birds are the selective agents, and there is very strong selective pressure to eliminate the non-cryptic form. More recent experiments in the Liverpool area have also demonstrated selective elimination of the more conspicuous morph (Bishop, 1972).

The situation is actually more complex than this since even in the most highly polluted areas, the melanics have never completely displaced the typical insects, and the population remains in balance with 90 or 95 per cent *carbonaria* and the remainder typicals or *insularia*. There is some evidence that the heterozygote has a slight selective advantage over the homozygous *carbonaria*, but the reason for this is unknown (Clarke and Sheppard, 1966). There is also evidence for different habits and behaviour of the caterpillars of the two morphs, and different emergence times for the adults, but the effect of these

factors on the frequencies of the melanics and the typicals is not known
(Ford, 1964). Finally, although the melanics differ from the typicals in
a single allele, there is evidence that local populations have evolved
different gene complexes. Melanic insects captured today are much
blacker than those which were first found over 100 years ago,
indicating that selection for modifiers which produce blacker insects
has occurred. If *carbonaria* from an industrial population are crossed
with typicals from Cornwall (an area where melanics are completely
absent apart from occasional mutations), and if the progeny are allowed
to mate amongst themselves, in the F_2 generation there is great variation in the expression of the *carbonaria* gene. In other words the
melanic insects come from a population in which there are many genes
present which accentuate the blackness produced by the *carbonaria*
allele, but the Cornish insects come from a population in which these
particular genes are absent since selection has never favoured their
retention. On crossing insects from the two areas, the locally adapted
gene complexes of the parents are mixed up, and the progeny in the F_2
generation are therefore very variable. It is probable that it is largely
predator selection that has been responsible for intensifying the blackness of the melanic insects over the past 100 years.

The peppered moth is not the only species which has adapted to
industrial conditions in this way. Melanic forms of more than seventy
species of moth are now known in Britain, and these are mostly
commoner in industrial regions than elsewhere. Industrial melanism is
also known from the industrial regions of Europe and North America.
In all of these moths the melanics have become common because of the
protective value of their coloration in a smoke-polluted environment.
There are, however, a few species of moths with melanic forms which
occur in non-industrial regions. In some of these the black forms may
also be cryptic in certain situations (see Section 4), but it is possible
that the black colour may have some other function not related to
predation. For example, populations of *Colias* butterflies from cold
climates contain a higher frequency of dark insects (partial melanics)
than do populations from warmer places. Watt (1968) showed that dark
Colias warm up quicker than do pale insects and that therefore they are
more active in cold climates. Conversely it is possible that in hot
climates they are more likely to suffer from heat stress than are pale
insects. Similar considerations may apply to some of the non-industrial
melanic moths which are active by day, but none has so far been
studied with this in mind.

The banded snail *Cepaea nemoralis*

Cepaea nemoralis is a snail that is widely distributed on basic soils
throughout western Europe. The ground colour of the shell may be
yellow, pink or brown, and there may be from zero (unbanded) to five

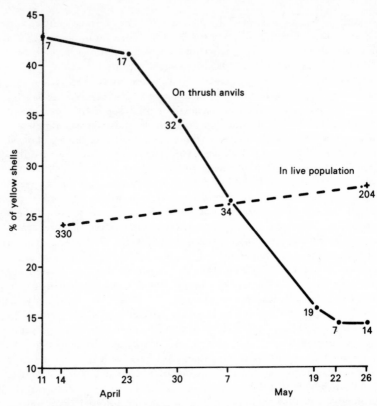

Fig. 2.32   Frequency of yellows in samples of *Cepaea nemoralis* collected at thrush anvils (solid line) and from the population of live snails in the same area (broken line). The small figures refer to the sizes of the samples. (*Data from* Sheppard, 1951.)

bands running round the whorls of the shell, either distinct from one another or with various degrees of fusion. The genetics of this system have been worked out by Cain and Sheppard and their colleagues and are summarized by Ford (1964). Cain and Sheppard found that in natural populations of *Cepaea* the frequencies of the various colour morphs varied in different habitats, and it seemed likely that this might be due to visual predation. In some areas the song thrush is an important predator, particularly in dry weather when other food is not available. Since this bird breaks snail shells on specific stones or anvils, it is easy to determine the proportions of the various morphs preyed on by thrushes by means of regular collection of broken shells from these anvils. In one woodland area in April, Sheppard (1951) found that over 40 per cent of shells found at thrush anvils were yellow, but a random collection showed that yellows formed only 24 per cent of the snails in the area (Fig. 2.32). In April the ground is largely brown with no vegetation, so the yellow shelled snails are the most conspicuous snails at this time, and the thrushes were therefore selectively preying more on yellows than on pinks or browns. By late May, however, the herbs in the wood had grown so that the background was predominantly green, and the yellows were the best camouflaged form (the soft parts of the snail are greenish grey so that the appearance of the live animal in its yellow shell is yellowish green). At this time of the year the proportion of yellows at the thrush anvils was only 15 per cent. So the direction in which selection (in the form of the song thrush) acts varies during the course of the year: in early spring the browns and pinks are favoured, but in summer the yellows are at a selective advantage. Hence the proportion of yellows surviving in the population increases from spring to summer (Fig. 2.32).

The situation is actually more complex than this since Allen has shown that thrushes can form a searching image for particular forms, and Arnold has evidence suggesting that in some localities thrushes selectively prey on the commonest morph present (apostatic selection). In some places there are physiological differences between the various colour morphs that are more important than predation in determining their frequencies. In the Pyrenees, for example, there is little visual predation, but Arnold (1969) was able to show that on hillsides, where vegetation is sparse and there is much exposure to sunlight, there is a higher frequency of unbanded snails than in the more wooded areas of the valleys. Lamotte had earlier shown in the laboratory that unbanded snails are more resistant to heat than are banded snails, so this explains why, in the Pyrenees, they are more favoured on hot dry hillsides than in the valleys. There is also evidence from subfossil *Cepaea* in southern England that 6 000–7 000 years ago, when the climate was warmer and drier than it is now, the frequencies of some of the morphs were significantly different to those found in the same areas today (Cain, 1971). This emphasizes that the protective value of coloration is not necessarily related to the impact it has on visual predators.

There are also behavioural differences associated with different morphs of *Cepaea*: unbanded snails migrate more than do banded snails; yellow snails rest in trees less frequently than do non-yellow snails; and yellow unbanded snails produce more clutches of eggs (at 20°C) than do yellow banded, pink banded, or pink unbanded snails (Wolda, 1963, 1965, 1967). There is also evidence that in some areas another predator, the glow-worm (*Lampyris noctiluca*) exerts selective predation on *Cepaea* (O'Donald, 1968). When given a choice between brown and yellow banded snails, these beetles showed a significant preference for attacking the browns (Fig. 2.33). In one population in Anglesey (Wales) there was a cline in the frequency of browns from 10 per cent in the north to 40 per cent of the population 80 m further south. Glow-worms were very common in the north but absent in the south, suggesting that they were responsible for the low frequency of

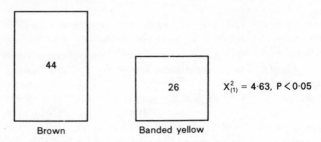

Fig. 2.33   Frequency of attacks by glow-worms on brown and banded yellow *Cepaea nemoralis* when given a choice between the two morphs. (*Data from* O'Donald, 1968.)

browns in the north. This predator is certainly not a visual hunter, so clearly colour of the snail is correlated with other physiological characteristics which are relevant to the beetle. The interaction of all these factors in determining the frequencies of the different morphs have not been worked out for any one population, but it is clear that camouflage is an important factor in those populations which are exposed to visual predation.

A further complicating factor is the presence of a second species of *Cepaea* in some populations — in Britain this second species is *C. hortensis*. The frequencies of the various morphs of *Cepaea hortensis* vary in different habitats, just as do those of *C. nemoralis*, and in some areas thrushes also prey on this species of snail. However, in any one habitat the morph frequencies of *C. hortensis* are not usually the same as the frequencies of the corresponding morphs of *C. nemoralis* (Clarke, 1960, 1962b; and see also Carter, 1967, and Clarke, 1969). For example, in beechwoods, populations of both species are composed very largely of brown shelled snails which are the most cryptic colour on the bare earth and leaf litter with sparse vegetation characteristic of this habitat. But whereas in *C. nemoralis* the shells are brown without bands, in *C. hortensis* they are usually yellow with brown bands (Fig. 2.34). Furthermore, the two bands on the upper surface of the shell in *C. hortensis* are usually fused so that the general appearance of the shell when viewed from above is brown. Closely related species in the same area are liable to suffer very heavy predation due to predators incorporating both species in a single searching image; but if the two species are different in appearance where they are sympatric the predators will have to learn a searching image for each species separately so that the intensity of predation will be reduced (see also p. 43 and Fig. 2.30). Hence selection favours diversity of appearance ('aspect diversity') in closely related sympatric species (Clarke, 1962a; Rand, 1967). An example of aspect diversity is the five or six species of caterpillars commonly found on pine trees in North-West Europe, all cryptically coloured amongst the fine needle-like leaves, but all characteristically different in appearance from one another (Herrebout *et al.*, 1963;

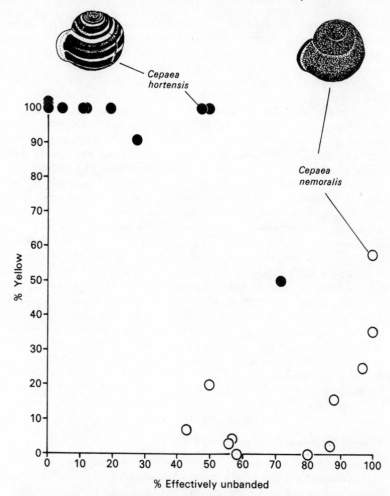

Fig. 2.34 Scatter diagram of the percentage of yellows against the percentage of effectively unbandeds in the snail population of beechwoods in southern England. Each population comprises both *Cepaea nemoralis* (open circles) and *Cepaea hortensis* (closed circles). Note how populations of both species consist largely of brown snails, but whilst this colour is achieved by brown pigment on unbanded shells in *C. nemoralis*, in *C. hortensis* the shells are mostly yellow and banded. (*Data from* Clarke, 1962b.)

Rand, 1967). Selection for aspect diversity may similarly account for the different ways in which the two species of *Cepaea* adapt to beechwoods.

The limpet *Acmaea digitalis*

Acmaea digitalis is a Pacific intertidal limpet that is found both on bare rock faces and on colonies of the goose-neck barnacle *Pollicipes polymerus*. On the barnacles the most cryptic limpets are pale with faint stripes, but on the rocks the most cryptic pattern is dark with prominent stripes. Giesel (1970) found that young larvae (veligers) settle on the rock and as they grow some migrate towards the barnacles. The paler limpets migrate faster than the dark ones and so reach the barnacles first, but Giesel found young limpets (3–3.9 mm length) of all colours both on the rocks and on the barnacles. However, old limpets (6–6.9 mm) are clearly separated by colour and habitat — pale ones are almost entirely confined to the barnacles and dark ones to the rocks. It is known that oystercatchers and other birds prey on limpets, and Giesel concludes that birds are responsible for this disruptive selection on the limpet populations in the two habitats. Since there is a tremendous mortality of inappropriately coloured limpets, selection will favour any behavioural mechanism which reduces this predation. When limpets are removed from their 'home' they soon find their way back to it. Pale limpets 'home' much more quickly than do dark ones, and limpets from exposed rocks, whatever their colour, home more quickly than do limpets from sheltered rocks. It is probable that in exposed areas, where there is very heavy predation, selection favours those limpets which return the quickest to their home, since they are less easy to prise off from their home than from other rock surfaces. But in sheltered areas predation is less intense, and selection for speed of homing is less important than selection for colour. The frequency of the different colour and behavioural morphs in the population is obviously determined by the nature of the habitat and the intensity of visual predation.

The citrus swallowtail *Papilio demodocus*

The young caterpillars of *Papilio demodocus* sit on the leaves of their food plant and are black, yellow and white with a close resemblance to bird droppings. The final instar is probably too large to be efficiently protected as a bird-dropping mimic and it is bright green with disruptive brown and white marks, and rests on leaves, on green stems and petioles (Plate 2*d*). The caterpillars are also protected by glandular secretions from a pair of orange processes just behind the head which can be extruded, but it is not clear if these are directed against birds or against invertebrate predators such as ants (see p. 244). Throughout most of tropical Africa the caterpillars feed on *Citrus* and on other Rutaceae, but in South Africa they also feed on umbelliferous plants. In South Africa, but not elsewhere, there is an alternative morph of the final instar caterpillar which has a colour pattern of black and green that is apparently more cryptic on the finely divided leaves of fennel and other umbelliferous plants than on the smooth shiny leaves of Rutaceae. Clarke, Dickson and Sheppard (1963) were able to show, first, that there is predation by birds on caterpillars; and second, that there is selective elimination of inappropriate larvae on the two types of host plant. They collected caterpillars at random from umbellifers and from *Citrus* trees. Young instars (bird-dropping mimics) were reared in the laboratory so that it was possible to determine which larval type they belonged to. It was found that there is a significantly higher proportion of umbellifer-type caterpillars on the fennel than on *Citrus*, but there were also several caterpillars intermediate in appearance between the two types. It was also found that the proportions of the two types of caterpillars on *Citrus* differed for early and last instars (Table 2.4). The figures strongly suggest that there was selective predation eliminating the intermediate and umbellifer-type caterpillars when on the wrong host plant so that only the citrus-type caterpillars survived.

Table 2.4   Numbers of citrus-type, intermediate, and umbellifer-type caterpillars of *Papilio demodocus* on *Citrus* plants in South Africa. (*Data from* Clarke, Dickson and Sheppard, 1963)

Age of caterpillar	Type of caterpillar		
	citrus	*intermediate*	*umbellifer*
Early instars	9	10	13
Last instar	5	0	0

Hawkmoth caterpillars

Caterpillars of many species of hawkmoths are polymorphic for colour. Usually there are green and brown morphs (for example in the deaths-head hawkmoth *Acherontia atropos*, and in *Coelonia mauritii*), but very few species have been studied beyond simple description of the colour morphs which exist. In the convolvulus hawkmoth (*Herse convolvuli*) there are several different colour morphs (Poulton, 1888; Bell and Scott, 1937; Pinhey, 1960). At Legon in Ghana one of the food plants of this insect is *Ipomaea aquatica*. The third and fourth instar larvae are either green with faint diagonal stripes or yellow with longitudinal purple stripes (Plate 1*b*, *c*). The green caterpillars rest on the under sides of green leaves whilst the striped caterpillars rest on the prostrate stems of the plant. Since the stems are yellowish green tinged with purple, each morph is cryptic when in its chosen resting place. The final instar caterpillars are all brown (Plate 1*d*) and hence are not cryptic when feeding on the plant. However, about forty-eight hours after moulting to the final instar, the caterpillars burrow into the earth during the day and only emerge to feed at night. On the earth of course they are cryptic. Hence when feeding on *Ipomaea aquatica* each morph of this species is well adapted in colour and behaviour so as to be cryptic.

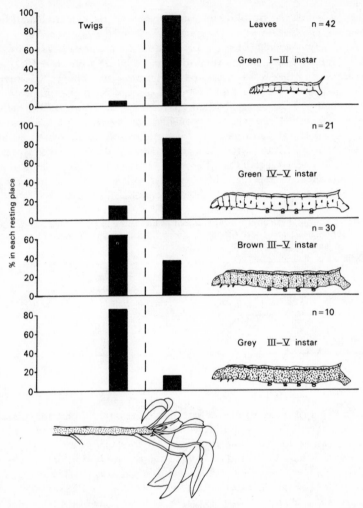

Fig. 2.35   Resting positions of green, brown and grey caterpillars of the hawk-moth *Errinyis ello* on *Poinsettia pulcherrima*. (*Redrawn with modifications from* Curio, 1965, Fig. 2.)

Similar studies on populations on other host plants would be of very great interest, particularly if quantitative.

The New World hawkmoth *Errinyis ello* has been studied in rather more detail. In the Galapagos Islands the young caterpillars are green and rest upside-down on green leaves. The last three instar caterpillars can be green, brown or grey, and whilst the green ones continue to rest on the green leaves, the brown and grey ones prefer to rest, dorsal side uppermost, on brown twigs (Fig. 2.35) (Curio, 1965). In Jamaica there are four morphs of the last instar of this species: green, blue, green-grey, and brown (Curio, 1970*a*, *b*, *c*). The first three of these feed and rest during the day on the leaves of *Poinsettia pulcherrima* where they are cryptic. The brown caterpillars rest during the day on the trunk where they too are cryptic, and they only come up to the leaves to feed at night. The principal predator is the wasp *Polistes crinitus* which hunts only amongst the foliage. Curio released 916 caterpillars (including some of each colour morph) on the foliage, and 131 on the trunks. From an inspection several hours later he was able to calculate that on the foliage the caterpillars suffered a mortality of 5.5 per cent per hour whilst on the trunk the mortality was only 0.7 per cent per hour. It appears that there is a tremendous selective advantage of the order of 88 per cent to the caterpillars which rest on the trunk. This advantage is not related to colour but to the trunk resting position since the wasps do not recognize prey by sight. Why then are the caterpillars poly-morphic for colour? Part of the answer may be because another predator, the lizard *Anolis lineatopus*, searches only on the trunk and main branches, but not on the leaves. *Anolis lineatopus* also appears unable to swallow large brown caterpillars, although it will eat equally large larvae of other colours as well as small brown caterpillars. It is likely too that caterpillars grow slower on the trunk than on the leaves since they cannot feed during the day-time, and hence they will be exposed to predation for a longer period of time than the quicker grow-ing foliage-resting caterpillars. Finally, there may be other predators (such as birds) which normally prey apostatically on the caterpillars, although possibly at the University campus in Jamaica, where Curio worked, these were absent. It is clear that the polymorphic caterpillars of hawkmoths require much more detailed study before we can claim to understand the selective factors favouring the different morphs in any one locality.

Pine looper caterpillars (*Bupalus piniarus*)

Caterpillars of the pine looper or bordered white moth are dimorphic in colour: the normal green larvae are homozygous whilst the rare yellow ones are heterozygous and comprise from 0.06–2.5 per cent of the population. (The rare homozygous blue larvae have never been seen in the wild.) Boer (1971) could find no differences between adult moths from green and from yellow larvae with respect to mating preferences, copulatory success, fecundity, egg hatching and growth rate of larvae, but he found that various predators take one or other morph of the caterpillars selectively. Great tits and coal tits show a significant prefer-ence for yellows when offered yellow and green caterpillars on pine needles, either in equal numbers or in the ratio one yellow: nineteen green (Fig. 2.36). Hence yellows are likely to be at a strong selective disadvantage relative to greens with respect to these birds presumably

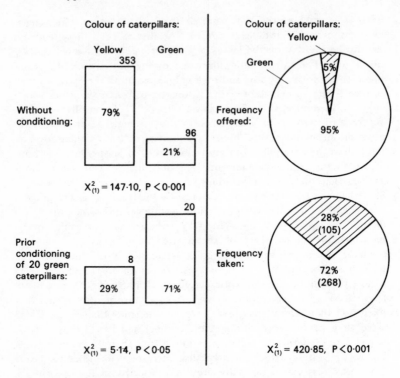

Fig. 2.36   Number of attacks by great tits and coal tits (*Parus major* and *P. ater*) on green and yellow pine looper caterpillars presented on green pine needles. The χ^2 values are calculated assuming that yellow and green should be taken in the same proportions that they were offered. Left: green and yellow caterpillars presented in equal numbers with no conditioning (above) and with prior conditioning of twenty green caterpillars (below). Right: green and yellow caterpillars presented in the ratio 19:1. (*Data from* Boer, 1971.)

because the greens are better camouflaged. However, when the tits were given twenty green caterpillars first, and then offered a choice between a yellow and a green, they took the green significantly more often than the yellow (Fig. 2.36). Presumably if the caterpillars are abundant in nature the birds can build up a searching image for the commoner green morph so that the rarer yellows are protected and at a selective advantage to the greens.

The crab spider *Xysticus* sp. is also a predator on caterpillars but it only takes small ones (principally the first three instars). All first instar caterpillars are of the same colour and it is only possible to determine if they are going to become green or yellow by rearing them. Nevertheless, when offered a choice between two first instar larvae, one potentially green and the other potentially yellow, *Xysticus* took the potentially yellow ones significantly more often than the green

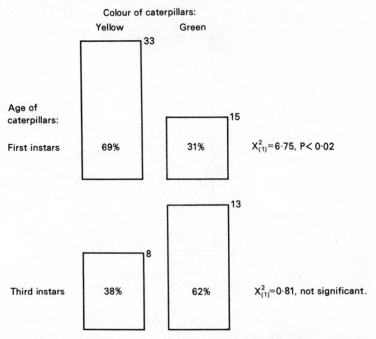

Fig. 2.37 Number of attacks by crab spiders (*Xysticus* sp.) when given a choice of green or yellow pine looper caterpillars. Above: first instar caterpillars; below: third instar caterpillars. Note that the colour of the caterpillars is not developed until the third instar so that the first instars are all of similar colour though different behaviour. (*Data from* Boer, 1971.)

(Fig. 2.37). By the third instar the caterpillars have developed their green or yellow colour, yet when these are offered to *Xysticus*, the spiders show no preference for either morph (Fig. 2.37). Boer found that the explanation for this anomaly is that the genetically yellow first instar caterpillars are much more active than are the greens, and that since the spiders attack moving objects they are therefore more likely to attack yellow than green caterpillars. By the third instar, however, there is no difference in the behaviour of yellow and green caterpillars, and hence they are likely to be attacked by *Xysticus* with equal frequency.

A third predator of *Bupalus* caterpillars is the wasp *Vespula vulgaris* which shows a significant preference for attacking green larvae rather than yellow ones (Fig. 2.38). By smearing twigs with haemolymph from green or from yellow caterpillars Boer showed that significantly more wasps landed on the twig smeared with green caterpillar haemolymph. Thus it is probable that the wasp is hunting by smell, not by sight. Hence in this species several conflicting selection pressures are operating: the colour of the caterpillars is important in protecting them

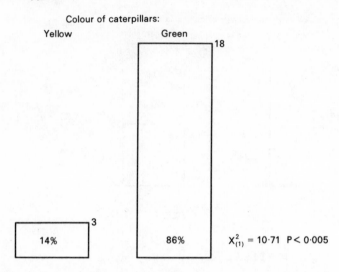

Colour of caterpillars:

Yellow   Green

18

3

14%   86% $x^2_{(1)} = 10 \cdot 71$ P < 0·005

Fig. 2.38   Number of attacks by the wasp *Vespula vulgaris* when offered a choice of a green or a yellow pine looper caterpillar. (*Data from* Boer, 1971.)

from predation by birds, but their behaviour and smell are also important with respect to other predators.

Summary

Crypsis is the resemblance of an animal to a part of its environment such that predators fail to distinguish it from the environment. We know most about colour crypsis because it is easy to study the responses of visually hunting predators to different coloured prey, but presumably animals can be cryptic with respect to predators hunting with the aid of other senses such as smell, hearing, pressure and electrical reception. Experiments with dimorphic prey have shown that predators take conspicuous prey more often than camouflaged prey, and predator selection has probably been responsible for improving the camouflage of many cryptic species. Many cryptic animals are counter shaded or have disruptive coloration, and some deep-sea fish are camouflaged from whichever direction they are viewed. Cryptic animals adopt characteristic resting postures which render then inconspicuous on their normal or chosen backgrounds, but a few species can change colour so that they can be cryptic on more than one type of background, whilst others modify their environment to conceal themselves more effectively. The principal disadvantages of crypsis are that it conflicts with other essential activities such as feeding and mating, and that predators can learn the signals produced by a cryptic animal and then quickly search out further animals with similar signals. Hunting by searching image occurs both with visually hunting predators searching for animals of particular colours or patterns, and also with predators hunting by smell for prey with particular scent characteristics. The best defence against predators hunting by the searching image method is for a species to be polymorphic.

Chapter 3
Aposematism

Animals which have dangerous or unpleasant attributes, and which advertise this fact by means of characteristic structures, colours, or other signals so that some predators avoid attacking them, are said to be *aposematic*, and the phenomenon is called *aposematism*.

For aposematism to be advantageous, *either* the predator must sample some of the prey, find them unpleasant, and learn from this to avoid animals of similar appearance in future; *or* the predator must have an innate avoidance response to the aposematic signals, which presumably would only result from a long phylogenetic process of selection for such built in recognition responses. In the former case, the predator must be capable of learning to associate colour with inedibility, as is known to occur with many vertebrate predators. In each generation of predators a few prey individuals must be sacrificed in order that the new predators can learn the aposematic patterns. It is also probable that in long-lived predators avoidance behaviour may require occasional reinforcement, so that an aposematic prey species which has been experienced in the past may be sampled again at intervals.

Aposematic animals all have secondary defence mechanisms in the form of a nasty taste, a sting, or some other noxious quality. They also have a conspicuous attribute to advertise themselves such as bright colours, contrasting pattern, strong smell or characteristic noise. The bright black and yellow wasps (*Vespa* and *Vespula* spp.) of Europe, which have painful stings, are probably the best known aposematic animals (Fig. 4.1). Other aposematic insects are shown in Plates 2, 4 and 5. Of course not all conspicuous colours in animals are warning colours: some may be cryptic on brightly coloured objects (e.g. the red eolid mollusc *Catriona aurantia* on the red hydroid *Tubularia* (Edmunds, 1966*a*)); some may be mimetic (e.g. the palatable black and yellow banded hover-flies (Fig. 4.1)); and some animals use bright colours as signals for conspecifics (e.g. in territorial and courtship displays of birds such as the robin).

Aposematic animals also usually have behavioural characteristics which make them conspicuous. Many move slowly and are comparatively easy for a predator to catch. This is important for two reasons: (1) because the predator has to catch them in order to learn — if it regularly sees the prey the message will be frequently noticed and this may assist in retention of the learned avoidance; and (2) because it is likely that slow movement does not activate prey-catching behaviour as readily as fast movement. An unpleasant slow-moving animal that a predator has already sampled is less likely to be attacked again than is a fast-moving animal. Aposematic animals are also often tough so that they can sometimes survive being caught and being tasted by a predator — provided the bite is not on some vital part of the body. Finally aposematic animals are not normally polymorphic since predators would have to kill twice as many prey individuals of a dimorphic species as they would of a monomorphic species in order to learn its colour patterns. Exceptions to some of these generalizations are discussed later.

Survival value of aposematism

In an experiment to show that brightly coloured insects are usually not eaten, Carpenter (1921) offered a large number of insects to monkeys. He found that of 220 which had bright colours only 20 per cent were eaten, but of 155 apparently cryptic insects 73 per cent were eaten. There is much further evidence indicating that brightly coloured insects are rejected as food by insectivorous predators. Thus the red and black ladybird beetle *Adalia bipunctata* is not eaten by a variety of birds and small mammals (summarized by Creed, 1966), and the black and yellow caterpillars of the cinnabar moth *Callimorpha jacobaeae* are also avoided by various predators in addition to which they are gregarious (summarized by Dempster, 1971, and Ford, 1955). The advantage of gregariousness is that if a predator tastes one caterpillar the visual impact of many others may speed up the process of learning to associate colour with inedibility.

There is, however, only one experimental demonstration in the wild of the advantage to a distasteful animal of being aposematically coloured. *Heliconius erato* is an unpalatable butterfly from South and Central America which roosts in large colonies that can be regularly and easily sampled. The Costa Rican race of this insect has black forewings with a bright red diagonal band. Benson (1972) painted the red bands of a number of insects black so that the entire forewing was black. These experimental insects were much less conspicuous to man, and so presumably to birds, than were the typical red banded insects. As controls (to check against the possible effect of paint on the insects' behaviour) he painted the black tip of the wings of some typical insects with black paint so that they had similar paint to the experimentals but similar appearance to normal insects. In 1968 he found that seven experimental insects survived in a wild population from 10½ to 52½ days (mean 31.7) whilst nine control insects survived between twenty-one and seventy-one days (mean 52.4). This difference was significant

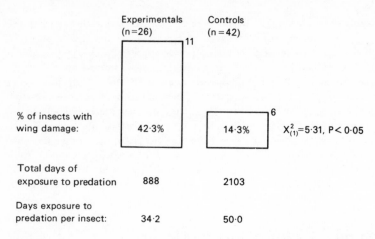

Fig. 3.1   Wing damage to experimental and control aposematic butterflies (*Heliconius erato*) in Costa Rica. Control insects have the forewing red and black (insect *c* in Fig. 3.4); experimentals have the forewing entirely black. (*Data from* Benson, 1972.)

(Mann-Whitney U test : $U = 13$; $p\ (U \leqslant 13) \approx 0.034$, one tailed test) showing that the experimental insects survived significantly less well than the controls. In 1969 the experiment was repeated, but in that year no significant difference was detected between the survival of the experimental and the control insects.

Benson also examined recaptured insects for evidence of wing damage indicating that they had been seized by birds but had then escaped. There were significantly more damaged wings amongst the experimental than amongst the control insects (Fig. 3.1), and this was even more striking if allowance is made for the longer period of exposure to attack by birds of the control butterflies. Hence in *H. erato* selection favours uniformity of appearance and tends to eliminate a rare morph of different appearance to the normal, even though this morph is actually more cryptic than the typical morph. Presumably for such unpalatable species as *H. erato* the advantages of being conspicuous outweigh the advantages of being cryptic, but this may not be so for all species of unpalatable animals.

Responses of predators to aposematic prey

If aposematism is to be of advantage to a prey species then it is important that predators should be able to learn to avoid it after only a few trials, or that they should have evolved an innate inhibition against attacking it.

Conditioned avoidance responses

There is now considerable evidence that vertebrate predators can learn to associate unpalatability with colour and to avoid prey with similar colours in future encounters. Windecker (1939) found that young birds will attack the yellow and black caterpillars of the cinnabar moth, take them into the beak, and then reject them. This was followed by vigorous wiping of the bill. He showed that the insides and skin of the caterpillars are both palatable, but it is the hairs that cause birds to reject them (shaved caterpillars were eaten). Nevertheless in rejecting further caterpillars the birds responded to the colour, not the hairs, since they refused to eat other insects with a black and yellow pattern as well as cinnabar caterpillars. Brower (1958a, b, c) has shown that Florida scrub jays quickly learn after one or a few presentations to avoid attacking the butterflies *Danaus plexippus, D. gilippus* and *Battus philenor*. Toads will attack any suitable sized prey but soon learn to avoid bees and bumblebees (Eibl-Eibesfeldt, 1952; Brower and Brower, 1962a; Brower, Brower and Westcott, 1960). In all of these cases the predator would still take other, palatable prey, so the inhibition against attacking was specific to prey of one particular colour pattern. Other examples of terrestrial vertebrates learning to avoid noxious prey are given elsewhere in this book, particularly in Chapter 4 on batesian mimicry.

Another important group of predators are the marine cephalopods. These have highly developed brain and eyes, and they are capable of learning (Wells, 1962). The common octopus (*Octopus vulgaris*) quickly learns to avoid hermit crabs which have the anemone *Calliactis* on the shell, but it continues to attack hermit crabs without an anemone (Boycott and Young, 1950; Ross, 1971). Unfortunately it is not known whether the octopus recognizes the anemone by shape, colour, or some other visual characteristic, but it is clear that aposematism can be an effective defence against cephalopod predators.

In most species of animal it is probably necessary for each individual to sample an aposematic animal in order to learn to avoid it. However, it is possible that in some species animals may observe one individual sampling and rejecting an aposematic animal and then themselves learn to avoid it in future. This is 'empathic' or observational learning (Klopfer, 1959; Gans, 1964). Fork-tailed flycatchers (*Muscivora tyrannus*) quickly learn to avoid the aposematic butterfly *Heliconius erato* and they then also avoid the palatable but similarly coloured *Anartia amalthea*. Untrained birds eat *Anartia*. However, if a bird trained to avoid *Heliconius* and *Anartia* sees an untrained bird eating an *Anartia*, then it may itself also sample *Anartia* (Alcock, 1969). Rats can also learn by imitating other rats (Zentall and Levine, 1972). It is not known, however, if observational learning of either palatable or of unpalatable food occurs at all commonly in nature.

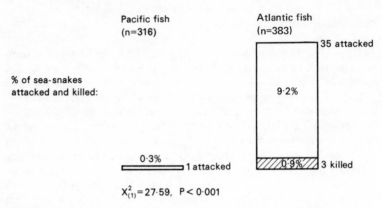

Fig. 3.2   Attacks by Pacific and Atlantic fish on sea snakes (*Pelamis platurus*). (*Data from* Rubinoff and Kropach, 1970.)

Innate avoidance responses

In some predators inhibition to attack certain signals can become built in to the genetic system so that it is now endogenous (i.e. innate), for example the response of birds to two large eyes. If a small bird or mammal searching for food in a tree comes across two large eyes this implies the presence of an owl, hawk or large mammal, any of which could attack. Innate withdrawal or flight in response to such a stimulus will be of very great survival value. The importance of this innate response to eyes can be judged by the large number of animals which have startle displays involving false eyespots, and Blest (1957a) has shown that eyespots do release fright behaviour in small birds. False eyespots are discussed further on p. 166.

Avoidance responses to bright colours are also probably innate, but they are responses to novel stimuli rather than to specific signals (Coppinger, 1969). These are further discussed on p. 152, but of relevance here is the demonstration that ten-day old domestic chicks develop a learned avoidance response to a novel stimulus more quickly than they do to a previously known stimulus (Shettleworth, 1972). Hence if an animal is evolving aposematic coloration there will be fewer individuals lost by predators sampling them if they have a novel (and hence conspicuous) colour than if they have a colour with which the predators are familiar (such as green or brown).

It has been shown that twenty-one Pacific fish belonging to ten different species would not attack either dead or live sea snakes (*Pelamis platurus*), but that twenty-one Atlantic fish of nine different species would occasionally attack them (Fig. 3.2) (Rubinoff and Kropach, 1970). The snakes are venomous and occasionally kill the fish which attacks them, but Atlantic fish which eat snakes and survive continue to eat them. It is unlikely that all of the twenty-one Pacific fish used in the experiments had learned to avoid *Pelamis* before capture, and it was found that two very young Pacific nurse sharks (*Ginglymostoma cirratum*), which certainly could not have encountered *Pelamis* before, refused to eat pieces of fresh sea snake although they would take similar pieces of squid. It appears that Pacific fish have an endogenous recognition pattern and avoidance response for *Pelamis*, probably involving chemosensory rather than visual cues, but Atlantic fish have no such endogenous behaviour. Presumably conditioned avoidance responses evolved first in the Pacific fish, but that since many of the trials with *Pelamis* resulted in death of the fish, selection favoured any individuals in which the avoidance to certain specific signals was endogenous. One reason why endogenous avoidance of aposematic species is not more widespread is that the majority of such species are not in fact lethal to their predators. Unless a predator is killed by the experience, natural selection is unlikely to produce an innate avoidance response. The most successful predators are likely to be opportunist, inquisitive, and prepared to tackle novel prey, rather than specialists programmed to attack only a few different sorts of prey and to avoid others, since the latter will be unable to survive if the favoured prey becomes scarce. These successful, more inquisitive predators are unlikely to evolve innate avoidance responses.

Some lizards also have innate aversions to the smells of unpalatable prey (Loop and Scoville, 1972), and it is probable that mammals do as well since, although no examples are known, some mammals do have innate avoidance responses to signals produced by predators. For example hand reared black-tailed deer refused to eat food in which there were droppings from certain predators (Müller-Schwarze, 1972), and hand reared hyaenas apparently have an innate avoidance response to the smell of lions (Kruuk, 1972). The avoidance responses of many small mammals and birds to hissing noises (as of a snake) are also probably innate. The aversions of monkeys and men to seeing a snake are perhaps also innate (Morris, 1967) although they could equally well be due to conditioning at an early age (for example seeing the agitated behaviour of a parent when presented with a snake or with a string moved in a snake-like manner).

Some unsolved cases

The most important predators in aquatic environments are fish, and it is therefore important to know whether or not fish are capable of learning to recognize and to avoid aposematic prey. If they cannot do this, then obviously there is no advantage to a prey animal in being conspicuously coloured, and so warning coloration will not evolve (Edmunds, 1966*a*). There is no doubt that fish are as capable of learning as are other vertebrates (literature reviewed by Welty, 1934; see also Lissmann, 1958, and Eibl-Eibesfeldt, 1970). In one experiment Triplett (1901) separated predators (perch) from their prey (minnows) by a glass partition. The perch soon learned to avoid hitting the glass in their attempts to catch the minnows which they then ignored. When the partition was removed from the tank, the minnows swam freely amongst the perch which continued to refrain from attacking them. This learned avoidance was found to be specific to a particular type of prey: perch which had learned to avoid hitting the glass partition when there were minnows on the far side immediately bumped into it when a worm was dropped amongst the minnows. There is also evidence that predatory fish can learn by experience not to attack distasteful brightly coloured fish such as the poison-fang blenny (*Meiacanthus*) (see p. 127).

However, in spite of extensive laboratory studies on learning in fish, there is little convincing evidence that fish in nature learn to avoid sampling distinctively coloured unpalatable invertebrates, such as many species of nudibranch molluscs (sea slugs). Several workers have thrown sea slugs into fish tanks or into the sea and then watched a variety of species of fish take them into the mouth and reject them (Thompson, 1960*b*). Often a slug will be sampled in this way by several fish before it finally reaches the bottom of the tank where it is then ignored. Such experiments show that the fish are not learning to recognize the prey and to reject it on sight alone, but that they are continuing to sample them. However, this experimental design is not likely to demonstrate learning in fish since sea slugs falling through the water column do not form a normal part of the experience of fishes in nature. Fish in aquaria are often fed in this way and may be expected to learn to associate objects falling through the water with food. The stimulus to snap at any object falling through the water is likely to be more powerful than any inhibition due to learning from a previous experience. When Brower demonstrated that Florida scrub jays can learn the colour pattern of aposematic butterflies she presented the insects motionless with the wings pinioned. Probably if they had been able to move the movements would have elicited many more pecks than occurred with the motionless insects. Clearly more critical experiments under more naturalistic conditions are required to find out how fish react to unpalatable prey.

Preliminary observations in Ghana using young trigger fish (*Balistes* sp.) as predators in a small aquarium and placing sea slugs on the floor of the tank have produced encouraging results. Four fish inspected and snapped at the dorid *Doriopsilla albolineata*, but then left it alone. When removed from the tank a few minutes later the dorid had two chunks from the mantle and one rhinophore missing. Similarly five out of the eight *Balistes* bit at a *Discodoris tema* crawling or resting on the floor of the tank, and one fish even carried it off for a few seconds in its mouth before dropping it. The dorid was then left alone and not attacked again, although it had several chunks from the mantle and one gill plume missing. *D. tema* has sulphuric acid glands, and *D. albolineata* has many small epidermal glands which may have a defensive function.

In further experiments the red eolid sea slug *Coryphella pellucida* was placed in a tank into which a grey mullet (*Mugil labrosus*) was introduced (Edmunds, Potts, Swinfen and Waters, unpublished). After once or twice taking the slug into the mouth, the fish quickly learned to avoid it either by inspection from a few centimetres distance or by briefly touching it with the lips. In every case the slug was undamaged after being sampled by the fish.* Another mullet was found to attack a

white cardboard model — ⌒ — but to avoid a similar model with

papillae on it — ⌒ —, suggesting that the papillae may be one

of the visual cues by which fish recognize, and then learn to avoid, sea slugs and sea anemones. These observations, though very preliminary, suggest that fish will sample any potential food, and they can learn to refrain from attacking a prey animal that they have a short time earlier found to be nasty. If further work confirms these results, then it will be clear that aposematism may be of advantage to aquatic invertebrates with respect to predatory fish just as it can be with respect to predatory cephalopods.

Another important group of predators belong to the phylum Arthropoda — insects, crustaceans, spiders and related forms. Arthropods are generally considered to be less capable of learning than are vertebrates, but there is now evidence that some arthropods can learn to avoid unpalatable prey on the basis of its colour. Tyshchenko (1961) studied the feeding reactions of the crab spiders *Xysticus ulmi* and *Misumena vatia* to inedible hymenopterans and their edible dipteran mimics. He found that, after experiencing a wasp, the spiders were much more restrained in their behaviour towards further wasps. They were similarly restrained in their behaviour towards dipteran mimics of the wasps, but they still attacked dipterans of different appearance to wasps without signs of inhibition. Hence they appeared to be discriminating between different species of prey and to be learning to avoid (or at least to treat with caution) insects of one pattern while still attacking insects of another pattern.

It has also been shown that mantids can learn to associate the colour of the milkweed bug (*Oncopeltus fasciatus*) with inedibility, but only when the mantid is well fed (Gelperin, 1968). A starved mantid will eat *Oncopeltus*. Gelperin also showed that a mantid can learn not to strike at a fly presented on a red background when this is coupled with an

---

* The three sea slugs used in these experiments may not actually have warning coloration: none of them is very abundant, and they occur most commonly underneath stones and rocks rather than in exposed places. Furthermore, *D. tema* is very cryptic when near the red sponge on which it probably feeds (Plate 4a), and *C. pellucida* is very cryptic when on the hydroid *Tubularia* which is one of its foods.

electric shock, but that it would still strike at a fly on a white background. However, it required thirteen presentations to produce strike inhibition, and such conditions are unlikely to occur in nature unless a noxious animal is exceptionally abundant or gregarious.

My own experience of *Sphodromantis lineola* suggests that even with a very unpalatable insect (the black and orange butterfly *Pyrrochalcia iphis*) the flapping wings of the prey provide such a strong stimulus to strike that this overrides any inhibition due to previous experience. On the few occasions when, after having tasted a *Pyrrochalcia*, the mantid did not strike, it also refrained from striking when presented with a palatable species of butterfly or with a grasshopper. There was no indication that failure to strike was associated with the colour of the prey. Thus although mantids and other predaceous insects may be capable of learning to avoid noxious prey on sight whilst still attacking other prey, this is unlikely to occur at all frequently in nature. Rilling *et al.* (1959) also concluded that colour is of no importance in determining whether or not a mantid will strike at prey — movement, and to a lesser extent size and shape, are the critical factors which elicit a strike.

Thus it is probable that aposematism will be found to occur only in animals which have vertebrate or cephalopod predators since only these animals are likely to learn quickly to avoid distasteful or dangerous prey species.

Further probable examples of aposematism

Many of the animals that are models for batesian mimetic associations are aposematic, and the evidence for this is discussed in Chapter 4 on batesian mimicry. Here I shall consider a number of other animals which are believed to be aposematic, but in some cases the evidence is very inconclusive.

Aposematism in the sea

Some eolid and other nudibranch molluscs are brightly coloured and may be aposematic, but this has not been proven. It has been discussed on p. 68 and is reviewed by Edmunds (1966a).

The sea urchin *Diadema* is also perhaps aposematic. It has long spines which are toxic and very irritant if they pierce the skin, and it is black and very conspicuous on exposed reefs (for example in the Caribbean). However, there is at present no evidence that predators which attack *Diadema* learn to associate its unpleasantness with colour or shape and so avoid it on sight in future encounters.

Species of scorpion fish (*Pterois* spp.) and gurnard (*Dactylopterus* spp.) also have poisonous glandular spines, and many of them are brightly coloured, conspicuous, and swim slowly. If molested they raise the pectoral fins in what appears to be a display and they may even

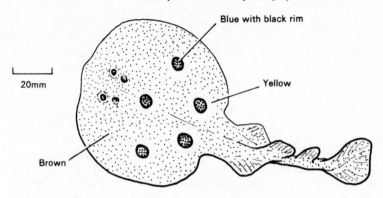

Fig. 3.3 The electric ray *Torpedo narke* in dorsal view. The body is brown with five blue spots each with a black rim and lying in a yellow area.

attack an intruder (Breder, 1963) (this is further discussed under secondary defence on p. 186). It is probable that predatory fish learn through experience to refrain from attacking scorpion fish, just as they do with the poison-fang blenny (see p. 127).

Another fish which may be aposematic is the electric ray *Torpedo narke*. Most electric rays are sandy brown in colour and hence cryptic, but *Torpedo narke* has five blue spots on the dorsal surface which make it conspicuous when resting on the sea-bed (Fig. 3.3) — at least it is conspicuous to aqualung divers and hence presumably also to predatory fish. It is likely (but not proven) that these eye-spots are aposematic marks since the electric ray can give a powerful electric discharge capable of stunning a potential predator.

Aposematism in tetrapods

There are very few examples of aposematism amongst mammals. The best known are all black and white: the striped and spotted skunks (*Mephitis* and *Spilogale*) from America, and the zorilla, polecat and ratel (*Ictonyx*, *Poecilogale* and *Mellivora*) from Africa. Some porcupines (*Hystrix* spp.) are also black and white, perhaps as a warning to predators, and all of these animals have dramatic secondary defences described on p. 160.

There are a number of aposematic amphibia and reptiles besides the well known coral snakes discussed on pp. 77 and 132, but few have been carefully studied. Most of them are red and black or yellow and black (see Cott, 1940), but some may be entirely black. The desert elapid snake *Walterinnesia aegyptia* is black and nocturnal in its habits. It is said to be very conspicuous when hunting on the sand by starlight (Zinner, 1971), so may perhaps be aposematic.

Many species of birds are conspicuously coloured, and Cott (1947) has shown that conspicuous species are usually less palatable than cryptic species. Kingfishers and drongos, for example, are relatively unpalatable and conspicuous birds. Colour in birds is important for intraspecific signals, but it is probably true that only those species that are slightly unpalatable have been able to evolve conspicuous signalling colours. Exceptions are the large birds such as pelicans, swans and ostriches which have few enemies, are conspicuous and edible. Most small birds, however, are either cryptic and palatable with few intraspecific visual signals; or they are conspicuously coloured, relatively unpalatable, and have many intraspecific visual signals. It is surely no coincidence that many of the best songsters (as judged by human ears) are cryptically coloured and often sing from perches well hidden in vegetation — they communicate with one another aurally — whilst the more conspicuous species often have rather poor songs and communicate with each other visually. Compare, for example, cryptic songsters such as nightingales, warblers and thrushes (even the robin is cryptic when viewed from above) with the conspicuous stonechat, finches and pied wagtail. However, it has not yet been shown that a predator, having sampled one or two individuals of a slightly distasteful species, refrains from further attacks.

Aposematism in insects

The best known examples of aposematism from insects are discussed elsewhere in this chapter or under batesian mimicry. As usual, the more a particular animal is studied, the more complicated are its defences found to be. Ladybird beetles are mostly brightly coloured insects which emit a nauseous smelling fluid when they are handled, so they appear to be aposematic. The two-spot ladybird (*Adalia bipunctata*) is normally red with one black spot on each elytron, and it is rejected as food by various species of birds and mammals (Frazer and Rothschild, 1962). It is found to be polymorphic in some places, which is unexpected since polymorphism does not normally occur in aposematic animals. The other morphs are black with two or three red spots on each elytron, and they are therefore partial melanics. These melanic ladybirds are most numerous in industrial areas, just as are melanic, cryptic, moths (Creed, 1966). This finding is unexpected since if they are aposematic one would expect selection to favour the conspicuously coloured forms in a sooty environment rather than the less conspicuous melanic forms. Furthermore, in a part of Birmingham, Creed (1971) found that as pollution and smoke declined (as a result of anti-pollution legislation), the percentage of melanics in the population declined from about forty-five to twenty-five, indicating that the melanics survive best in polluted environments whilst the reds survive best in the non-polluted environments. It is possible that the melanics are physiologically better adapted to live in polluted conditions, but it is not clear why physiological superiority should be correlated with coloration. A more likely explanation is that ladybirds are normally aposematic but that in certain environments they are better protected if they are cryptic. Aposematism inevitably results in some individuals being killed before all of the predators have learned to avoid the species. Hence it is possible that the normally rare melanic morph has proved to be advantageous in industrial areas because it is cryptic and so rarely found and sampled by predators. The red spots on the melanics are still valuable in defence since if the insects are found, the predator may be able to learn to recognize and to avoid further ladybirds because of them. If this explanation is correct, the red ladybirds rely for primary defence on aposematism whilst the black ones are cryptic (in a polluted habitat) but have warning colours as a secondary defence once they are discovered.

Aposematism is normally thought of as applying to animals with conspicuous colours whose predators hunt by sight. But it can equally apply to animals whose predators hunt by hearing or smell. Bees buzz continuously when flying, and this is probably a warning sound to predatory birds which have already sampled a buzzing, stinging bee. The caterpillars of the buff tip moth (*Phalera bucephala*) are gregarious and have a powerful characteristic smell, possibly as a warning to predators (Ford, 1955). Other animals that have warning sounds and smells usually emit them only after they have been alerted to the presence of a predator. These are therefore secondary defences and they are discussed in Section 2.

Limitations of aposematism

1. Some prey animals must be sacrificed in every generation of predators

One disadvantage of aposematism is that some prey individuals must be killed in every generation of predators, so protection can never be 100 per cent. If the species is common, this sacrifice will not significantly affect the population, but if it is rare it may result in a substantial proportion being killed before they have had a chance to reproduce. So warning coloration is unlikely to occur in rare species unless they are particularly tough and able to survive the experience of tasting by a predator. For an unpalatable species that is rare it may be of greater protective value to be cryptic, and hence rarely found, than to be conspicuous.

As a result of this sacrifice of aposematic animals in every predator generation, four further adaptations have sometimes evolved in aposematic species:

(i) *Post-reproductive longevity*. Blest (1963*b*) has pointed out that selection in aposematic species should lead to the prolongation of life after reproduction has ceased since a post-reproductive animal can still form the prey of a naive predator and help it to learn the prey colour pattern. If predators kill animals that have already reproduced rather than younger animals, this will benefit the prey species. Selection will favour a long post-reproductive life as a result of kin-selection. The killing of a post-reproductive individual reduces the chances of that individual's offspring and sibs being taken by a predator, and thus increases the probability of the aged animal's own genes continuing to exist in subsequent generations. Conversely, it will be advantageous in a cryptic species for death to follow closely on cessation of reproduction since if cryptic animals stay alive the density of the species will be increased, and so predators will be more likely to find them and hence to build up a searching image for the species.

A further advantage of longevity in aposematic species is that it increases the likelihood of predators seeing the prey, and, as already mentioned, this probably aids in retention of the learned avoidance response.

As evidence that cryptic species live shorter post-reproductively than aposematic species, Blest showed that, after mating, male saturniid moths of the cryptic genus *Lonomia* live on average five days in the laboratory whilst males of the related, but aposematic, genus *Dirphia* live on average ten days. He also showed that females of a variety of species of cryptic insects lived only one or two days after laying their final batch of eggs whilst females of aposematic species lived from two to nine days. Hence there is some evidence to suggest that cryptic species have short post-reproductive lives whilst aposematic species have long post-reproductive lives.

(ii) *Müllerian mimicry.* Some groups of aposematic species have come to share the same colour pattern, and this is known as müllerian mimicry. The advantage of it to the prey species is that predators have only to learn one colour pattern for several different species to gain protection. Hence the number of prey individuals killed in each generation of predators is much less than if each prey species had a different colour pattern. Examples of müllerian mimicry are the various species of yellow and black social wasps (*Vespa* and *Vespula* spp.) together with several genera of similarly coloured solitary wasps. The cinnabar caterpillar also belongs to this array of müllerian mimics since Windecker (1939) has shown that birds which learn to avoid cinnabar caterpillars also avoid wasps, whilst control birds with no experience of cinnabar caterpillars attack wasps.

In the New World there are many müllerian mimics amongst the heliconiid butterflies. For example, *Heliconius erato* and *H. melpomene* are very similar in colour pattern over much of their geographical range, and both are unpalatable to birds (Brower *et al.*, 1963; Benson, 1971; Turner, 1971*a*, *b*, 1973). In Benson's experiments in Costa Rica, described on p. 63, it was found that selection will tend to eliminate insects with a different colour pattern to that of the rest of the population in the area; in this case the population comprised both *H. erato* and *H. melpomene*. However, elsewhere in the New World both of these species have different colour patterns to that found in Costa Rica (Fig. 3.4). Since the experience of a predator sampling and killing a *Heliconius* and then learning to avoid similar coloured insects is of advantage to the butterfly population as a whole, but not to the individual killed, selection can only favour increased unpalatability of the insect provided that the population contains similar genes to the sacrificed individuals. This is 'kin selection' which can only occur if there is much inbreeding in the population (see p. 74). The more unpalatable species of *Heliconius* live in discrete colonies with large roosting aggregations, and the adult butterflies may live for six months or more in the same area, unlike typical butterflies which live only a few days or weeks. These habits have resulted in inbreeding in each population and this has led to the evolution of distinct geographical races. In *Heliconius erato* and *H. melpomene* each geographical race is still capable of breeding with other races although this occurs only rarely in the wild. No doubt in time these races will evolve into distinct species, as has apparently already occurred in other species of *Heliconius*. The evolution of parallel geographical races with similar colour patterns in *H. erato* and *H. melpomene* is a remarkable testimony to the advantages for distasteful animals of müllerian mimicry. These two species belong to different sections of the genus *Heliconius* and are not closely related. Their larvae have different species of food plant so do not come into competition with one another, and the adults of both species presumably benefit from their resemblance to one another since their colour patterns have evolved in parallel. There are several other species of *Heliconius*, including *H. elevatus* and *H. aoede*, which also share the same colour patterns as *H. erato* and *H. melpomene* over part of their geographical range. Many populations of *H. doris* are polymorphic with one form resembling *H. erato*, another form resembling the blue and yellow *H. sara*, and a third form being green and apparently nonmimetic. *H. doris* is much less nasty to birds than are *H. erato* or *H. melpomene* (Brower *et al.*, 1963), so it is possible that *H. doris* is a batesian mimic with relation to some predators. The situation is obviously very complex and requires much further investigation in the field.

Another müllerian mimetic assemblage is that of the burnet moth *Zygaena ephialtes* which resembles *Z. filipendulae* in central Europe and *Amata phegea* in southern Europe (summarized by Turner, 1971*b*). The African butterflies *Danaus chrysippus* and *Acraea encedon* are also probably müllerian mimics, but since this assemblage includes some palatable batesian mimics it is discussed in the section on batesian mimicry (p. 108).

Another very complex example of müllerian mimicry is that of the

Fig. 3.4   Geographical variation in *Heliconius erato* and *H. melpomene* in Central and South America. Insects of the two species from each locality have closely similar colour patterns. In each pair of insects *H. melpomene* is drawn on the left, *H. erato* on the right. Colour key: black = black; white = yellow (except in insects from (*a*) and (*b*) in which white = white); black stipple = red; white stipple (in insects from (*a*)) = blue. (*After* Turner, 1971*b*.)

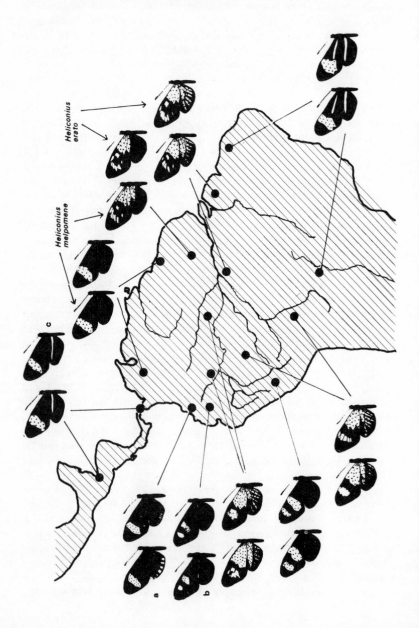

pyrrhocorid cotton stainer bugs of the genus *Dysdercus* and reduviid bugs of the genus *Phonoctonus* (Plate 5*a*). In both tropical America and Africa several species of *Dysdercus* occur in the same locality and they have similar colour patterns, usually of red and black (Doesburg, 1968; Stride, 1956*a*; Fuseini, 1972). Some species of *Dysdercus* show geographical variation in colour such that the colour of each race matches that of other common species of *Dysdercus* in the area (Doesburg, 1968). The situation is evidently very similar to that of the *Heliconius* butterflies, but it has not been investigated. It is known that species of *Dysdercus* and *Phonoctonus* are distasteful to a variety of species of birds and mammals, although the African lizard *Agama agama* will eat both types of bug. Adult *Dysdercus* from Ghana are not conspicuously coloured with much grey and black on the dorsal surface, but the larvae are bright red and very conspicuous. The larvae occur in dense aggregations where a predator that has sampled one can quickly learn to avoid others of similar colour, but many of the adult insects fly off to found new colonies alone. It may be more advantageous for a single *Dysdercus* to be cryptic and not easily found by a predator than to be conspicuous and so easily seen and killed.

In southern Ghana *Dysdercus voelkeri* is the commonest savanna species, but the almost identically coloured *D. superstitiosus* and *D. fasciatus* may occur with it in mixed colonies. *Phonoctonus fasciatus* usually occurs in these colonies and has an almost identical colour pattern to the *Dysdercus*. There are in fact species pairs of pyrrhocorid and reduviid which characteristically occur together and have the same colour pattern: *D. voelkeri* (and others) with *P. fasciatus*; *D. melanoderes* with *P. subimpictus*; and *Odontopus sexpunctatus* with *P. lutescens* (Fig. 3.5). The pairs are not invariably associated with one another, but they very often are, and it is found that the *Phonoctonus* preys heavily on the *Dysdercus* with which it lives. Since *Phonoctonus* is comparatively rare in colonies of *Dysdercus* it is likely that predators learn to avoid the colour pattern of the insects by sampling the much more numerous *Dysdercus* so that *Phonoctonus* gains protection from the resemblance. However, the situation is probably much more complex than outlined here, and much more information is required on the predators of these insects before we can claim to understand the significance of their coloration.

An example of müllerian mimicry from the vertebrates is provided by the coral snakes of the New World studied by Mertens (1966), whose findings are summarized by Wickler (1968). There are about seventy-five species of coral snakes: some are poisonous müllerian mimics and others are non-poisonous batesian mimics. The situation is very complex and is discussed further on p. 132

(iii) *Innate avoidance of aposematic patterns.* The sacrifice of aposematic prey in each generation of predators can be eliminated if the inhibition to attack a particular pattern becomes innate. This is what

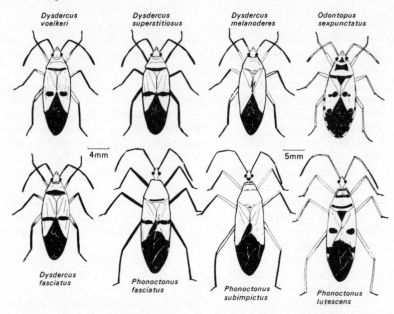

Fig. 3.5   Müllerian mimicry amongst West African Heteroptera. The reduviid *Phonoctonus fasciatus* normally lives in colonies of one or more of the pyrrhocorids *Dysdercus fasciatus*, *D. superstitiosus* and *D. voelkeri*; *Phonoctonus lutescens* normally lives in colonies of *Odontopus*, sometimes with *D. voelkeri* as well. All of these bugs are black, red (shown stippled) and greyish orange (shown white). *P. subimpictus* normally lives in colonies of *Dysdercus melanoderes*. Both are cream or greyish yellow (shown white) with grey on the wings (shown black) and with black on the antennae and legs.

appears to have happened in the case of the Pacific fish which encounter sea snakes. As already mentioned, for a predator there is no advantage to be gained by developing an innate inhibition against attacking a particular pattern as compared with a learned inhibition, provided that the prey species is noxious but not dangerous. However, if the defence of the prey can be fatal to the predator, then obviously it will be of advantage to the predator to develop an innate inhibition against attacking that particular stimulus. It is possible that this also accounts, in part, for the occurrence of deadly snakes in the assemblage of mimetic coral snakes (see p. 132), but as yet it has not been demonstrated that any terrestrial snake-eating predator in South America shows an innate avoidance of deadly species.

(iv) *Emetic properties*. Brower (1969) and Rozin and Kalat (1971) have shown that where a noxious food causes illness of the digestive tract (e.g. vomiting) a predator is not likely to sample that food again. On the other hand, if rejection occurs at the mouth, without swallowing, the predator may 'forget' and attack that species of prey occasionally in future. If an aposematic prey species is common it may be

a | b
---
c | d

Plate 1

(a) Caterpillar of an unidentified lymantriid. Note the flattened body with laterally directed hairs to minimize shadow. The caterpillar has been disturbed and has responded by raising its head and thorax and erecting two tufts of black irritant hairs in a deimatic display. Legon, Ghana.
(b) Convolvulus hawkmoth caterpillar (*Herse convolvuli*), fourth instar striped form resting on stem of *Ipomaea aquatica*. Legon, Ghana.
(c) Convolvulus hawkmoth caterpillar, fourth instar green form which normally rests on leaves of *Ipomaea aquatica*. Legon, Ghana.
(d) Convolvulus hawkmoth caterpillar, final instar from *Ipomaea aquatica* in Ghana. It feeds at night and spends most of the day buried in earth.

a	b	c
d	e	f

Plate 2

(a) Caterpillar of the ctenuchid *Euchromia lethe*. Note aposematic coloration and irritant tufts of hairs. Legon, Ghana.

(b) Cocoons of *Euchromia lethe*. The caterpillar on the right is pulling out hairs from its body and sewing them into the wall of its cocoon and also onto the nearby branch. A completed cocoon is on the left.

(c) Adult of the aposematic moth *Euchromia lethe*.

(d) Caterpillars of the citrus swallowtail, *Papilio demodocus*. Right, young caterpillar resting exposed on a leaf of *Citrus* and resembling a bird's dropping. Left, final instar feeding on a leaf, showing cryptic coloration with disruptive black marks. Legon, Ghana.

(e) Young caterpillars of the moth *Trilocha kolga* resting on a leaf of the food plant *Grossera vignei* and resembling bird droppings. Mount Atewa, Ghana.

(f) Final instar of *Trilocha kolga* on *Grossera vignei*, apparently resembling a large lizard or bird dropping.

advantageous to evolve emetic properties even though this means that the individuals attacked are swallowed and hence killed. But if an aposematic prey species is rare it may be more advantageous to be tough so that individuals are likely to survive one or several attacks by predators.

2. **Conspicuous colours may attract a predator to prey it had not previously seen**

A second disadvantage of aposematism is that whilst some predators are capable of being conditioned to associate colour with inedibility, others (particularly invertebrate predators) are not. The bright colours and ineffectual escape movements of an aposematic animal may attract these predators, and even though they may not eat the prey, they may kill or seriously injure many more individuals than they would of an equally abundant but cryptic species. Some predators, too, reject aposematic prey if they are well fed but take them if they are starved (e.g. mantids and the milkweed bug studied by Gelperin, 1968, and described on p. 69). Sexton, Hoger and Ortleb (1966) trained lizards (*Anolis carolinensis*) to avoid the distasteful firefly *Photinus pyralis*. Groups of lizards were then kept on four different nutritional levels and again presented with fireflies seven days later. It was found that lizards which had been on a low ration diet captured more *Photinus* than did lizards on a high ration diet. Similar results using birds and aposematic butterflies are summarized by Swynnerton (1915a). Thus aposematism can be of advantage to a prey species only if the relevant predators are in a physiological condition such that they will reject them. A starved predator would probably be attracted by these same colours. Only when it had fed again would the impulse to attack and eat become subordinate to the conditioned rejection response. However, if the prey is very distasteful, then it may be avoided by even a starved predator — lizards on a low ration diet continued to refuse to eat *Oncopeltus* Sexton *et al.*, 1966).

There are also some predators which have developed tolerance for the noxious qualities of some aposematic animals (see, for example, Rothschild and Kellett, 1972), and for these the colours must make the prey easier to find than if it were cryptic. Cuckoos regularly take hairy, brightly coloured caterpillars that are rejected by most other species of birds. The black redstart has also been recorded as feeding considerable numbers of cinnabar caterpillars to its young (Hosking, 1970) despite the fact that these caterpillars are aposematically coloured, gregarious, and avoided by most birds. Obviously there must be a balance between the advantages of aposematic coloration which makes a species easily seen by predators, and the degree of unpalatability in relation to the various species of predator in the environment. One solution to this problem is for an animal to be cryptic from a distance but aposematic when seen from nearby. This may apply to red and blackish green burnet moths (*Zygaena* spp., Plate 5*d*) (Rothschild, 1964), to adult *Dysdercus* (see p. 77), to melanic ladybirds (see p. 72) and to the caterpillars of *Danaus chrysippus* (Plate 4*d* and see p. 263) and the cinnabar moth. Black and yellow cinnabar caterpillars are always conspicuous on the leaves of their food plant (*Senecio jacobaea*), but they are not conspicuous from a distance when on the bright yellow flowers (Plate 4*f*). Since they often feed and rest on the flowers they are sometimes cryptic and sometimes aposematic. Thus an animal may rely on crypsis for primary defence, but if it is discovered the first secondary defence is a warning coloration or a warning display (further considered on p. 263).

Summary

Aposematic animals are distasteful and have bright colours or other signals which cause predators to recognize and avoid attacking them. Vertebrates and cephalopod molluscs can learn through experience to avoid attacking aposematic animals, but there is no evidence that any arthropod predator learns to avoid aposematic prey in nature. Some vertebrate predators have evolved an innate avoidance response to signals produced by certain distasteful or dangerous animals. Nevertheless, starved predators as well as some specialist predators do eat certain aposematic animals. The conspicuous signals of the aposematic animals will attract these predators and be disadvantageous so that selection may lead to an animal having colours that are cryptic from a distance but conspicuous when seen from nearby.

Aposematic animals are often slow moving, gregarious, and they tend to have longer post-reproductive lives than do related cryptic animals. Several species of aposematic animals have evolved similar colour patterns, presumably so that predators can include all of them in a single learned avoidance response (müllerian mimicry).

Chapter 4
Batesian mimicry

Mimicry is commonly understood to imply the resemblance of one animal (the mimic) to another (the model) such that a third animal is deceived by the similarity into confusing the two. Wickler (1968) gives a thorough discussion of mimicry throughout the animal and plant kingdoms and concludes that mimicry is the imitation of a signal which is of interest to a signal receiver. The signal receiver is of course usually an animal; the signal itself may be produced by another animal of the same or of a different species, or by a plant. Wickler's definition of mimicry excludes müllerian mimicry since here no deception is involved: the signals of all the animals involved are genuine warning signals, not counterfeit ones. But it includes cases of intraspecific mimicry such as the dummy eggs on the anal fin of the male cichlid fish *Haplochromis burtoni* which ensure that fertilization of eggs in the female's mouth occurs; the swollen, coloured anal regions of some male monkeys which mimic the genital regions of females in oestrous and are used in submissive displays; or the ears and eye-stripes of some artiodactyls which are said to mimic horns and are also used in intraspecific social contexts (Wickler, 1968; Guthrie and Petocz, 1970).

In this book I am concerned with mimicry only as a defence against predators. Batesian mimicry is a form of defensive mimicry which can be defined as the 'close resemblance of one organism to another which, because it is unpalatable and conspicuous, is recognised and avoided by some predators at some times' (Remington, 1963). To put this in Wickler's terminology, batesian mimicry occurs when a predaceous animal, which avoids eating one animal producing a particular signal, is deceived into avoiding a second animal which produces a similar but counterfeit signal. The other important form of defensive mimicry, müllerian mimicry, involves no counterfeit signals and has already been considered under the section on aposematism, but it will be discussed again here in the context of the evolution of mimicry.

Batesian mimicry can be regarded as a special case of camouflage (crypsis), but if differs in that the predator normally detects the prey animal because it produces clear signals, whereas in crypsis the predator fails to pick up any signals from the prey animal. Cases of special resemblance to leaves or bird droppings fall between the two categories since they may have evolved from typical crypsis, but the predator may distinguish the animals from their surroundings and then fail to recognize them as being edible (Robinson, 1969a, b). For convenience, these also will be considered here since, like typical cases of batesian mimicry, they involve the production of counterfeit signals.

Survival value of batesian mimicry

As a demonstration that mimicry can be of protective value to an animal, Brower (1958a, b, c) presented a number of mimetic butterflies together with their models to caged birds (Florida scrub jays). She was able to show that birds which had been conditioned to avoid the aposematic monarch (*Danaus plexippus*) subsequently avoided the very similarly coloured viceroy (*Limenitis archippus archippus*). Control birds which had not experienced the monarch ate the viceroy readily. Similarly the Florida variety of the viceroy (*L. archippus floridensis*) was not eaten by birds that had been conditioned to avoid the unpalatable queen (*Danaus gilippus berenice*); and three species of *Papilio* swallowtail (*P. troilus, P. polyxenes* and black *P. glaucus*) were all avoided by birds that had been conditioned to avoid the aposematic, black *Battus philenor* (Fig. 4.10).

Mostler (1935) showed that bees, bumblebees and wasps are distasteful to a variety of species of birds. The sting is of course the initial source of unpalatability, but the toughness of the bumblebee and the unpalatable flesh of bees and wasps also contribute to their rejection as food by birds. He showed that mimetic Diptera were avoided by birds which had previously learned to avoid the hymenopteran models (see Fig. 4.1), but they were eaten by young birds which had had no experience of the models.

Toads (*Bufo terrestris*) can also be deceived by a mimetic resemblance. They quickly learn to avoid bumblebees (*Bombus americanorum*), and then subsequently refuse to eat the mimetic but edible asilid fly *Mallophora bomboides* (Brower *et al.*, 1960). Similarly, the dronefly (*Eristalis vinetorum*) is edible to toads, but after they have learned to avoid bees (*Apis mellifera*) they also avoid eating droneflies (Brower and Brower, 1962a). Figure 4.2 shows that toads eat a higher proportion of wingless bees (which cannot buzz) than of winged bees (which can). They also eat a higher proportion of wingless droneflies than of winged ones: both are equally edible, but only the winged ones buzz (Brower and Brower, 1965). This suggests (but does not conclusively prove) that the buzzing of both bees and droneflies contributes to their protection, and that the buzzing of droneflies is an example of sound mimicry.

Other demonstrations of the effectiveness of mimicry under experimental conditions in the laboratory as well as in natural situations in the field are discussed elsewhere in this chapter.

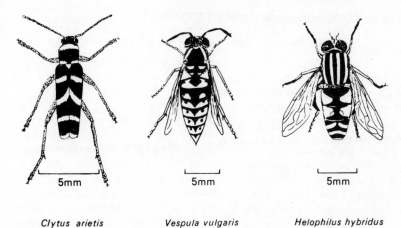

Clytus arietis   Vespula vulgaris   Helophilus hybridus

Fig. 4.1   The aposematic wasp *Vespula vulgaris* and two batesian mimics, the beetle *Clytus arietis* and the hoverfly *Helophilus hybridus*. In all three the body is dark brown or black with yellow bands.

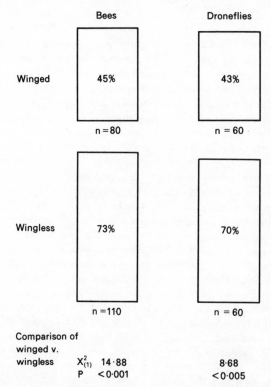

Fig. 4.2   Percentages of winged and wingless bees (models) and droneflies (mimics) eaten by toads. (*Data from* Brower and Brower, 1965.)

Limitations of batesian mimicry

1.   The mimic cannot become too common relative to the model

It is obvious that if the mimic gains protection because of its resemblance to a distasteful model then it can only receive protection if the
model is present in the same environment. Predators cannot otherwise
learn to avoid the colour pattern. It is obvious too that if the model is
very rare then predators are more likely to discover, and to develop a
searching image for, the edible mimic than they are to sample and to
develop a conditioned avoidance response towards the noxious model.

Brower (1960) performed a series of experiments using starlings as
predators and mealworms coloured with cellulose paint as models and
mimics. The models were banded with green and dipped in quinine
solution; mimics were also banded green but were dipped in distilled
water; and edible controls were banded orange. The starlings soon
learned to avoid the green models and to eat the orange controls
(despite orange normally being associated with warning coloration in
nature). The mimics and models were then presented to the birds in
random order at various frequencies, together with alternate edible
mealworms. The results (see Fig. 4.3) show that with 90 per cent
models and 10 per cent mimics presented, only 6 per cent of the
mimics (one out of sixteen) were eaten, the rest being rejected. Hence
the mimicry can be said to be 94 per cent effective at protecting the
mimics. Two perhaps surprising conclusions are also apparent: first, the
mimic appears to receive equal protection when there are 70 per cent
models in the population and when there are only 40 per cent models
in the population (about 80 per cent protection); and second, even at a
model frequency of only 10 per cent, with the mimics outnumbering
the models by nine to one, 17 per cent of the mimics (fifteen out of
ninety) were not eaten. In other words, there is some protective value
for a mimic even if it is much commoner than its model. Emlen (1968*a*)
supports this conclusion on theoretical grounds but also shows that the
advantage to the mimic declines at very high mimic frequencies.

Many animal populations fluctuate seasonally. If the population of a
model species fluctuates during the year this will affect the level of
protection given to its mimics. The buff ermine moth (*Spilosoma
lutea*), although it is unpalatable to some predators, is considered to be
a mimic of the much more distasteful aposematic white ermine (*S.
lubricipeda*) (Fig. 4.4), but the two fly at different seasons (Rothschild,
1963). Since the white ermine flies first, potential predators are likely
to sample this species first and then learn to avoid it. Later on, the buff
ermines emerge, and they receive a high level of protection because
many of the predators in the area have already learned to avoid the
white ermine. In addition predators themselves often have distinct
breeding seasons, and the period when fledglings are starting to find
their own food and are sampling models and mimics for the first time

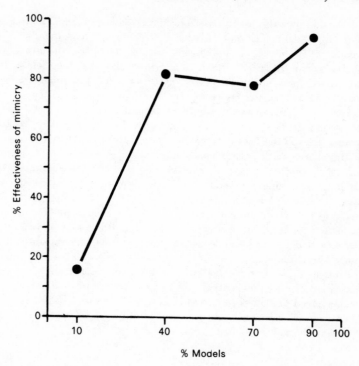

Fig. 4.3   The effectiveness of mimicry using starlings as predators and painted mealworms with or without quinine as models and mimics. The mealworms were offered at four different frequencies of models and mimics (horizontal axis), and the vertical axis shows the percentage out of all mimics offered which survived presentation to a bird. (*Redrawn from* Brower, 1960, Fig. 2. © by the University of Chicago.)

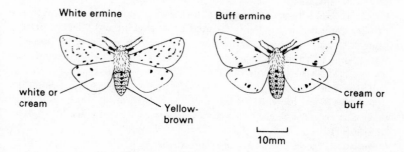

Fig. 4.4   The white ermine moth (*Spilosoma lubricipeda*) and the buff ermine (*S. lutea*). The abdomen of both is yellow or brownish yellow with darker markings, and the wings are white or cream with black spots.

will be particularly critical for the mimics unless their population is very large. Waldbauer and Sheldon (1971) found that the dipteran mimics of bumblebees and wasps in Illinois have two peaks of abundance in the year. The first is in late May and June when the models are present but not very numerous. Possibly the predators involved remember the Hymenoptera from unpleasant experiences the previous autumn when they were very abundant – Mostler (1935) and Rothschild (1964) have shown that some birds can retain learned avoidance responses for over a year. In July and August there are few or no dipteran mimics flying. This is the season when most birds in the area have finished breeding and there is a peak in the number of fledglings searching for food. The hymenopteran populations are by then building up, so the chances are high that these inexperienced predators will sample unpalatable models rather than palatable mimics. By September the hymenopteran population is at its maximum and the surviving fledglings have had considerable experience of various insect foods. There is a second peak of mimics at this time.

Thus in these dipteran mimics, the life cycles are adjusted to the population fluctuations of the model and to the seasonal abundance of inexperienced predators so that they receive a high level of protection. When the models are rare or when inexperienced predators are abundant, mimics are almost completely absent.

2.   If some predators eat the model they will also prey on the mimic

Under conditions of food shortage, predators which normally refuse to eat an unpalatable model may take it (see p. 79) and they will then prey on the palatable mimic as well. Similarly some species of predator may be unaffected by the noxious qualities of the model, and these too will be attracted rather than inhibited from preying on the mimic. Hence one might expect mimics of very nasty models to be better protected than mimics of slightly nasty models, and that therefore a mimic of a slightly nasty model will be better protected if it is rare than if it is common. This was investigated by O'Donald and Pilecki (1970) using artificial flour and lard prey with wild sparrows as predators. Unfortunately the numbers of rare mimics eaten in their experiments was very small so that the results do not give a decisive answer. However, they do seem to indicate that, relative to other prey, common mimics are eaten more when the model is slightly nasty than they are when it is very nasty (see discussion of these experiments by J. and M. Edmunds, 1974).

One way in which a species comprising rare mimics can still be numerous is to be polymorphic. Hence, with slightly nasty models which are often eaten by predators, selection may favour the evolution of polymorphism in the mimetic species with each morph mimicking a different species of model or else being non-mimetic. With

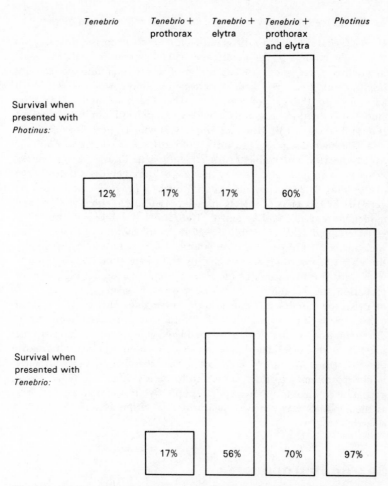

Fig. 4.5 Survival of artificially made *Tenebrio* mimics of *Photinus* using *Anolis* lizards as predators. Each test insect was presented together with either a *Photinus* (above) or a *Tenebrio* (below). Ninety-six per cent of these control *Photinus* and 21 per cent of the control *Tenebrio* survived. (*Data from* Sexton, 1960.)

very nasty models which are rarely eaten by predators there is no evidence that a common mimic is at a selective disadvantage to a rare mimic, so one would expect such mimetic species to be monomorphic. The best known polymorphic mimetic butterfly is *Papilio dardanus*. The males of this species are non-mimetic, but the females mimic a variety of different models in Africa. Unfortunately although the genetics of this species are well known, there is very little information on the palatability of the different models or on the protection afforded to the different mimics. Hence the suggestions made in this section must be regarded as very tentative.

3.   Some predators may learn to discriminate mimics from models

If any predators learn to distinguish the models from the mimics then the bright colours of the mimic may attract predators, and the mimic might then suffer more predation than if it were non-mimetic (and not brightly coloured). In these circumstances polymorphism will be of advantage to the species. One case where an individual predator (a chicken) learned to discriminate between model and mimic salamanders is described on p. 130. However, the way in which a predator discriminates is rather complex. Sexton (1960) offered aposematic *Photinus pyralis* beetles and edible *Tenebrio molitor* to lizards (*Anolis carolinensis*). To provide mimics he glued the coloured elytra and prothorax of *Photinus* to *Tenebrio*, and used untouched *Tenebrio* as controls. He gave the lizards pairs of insects in each presentation, either a mimic and model, or a mimic and a control. The results (see Fig. 4.5) show that when coupled with an edible *Tenebrio*, good mimics of *Photinus* are well protected (70 per cent not eaten). Less precise mimics, with only the elytra of *Photinus*, receive slightly less protection, and the mimics with only the prothorax of *Photinus* encounter a similar high level of predation as do control *Tenebrio*. However, when presented with models, the mimics are much more vulnerable, and only the good mimics are well protected (60 per cent not eaten). Hence the powers of discrimination of the predator depend on whether the mimic is seen together with the model or not, and also on whether there is alternative prey available. We should therefore expect batesian mimics which live in close proximity to their models to be almost identical with them in appearance, whereas mimics which simply live in the same general area as their models may receive equal protection from a much less precise resemblance.

Sex-limited mimicry

One solution to the problem of a mimetic butterfly whose model is either not particularly unpalatable or is periodically rather scarce is to develop a sex-limited mimicry. In every case known it is the female that is mimetic and the male non-mimetic. This means that if the species is as numerous as its model, the frequency of models to mimics will be 2 : 1 instead of 1 : 1, and so the females will receive greater protection than if both sexes were mimetic. The female requires greater protection than the male since she lays eggs and this is a more time-consuming process than the act of fertilizing a female. Further, a male can quickly mate several females, and each mating may suffice for several batches of eggs. It matters little if he is soon killed, but the female can continue to lay eggs for several days or weeks, and so the longer she survives the greater will be the number of progeny she may be able to leave. The female too is often heavier than the male (because of carrying eggs) so she is less agile in flight and more liable to be caught by a bird. For these reasons, if only one sex is to be mimetic it must obviously be the female.

It is known that in many butterflies visual signals play an important part in courtship, for example *Pieris napi* males are attracted to white females more than to yellow females (Petersen *et al.*, 1952). Hence it is possible that the demands of sexual selection might conflict with mimicry, and males might court preferentially with females having a non-mimetic colour pattern. Burns (1966) found some evidence for this in *Papilio glaucus* in North America. Here the male is non-mimetic, whilst the female is dimorphic, either resembling the male or mimicking the aposematic *Battus philenor* (Fig. 4.10). Burns found that mimetic females are mated only 81 per cent as frequently as non-mimetic females, and he confirmed this by counting the numbers of spermatophores in the spermathecae of the two morphs. In one population from Virginia each mimetic female, on average, had 1.69 spermatophores whilst each non-mimetic female had 2.08 spermatophores. Due to the small size of one of his samples, however, this difference is not significant, but Levin (1973) claims to have further data from Virginia which support Burns' figures. Pliske (1972) examined a much larger number of insects from Florida (110 yellow and 110 black females), and he found that both morphs had on average 1.15 spermatophores per female. Hence there is evidence for preferential mating of non-mimetic females in some populations of *P. glaucus*, but not in others. There do not appear to be any differences in fecundity or fertility between the two female morphs (Levin, 1973), so the selective advantage of the yellows relative to the blacks is likely to be very small.

Stride (1956*b*, 1957) found a somewhat similar situation in *Hypolimnas misippus* in Ghana. He showed that the black and white males court more actively with females whose hindwings are orange than they do with females whose hindwings are white — in Ghana white-hindwinged females bear a much closer resemblance to the aposematic *Danaus chrysippus* than do orange hindwinged females (see p. 111). However, these observations do not necessarily prove that sexual selection by the males for a particular patterned female will oppose predator selection for mimetic females. When females that have already been mated are courted by a male they give a characteristic display flight which continues until the male gives up the pursuit (Stride, 1957, 1958). There is no difference in the frequency of this display flight between orange and white hindwinged females (Edmunds, 1969*a*), so though the males court non-mimetic females more actively than mimetic ones, all females quickly get mated under natural conditions.

L. P. Brower (1963) studied the courtship of wild male queen butterflies with females that had been painted. Some were painted like normal females, others were painted white, but he found no difference in the frequency of successful matings. Hence although male butterflies do show a preference for courting females of a particular pattern, there is no evidence that this results in differential fertility of the females nor that it significantly affects the selective advantage which mimetic females possess over non-mimetic females. Nevertheless, in conditions of very low population density it is possible that mimetic females are at such a disadvantage relative to non-mimetic females that their fertility is reduced.

It is also possible that female butterflies mate preferentially with non-mimetic males rather than with mimetic males, but at present there is no evidence to support this (L. P. Brower, 1963; Magnus, 1963). It is usually the male that makes the initial moves in the sequence of court-ship behaviour patterns, and the female's responses more often involve scent than sight. In the Central American nymphalid *Anartia fatima* Emmel (1972) found that both white and yellow males were more attracted to white females than to yellow females, but the female will mate with whichever male comes first. Thus sexual selection in this species occurs by the male preferring a particular morph of the female, not by the female selecting a particular type of male. (In this particular example yellow males fly later in the morning than white ones so the female is more likely to mate with a white than with a yellow. The reason why yellows persist in the population is probably either because yellows are more active than white in the heat of the day, or because the yellows have a slight mimetic resemblance to species of *Heliconius*.)

Evolution of mimicry

At first sight one might wonder how two unrelated animals could ever come to resemble each other closely either as batesian or müllerian mimics. Several experiments with artificially made mimics and models show how even very slight resemblances may have survival value. Morrell and Turner (1970) used wild suburban English birds as predators and pastry rolls presented on coloured triangles of card as models, mimics and controls. They used green card for the edible controls, red card for the models (which were soaked in a quinine solution), and either red or yellow cards, with or without a black bar, for mimics of varying similarity to the models. Eight prey of each type of mimic were presented to the birds in each trial together with sixteen models and eight controls, and there were five repetitions of each trial. The average numbers of prey eaten per trial are given in Fig. 4.6. A statistical analysis of the results showed that significantly more yellow than red mimics were eaten (5.0 compared with 2.4), hence poor mimics are more heavily preyed upon than good mimics. Similarly a significantly higher proportion of barred mimics were eaten than of unbarred mimics (4.3 compared with 3.1), and in fact a significantly higher proportion of red barred mimics were eaten than of red unbarred mimics (perfect mimics). Nevertheless, even the poorest mimics (yellow

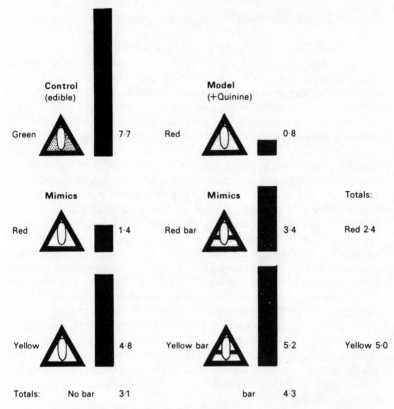

Fig. 4.6   Predation by wild birds on artificial models and mimics. The blocks indicate the average number of prey eaten during five series of trials out of a total of eight prey offered in each trial. The figure for models is out of sixteen offered in each trial. For further explanation see text. (*Data from* Morrell and Turner, 1970.)

barred) received more protection than the green controls. These results imply that birds can generalize, and hence even rather a poor mimic will receive some protection due to its resemblance. But birds can also discriminate good mimics from poor mimics, so selection will also favour the progressive evolution of a very close resemblance between mimic and model.

Schmidt (1958, 1960) arrived at very similar conclusions using chickens as predators and mash presented with various coloured cardboard butterflies as mimics and models. He found that the closer the similarity between mimic and model, the higher the number of mimics that are rejected; but also that even mimics with a very slight resemblance to the model received some protection compared with differently coloured controls. He also obtained evidence that individual birds respond to different visual cues when recognizing a cardboard insect as a model.

Ikin and Turner (1972) attempted to show that birds recognize prey by its overall appearance (*Gestalt* perception) rather than by certain specific visual cues only. They trained wild birds to avoid quinine impregnated bait presented on two backgrounds, green with black lines (GL) and brown with black triangles (BT). They then offered four types of mimics with palatable bait, two 'perfect' mimics (GL and BT) and two imperfect or hybrid mimics (GT and BL). In one series of experiments they showed that the perfect mimics were eaten less often than the imperfect mimics, hence they concluded that some of the birds were recognizing the mimics by overall appearance (colour and pattern combined) rather than by a single visual cue (e.g. pattern).

However, it is doubtful if two characters can legitimately be considered as constituting overall appearance — in a naturally occurring situation many features of colour and pattern together with behavioural characters would together contribute to the overall appearance perceived and recognized by a predator. Thus the experiment does not prove that birds recognize prey by *Gestalt* perception, but only that they may recognize it by two features together. To test for *Gestalt* perception in birds would require the use of more elaborate models such as those used by Schmidt (1958, 1960) or by Robinson (1973). Nevertheless, it is likely that a whole spectrum exists from birds which recognize prey by a single visual cue, and are hence easily deceived by even a poor mimic, to birds which recognize prey by overall appearance or by many visual cues and which can distinguish almost perfect mimics from their models.

Pilecki and O'Donald (1971), using coloured flour and lard bait with wild birds as predators, also showed that imperfect mimics (paler in colour than the models) derive some protection from their resemblance to the model. With a model : mimic ratio of 3 : 1 they found that the mimics were protected as well as if they had been identical in hue to the models. But with a 1 : 1 ratio, significantly more mimics were taken than models, indicating that the predators were then distinguishing the mimics from the models. Hence with a large number of mimics in the population selection will favour mimics which are much more similar to the model in appearance than it will with a smaller number of mimics in the population.

Ford (1971) also showed that even very imperfect mimics receive some protection from similarity to a model. Possibly the novelty of these imperfect mimics causes predators to avoid them and to concentrate on more familiar prey (see p. 152).

There is some evidence supporting these conclusions from experiments with caged predators and live prey. Thus J. V. Z. Brower (1963) showed that birds trained to avoid the monarch butterfly will also avoid not only the mimic *Limenitis archippus archippus* (Fig. 4.13), but also the differently coloured *L. a. floridensis* and its model *Danaus gilippus*. Birds conditioned to avoid the aposematic *Heliconius erato* also refused to eat two other butterflies (*Biblis hyperia* and *Anartia amalthea*) whose similarity to *Heliconius* is very imperfect, but they would attack a control species (*Anartia jatrophae*) of very different appearance (Brower, Alcock and Brower, 1971). Similarly, Windecker (1939) showed that birds conditioned to avoid the yellow and black caterpillars of the cinnabar moth also avoid wasps (*Vespula* spp.) which have a similar coloration but totally different shape. Hence these predatory birds were all able to generalize so that insects with only a slight resemblance to a nasty model were nevertheless protected.

In a rather different type of experiment Duncan and Sheppard (1965) trained chickens to avoid drinking a dark green solution when this was coupled with a low level electric shock. The birds were then offered equal numbers of model solutions (dark green with an electric shock) and mimic solutions (without a shock). The mimic solutions were diluted to varying degrees so that some were almost perfect mimics (82 per cent concentration of the model solution) whilst others were very poor mimics (10 per cent concentration of the model). Figure 4.7 shows that the more concentrated solutions were taken less

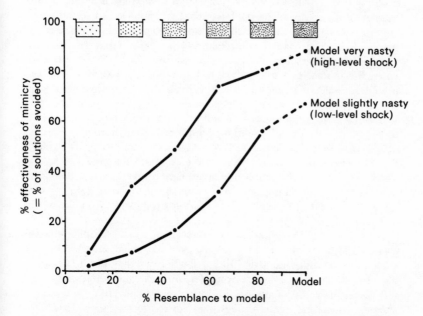

Fig. 4.7   Predation by chickens on mimetic solutions of green water. The degree of resemblance of the mimics to the model (extreme right) is indicated at the top of the figure. The vertical axis gives the frequency with which birds refused to drink solutions of each colour. For further explanation see text. (*Data from* Duncan and Sheppard, 1965.)

frequently than the less concentrated ones — in other words the more the mimic resembles its model, the better is it protected. Natural selection could therefore be expected to favour the evolution of batesian mimics which resemble their model very closely, since these would have a greater chance of survival than less accurate mimics. Duncan and Sheppard repeated their experiment with a higher level shock — in other words with a more distasteful model. Comparing the two series of experiments it is clear that the more distasteful the model, the greater is the degree of protection for any level of resemblance. It is also evident that both the 64 and 82 per cent perfect mimics are avoided almost as frequently as the very distasteful model, whilst with the low level shock they receive much less protection. In other words, one might expect selection to lead to a very close resemblance between mimic and model if the model is moderately nasty, but to a rather less precise resemblance if the model is very nasty since there is little advantage to be gained from increasing the similarity still further.

Evidence supporting these conclusions but using live prey is provided by a long series of experiments by Brower and Brower (1965). Toads quickly learn to avoid bumblebees and bees, but it was found that *Bombus* is avoided much more consistently than *Apis*. Presumably *Bombus* is nastier due to having a more virulent sting or because of its bristles. Both of these insects have dipteran mimics, but, using toads conditioned to avoid the appropriate model, 93 per cent of the *Bombus* mimics escaped predation whilst only 46 per cent of the *Apis* mimics escaped predation. This indicates that the mimicry is more effective when the model is very unpalatable (although it could also be interpreted as indicating that the toads were able to discriminate between *Apis* and its mimic better than between *Bombus* and its mimic).

Holling (1965), basing his calculations on the observed behaviour of predators in a variety of situations, supports the conclusions of Brower (1960), Morrell and Turner (1970) and Duncan and Sheppard (1965) on theoretical grounds. He shows that even when a model is rare a mimic should nevertheless receive some slight protection; that the more unpalatable is the model the greater is the protection given to the mimic; and that the degree of protection depends also on the availability of alternative prey.

Alcock (1970*b*) reached similar conclusions from experiments with caged birds (*Zonotrichia albicollis*) and prey concealed under sunflower seeds marked with four paint spots. He showed that the birds could discriminate between mimic and model when the mimetic resemblance was rather poor and the model was either slightly distasteful (salted) or emetic. When the resemblance was more precise, they failed to discriminate between model and mimic but regularly attacked the salted models and mimics whereas they avoided all emetic models and mimics. However, the birds were capable of discriminating between the mimic and model, even when the resemblance was close, since they regularly chose the mimic rather than the model when the model was simply an empty seed. Hence the protection of the mimic is a function both of the unpalatability of the model, and of the degree of resemblance of the mimic to the model. A slightly distasteful model gives protection only if alternative prey is available, but an emetic model can protect a mimic even if there is no alternative food. In both cases the degree of unpalatability of the model determines whether or not the bird discriminates model from mimic. The empty models with no reward and their mimics can be compared with sticks or leaves and stick or leaf-mimicking insects. These have considerable protective value provided that the models vastly outnumber the mimics and provided that the degree of resemblance is very close, since predators can discriminate much more finely between inedible models and mimics than between unpalatable models and mimics. Further, chickadees (*Parus atricapillus*) soon give up turning over sunflower seeds if none is rewarded whereas they persist much longer with inspecting salted prey or even emetic prey (Alcock, 1970a). So a mimic is best protected if it resembles either an emetic model which need not be common, or an inedible model which must be superabundant. But the degree of similarity to the model should be much greater with the inedible model than with the emetic model.

All of these experiments suggest how mimicry might evolve by selection favouring progressively closer resemblance between mimic and model with each stage having a slight protective advantage over the previous one. There is some circumstantial evidence from natural populations that this does in fact occur. *Papilio dardanus* is a large butterfly with a variety of different morphs in East Africa each of which mimics a different species of aposematic butterfly. Around Entebbe, Uganda, the models outnumber the mimics and the resemblance of the mimetic *P. dardanus* to their respective models is very close — only 4 per cent were 'imperfect' and differed significantly in coloration from their model (Fig. 4.8). At Nairobi, Kenya, on the other hand the mimics outnumbered the models, and a third of the *P. dardanus* were imperfect with no close resemblance to any model (Ford, 1964). Thus where the models are abundant selection eliminates edible insects which differ in colour from the model (i.e. 'imperfect' mimics); but where there are few models these imperfect mimics are not at such a disadvantage relative to the more perfect mimics (since fewer predators have learned to avoid the models) and so they occur at higher frequency. Clearly mimicry is more likely to evolve a very close resemblance between mimic and model at Entebbe than it is at Nairobi. It would be interesting to examine the situation in these two places again to see whether or not it has changed since Carpenter made these observations forty years ago.

A similar but rather more complex situation occurs in the butterfly *Pseudacraea eurytus* which has several colour forms mimicking different

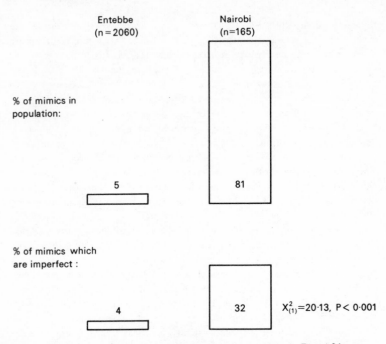

Fig. 4.8   Frequency of imperfect mimics of *Papilio dardanus* in East Africa compared with the frequency of mimics in the entire population (of mimics and models combined). (*Data from* Ford, 1964.)

species of *Bematistes*. Both in Sierra Leone and at Entebbe, the models outnumber the mimics and few mimics depart significantly in colour from that of one of the models. But on islands in Lake Victoria where the models are rarer, the mimics are very variable in coloration (Fig. 4.9) (Carpenter, 1949; Owen, 1971). These observations on *Papilio dardanus* and *Pseudacraea eurytus*, together with the experiments by Morrell and Turner, Schmidt, Duncan and Sheppard and the Browers, demonstrate the importance of predator selection in perfecting and maintaining a mimetic association.

The value of a very close resemblance between mimic and model is demonstrated by some data of Kirkpatrick (in Rettenmeyer, 1970). He examined the stomach contents of twenty-seven cattle egrets. One bird had eaten ninety *Syrphus corollae* (which is a dipteran mimic of wasps) another had eaten twenty-seven, eight more had eaten one or two, and the remaining seventeen birds had eaten none, yet all had been feeding in the same area. The most likely explanation of these figures is that most birds failed to discriminate between the syrphid mimic and its hymenopteran model, so they ate few or no syrphids. But two birds had learned to distinguish them and had developed a searching image for the syrphid so that they preyed on it very heavily. Under these

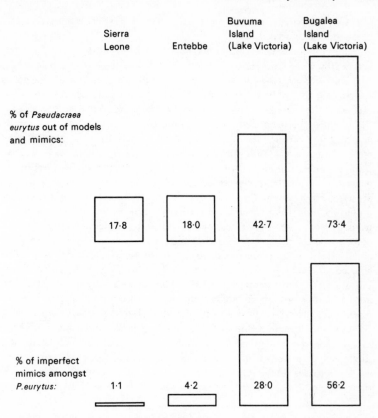

Fig. 4.9 Frequency of imperfect mimics of *Pseudacraea eurytus* in different populations related to the frequency of *P. eurytus* in the entire population of model and mimic butterflies. (*Data from* Owen, 1971.)

conditions any syrphid that resembles its model even more closely, so that the predators cannot distinguish it from the model, will have a greater chance of survival than less perfect mimics.

In West Africa *Hypolimnas misippus*, with orange hindwings, probably derives little if any protection from any resemblance to the aposematic *Danaus chrysippus* which has white hindwings. But individuals with even a small amount of white on the hindwings have an increased survival rate so long as the model (*Danaus*) is abundant (Edmunds, 1969b, and see p. 111). Hence even quite small similarities may be of protective value, and selection over the course of many generations would gradually cause the mimic to resemble its model more and more closely. There is no need to postulate some macro-mutation producing a more or less perfect mimic in one step; close resemblances have evolved gradually with each stage having a slight selective advantage over the previous one because of a closer resemblance to the model.

Where there is selection for a polymorphism, and where the characters which vary are controlled by many different genes, then selection will favour any translocations and inversions which bring the genes controlling these characters close together on the same chromosome. In this way the alternate sets of alleles will be inherited together so that the phenotypes produced clearly belong to one or other of the morphs and are not intermediate in appearance. In addition, the initial gene which controls the polymorphism may change its phenotypic expression by selection of various modifier genes so that the phenotype produced depends solely on which alleles of this 'switch' gene are present. Each population will evolve its own collection of modifiers under the control of one or more switch genes. But different populations of the same species may acquire different collections of modifiers controlled by the same switch genes. This is what appears to have occurred in *Papilio dardanus* and *Papilio polytes* which have been studied in detail by Clarke and Sheppard (see Sheppard, 1962; Ford, 1964; Clarke and Sheppard, 1972*b*). The males of these butterflies are not mimetic but there are a variety of different mimetic females resembling different species of models in different parts of their ranges. By crossing insects from different races, Clarke and Sheppard have shown that the control of the various morphs is usually by means of a single switch gene. But in each area different modifier genes are present so that crosses and backcrosses with other races cause breakdown of their interaction and produce very variable, non-mimetic progeny.

The question arises, what halts the process of evolution of a mimic? What stops it from becoming even more similar to the model? The answer must differ for every example of batesian mimicry since it will vary with the particular species of predators present, their powers of discrimination, and their degree of starvation; and also with the abundance and the degree of unpalatability of the model. Duncan and Sheppard's experiments suggest that with a very unpalatable model selection will not favour very perfect mimics since less perfect ones are equally well protected. But the palatability of the model may itself vary − in danaid butterflies it depends on the particular food plant on which the larva fed, which may be quite palatable or very unpalatable (see p. 104) so that there is a whole spectrum of palatability of models.

From the point of view of the model, having edible mimics is disadvantageous since predators may sample them and thus be encouraged to attack the models. If the model is only slightly nasty and has many mimics, selection will therefore tend to favour any models that are distinct from the edible mimics. The mimic will be evolving towards the model in its colour pattern and behaviour, but the model will be evolving away from its mimic. Selection may also be expected to favour aposematic insects which have more unpalatable qualities so that fewer have to be killed before the predator learns to avoid them in future encounters. This selection for nastiness may lead to tough unpalatable animals which can be sampled by a predator yet survive, and which produce repellent secretions or stings; or it may lead to palatable animals that have to be eaten but then cause the predator to vomit repeatedly so that it never touches another similar animal. The tough animal with repellent secretions is obviously the more effective if the species population is small, but predators may require regular reinforcement by attacking them from time to time. The palatable but emetic animal gives greater protection to conspecifics in future encounters since a single experience may cause a predator to refuse all such animals for many months or years (Brower *et al.*, 1970; Rozin and Kalat, 1971), but it can only be of value if the population is large enough to withstand the loss due to killing whilst the predators learn.

The distinction between batesian and müllerian mimicry is probably not always clear. Two müllerian mimics need not be equally unpalatable, and in conditions of starvation a predator might regularly take one such slightly unpalatable mimic which has thus become a batesian mimic. As de Ruiter (1959) points out, müllerian mimicry is really only an extreme form of batesian mimicry, and all intermediate stages of relative palatability may occur. Huheey (1961) even goes so far as to suggest that müllerian mimics have evolved from batesian ones by becoming progressively more unpalatable. But this appears to me unlikely except in special cases as, for example, when the mimic gradually becomes commoner relative to its model and so actually requires progressively greater protection. If it already has a high level of protection there is little to be gained from becoming distasteful as well since sampling by a predator (even of an unpalatable animal) is often fatal. Only if the model becomes scarce relative to the mimic, or if alternative prey becomes scarce, would the evolution of unpalatable attributes in the mimic be advantageous. But since a single species can contain both unpalatable models and palatable mimics (see p. 105), each individual case of mimicry must be looked at on its own merits — it is rash to generalize from one mimetic association to another unless the species concerned are closely related. The only generalization it is safe to make is that mimicry almost certainly evolves slowly over many generations by the accumulation of alleles into particular gene complexes and that these particular gene complexes become fixed in particular populations, though they may break down in the (unnatural) event of breeding with an individual from another population. Each population and each species of animal is affected by a different set of selection pressures whose complex interaction will result in different adjustments. Hence the actual path by which mimicry has evolved is different for every species of mimetic animal.

Some mimetic associations that have been intensively studied

In this section I shall describe some mimetic associations that have been intensively studied, and I shall try to point out the more important selection pressures which gave rise to and which now maintain these mimetic resemblances.

Mimicry in butterflies

Battus philenor *and its mimics*

One of the commonest aposematic butterflies in North America is the troidine papilionid *Battus philenor* (the pipe-vine swallowtail), whose colour is largely black. In the eastern part of the United States it is common from New York south to the Carolinas, but it is rare further south in Florida and Georgia, and completely absent from Canada. Several species of butterflies mimic *Battus philenor* including *Papilio troilus*, female *P. glaucus*, and the nymphalid *Limenitis arthemis astyanax* (Fig. 4.10). Birds quickly learn to avoid *Battus*, and they are then deceived into avoiding the edible mimics as well (Brower, 1958*b*; Platt *et al.*, 1971). We have seen that if mimicry is to be of protective value to the mimic then the mimic must not be too abundant relative to the model (see pp. 84–6, but see also the comments on p. 84), and it follows that the mimic will have no advantage in areas where the model does not occur at all. Brower and Brower (1962*b*) found that *Papilio troilus* is very uniform in colour in Tennessee and North Carolina where *Battus* outnumbers it by at least 8 : 1, but it is very variable in colour in Florida and Georgia, where *P. troilus* outnumbers *Battus* by at least 8 : 1. As with *Papilio dardanus* in East Africa (see p. 95), where the model is very rare relative to its mimics, selection does not eliminate imperfect mimics since birds rarely sample the model and there is therefore no advantage in resembling it closely.

In the tiger swallowtail, *P. glaucus*, only the female is mimetic, but it has two morphs, a black mimetic morph and a yellow non-mimetic morph similar in colour to the male (Fig. 4.10). In the Great Smoky Mountains the model is commoner than all of its various mimics, and 93 per cent of female *P. glaucus* are mimetic. Further north the frequency of *Battus* declines, and it does not occur at all in northern New York State (Fig. 4.11). There is a corresponding decline in the frequency of black female *P. glaucus* as one goes north, and in northern New York State the entire population is yellow. It is evidently of advantage to *P. glaucus* to be mimetic in areas where this can give even slight protection against predators, but where models are very rare or absent it is of greater advantage to be non-mimetic – possibly because males court yellow females more actively than they court black ones (see p. 89).

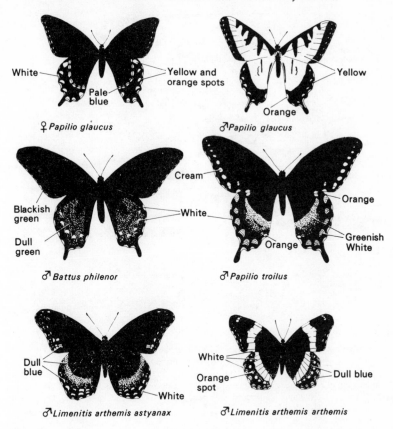

Fig. 4.10 The aposematic butterfly *Battus philenor* and its mimics. For further explanation see text.

The frequency of *Battus* also declines southwards and in Florida it is very rare relative to its various species of mimics. There is also a decline in the frequency of black female *P. glaucus* as one travels south, but in Georgia 96 per cent of female *P. glaucus* are black whilst only 7 per cent of the entire population are models, and in Central Florida the frequency of black females in the *P. glaucus* population has increased from 6 to 31 per cent in fifteen years, yet the model remains very rare. This unexpected high frequency of black *P. glaucus* may be caused by migration of insects southwards from the Tennessee and Virginia region (Pliske, 1972). If this is the correct explanation, then in Florida the blacks should be at a strong selective disadvantage relative to the yellows, whilst further north where the frequency of blacks is high and where *Battus* is abundant the two colour morphs should survive equally well. There is, however, no information on the survival of marked *P. glaucus* in different parts of its range, so this suggestion remains unproven.

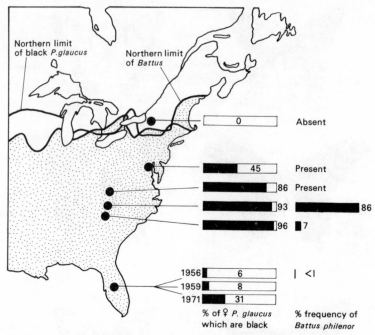

Fig. 4.11 Map showing the geographical area in which *Battus philenor* occurs (stippled) and the northern limit of black mimetic female *Papilio glaucus* (solid line). The proportions of black and of yellow female *P. glaucus* at six localities are shown on the right, and where available, the frequency of *Battus* out of the entire population of models and mimics is also shown. (*Data from* Burns, 1966, and Platt and Brower, 1968.)

The third mimic of *Battus philenor* that has been studied is *Limenitis arthemis astyanax*. This is black and mimetic in the range of its model but further north from Canada to Massachusetts, where *Battus* does not occur, it is non-mimetic and disruptively cryptic in colour (form *arthemis*) (Figs. 4.10 and 4.12). There is a narrow clinal zone where both forms hybridize freely (Platt and Brower, 1968). The mimetic *astyanax* form has no white band on the wings, and no orange-red spots on the edge of the hindwing, but there is a blue iridescence on the hindwing. The non-mimetic *arthemis* form has a white band and orange-red spots, but no iridescence. To the south of the clinal zone selection for mimicry is important since *Battus* is present, and here most of the insects have the *astyanax* characters fully developed (stage 6 on histograms in Fig. 4.12). As one passes north through the clinal zone insects with these characters at intermediate stages of development are common (stages 2 to 5). In the extreme north where *Battus* never occurs the white band is usually present (form *arthemis*), but iridescence may also be present in some individuals of the population whilst orange spots are occasionally absent. Evidently for a black

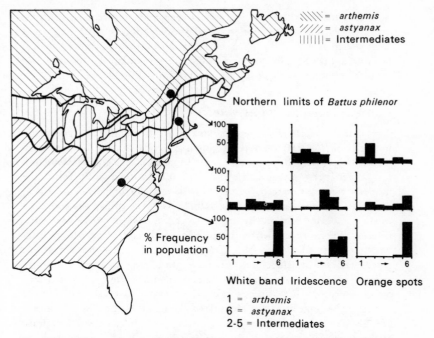

Fig. 4.12 Map to show the distribution of the *arthemis* and *astyanax* forms of the butterfly *Limenitis arthemis*. The area in which intermediates between these two types occur is indicated and can be compared with the northern limit of the model *Battus philenor*. For further explanation see text. (*After* Platt and Brower, 1968, Figs. 2, 4 and 5.)

butterfly, a broad white patch is likely to be advantageous in breaking up the outline of the insect, but whether or not the small spots and iridescence occur is of less importance. Hence selection in the north eliminates all insects without a white band, but is less stringent in eliminating insects with traces of iridescence or without orange spots.

It is clear that the mimetic form is selected in the south where *Battus* is present whilst the disruptively patterned *arthemis* is selected in the north where *Battus* is absent. The colour of the population in the clinal area probably varies in different years according to how numerous *Battus* is, so selection cannot lead to a very sharp discontinuity between the two forms. It may actually be advantageous here to have a variety of different forms so that predators cannot build up a searching image for any one type.

Thus in all three species of mimics predator selection has resulted in a high proportion of the population being mimetic in areas where *Battus philenor* is common. Where *Battus* is rare or absent, this selection pressure for a close resemblance is relaxed, and *P. troilus* is much more variable in coloration in such areas. The population of *P. glaucus* is almost entirely non-mimetic in such areas, possibly because selection here favours a similarity in colour between male and female, whilst in areas where *Battus* is common, predator selection is more important than sexual selection. In the case of *Limenitis arthemis*, where *Battus* is rare, predator selection for a close mimetic resemblance is reduced, but predator selection for a disruptive pattern is intensified.

Danaus plexippus *and its mimics*

The monarch butterfly, *Danaus plexippus*, is a common aposematic butterfly in the eastern part of North America, and in the Caribbean region (Fig. 4.13). Birds quickly learn to avoid it on sight after one or a few experiences of eating it, and such birds subsequently reject the mimetic viceroy *Limenitis archippus* (Brower, 1958a; Platt *et al.*, 1971) which is, however, slightly unpalatable. As a caterpillar the monarch feeds on species of plant belonging to the milkweed family (Asclepiadaceae) which are known to be poisonous to cattle and other vertebrates. Many plants of this family contain cardenolides (cardiac glycosides) which affect the rate and depth of heart beat of vertebrates and are lethal in heavy doses. Birds and mammals are caused to vomit by these drugs so that if taken orally they may in fact have no effect on the heart because they are eliminated. Monarch butterflies commonly feed on *Asclepias humistrata* and *A. curassavica* both of which contain cardenolides (Reichstein *et al.*, 1968; Brower, 1969). The caterpillars accumulate cardenolides and these are retained in the pupa and adult butterfly so that a bird that swallows a monarch is caused to vomit repeatedly over the next half-hour (Brower *et al.*, 1967), and subsequently it rejects further monarchs. However, if a monarch has been reared on a plant that does not contain cardenolides then the butterflies produced are edible and do not cause birds to vomit. Different species of milkweed contain different concentrations of cardenolides, and monarchs that have been reared on *Asclepias syriaca, tuberosa* or *incarnata* or on the related *Gonolobus rostratus* contain few or no

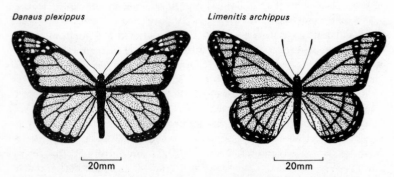

Danaus plexippus　　　　　*Limenitis archippus*

20mm　　　　　20mm

Fig. 4.13 The monarch butterfly *Danaus plexippus* and its mimic the viceroy, *Limenitis archippus*. The stippled areas are orange.

cardenolides and are palatable to birds. There is in fact a wide variation in the concentration of cardenolides in different species of milkweeds, and in the degree of palatability of monarchs reared on them. A single monarch that has been reared on *A. humistrata* contains such a high concentration of cardenolides that it can cause eight blue jays to vomit if shared between them. *A. curassavica* and *Calotropis* are less rich in cardenolides but have a sufficient quantity to cause one or a few birds to vomit, whilst *Gonolobus* reared monarchs are entirely palatable (Brower *et al.*, 1968; Brower, 1969).

The result of this variation in palatability of monarchs depending on which food plant they have been reared on is that natural populations contain some insects that are unpalatable (emetic) and some that are palatable. The palatable ones are of course mimics of the unpalatable models, and their resemblance is perfect since they are identical in appearance to the models. This intraspecific mimicry has been called *automimicry* (Brower, 1969). Fifty monarchs were collected in Massachusetts and fed to blue jays: twelve of the insects caused birds to vomit, so that in this population approximately 24 per cent were unpalatable and 76 per cent were automimics. One might suppose that with edibles outnumbering emetics by 3 : 1 predators would develop a searching image for the insects and prey upon them rather heavily. In fact as soon as a bird eats an emetic insect it subsequently rejects all insects of similar colour — in other words a single nasty experience is usually sufficient to cause the predator to reject similarly coloured insects in future for at least several months. The three edible: one emetic ratio is really an oversimplification of the true situation since it only applies to a population whose principal predator is the blue jay. Butterflies sampled in Massachusetts more recently have been shown to have very considerable variation in cardenolide content (Brower *et al.*, 1972). Some predators in the area are likely to be more sensitive to cardenolides than are blue jays whilst others are likely to be less sensitive, hence it is very difficult to assess what proportion of the butterfly population is emetic and what proportion is edible. Nevertheless, the three edible: one emetic is a useful ratio for considering the effects of different levels of predation on such a population.

Let us assume that a bird normally eats sixteen butterflies in a period of time (such as one butterfly generation), but that as soon as it eats an emetic one it will eat no more. Let there also be a butterfly population of 1 600 insects. It is then possible to calculate the numbers and percentages of butterflies that are likely to survive at various frequencies of model and mimic in the population, and at different levels of predation. With 100 predators in the population, if all the insects are palatable then all of them will be eaten (Fig. 4.14a). At the other extreme if all are unpalatable then 1 500 out of the 1 600 (93.75 per cent) will survive. Figure 4.14a shows that with 50 per cent or even 25 per cent of the population palatable automimics, the level of protec-

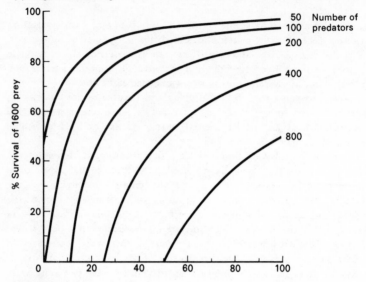

(a) Single trial learning

(b) Two consecutive trial learning

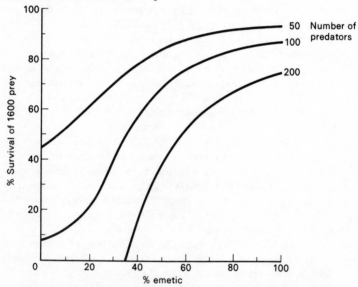

Fig. 4.14 Percentage survival of automimetic prey at different levels of predation with each predator taking a maximum of sixteen prey. The percentage of automimics in the population is 100 minus the percentage of emetics: (a) single trial learning, based on the formula derived by Brower et al., 1970, in which $P = (1 - (1 - k)^n)/k$. (b) two consecutive trial learning in which $P = 1 + (1 - (1 - k^2)^{n-1})/k^2$. P is the fraction of the prey population eaten, k is the frequency of unpalatables in the predator's sample of prey, and n is the maximum number of prey which a predator can eat assuming none is unpalatable (in this case $n = 16$). For further explanation see text.

tion achieved is almost as high as with the entire population unpalatable. If Brower's insects were subjected to this level of predation then, with nearly one-quarter of the population emetic, the insects would receive a 75 per cent level of protection. On the other hand with a higher level of predation (400 or 800 predators), the advantage gained by automimicry decreases sharply. Conversely with very slight predation (fifty or less predators) there is little advantage to be gained from being unpalatable since only a small proportion of the population is attacked. Automimicry is only effective with intermediate levels of predation. Figure 4.14*a* was constructed on the assumption that as soon as a predator takes one emetic insect it will never touch another of similar appearance (single trial learning). Figure 4.14*b* gives the survival of prey at different levels of predation assuming that a predator must take two consecutive emetic prey before it refuses to touch any more. It is clear that automimicry can also be of advantage under these conditions, but obviously for any frequency of emetics the advantage gained by automimicry is much less than with single trial learning (Pough *et al.*, 1973).

If predators with single trial learning can take more than sixteen edible prey (say twenty or fifty), then the advantage gained by automimicry is even greater (Brower *et al.*, 1970), so that there is even some advantage with only 1 per cent or fewer emetics in the population. With moderately high predation automimicry allows a population to increase beyond the limits imposed by the supply of emetic food plants but without significantly increasing losses due to predation. Let us suppose the emetic food plant can support a population of 1 600 insects, and there are 100 predators in the area. With a population composed entirely of emetics, 100 will be eaten and so 1 500 will survive. However, if there is an alternative, non-emetic food plant which can also support 1 600 insects, the predators will take twice as many insects as before (100 emetic and 100 edibles), but despite this twice as many (3 000) insects will survive. It should be noted that with the particular assumptions made here the relative frequencies of model and automimic are naturally self-replicating. If females lay eggs on the food plant on which they themselves were reared and the population consists of 25 per cent mimics, then this frequency will persist indefinitely because predation on the two forms is equal. Of course there may actually be other checks such as scarcity of one or other food, differential parasitization of the larvae on the different foods, etc., which may alter the ratio of mimics to models.

The situation is complicated by the occurrence of a batesian mimic of the monarch, the viceroy, so that the proportion of emetics in the entire population may fall below 25 per cent, but this situation has not been analysed. In Trinidad the monarch occurs with another similarly coloured danaid, the queen, *Danaus gilippus xanthippus*. Both feed on milkweeds, so both can be either edible or emetic. Here 65 per cent of the monarchs are emetic (feeding on *Asclepias curassavica*) whilst only 15 per cent of the queens are emetic (Brower, 1969). So although some queens are aposematic and hence müllerian mimics of the monarch, most of them are palatable batesian mimics. In other parts of America where both these species occur together they are very different in appearance, and the queens are probably straightforward automimics and models. Only in Trinidad, where the monarch displaces the queen from its principal emetic food plant, has the queen evolved a resemblance to the monarch. However both in Trinidad and on the mainland the larvae of the monarch and the queen probably come into competition for food plants (Brower, 1962), so the relation between these species is very complex.

Danaus chrysippus *and its mimics*

In Africa there is a closely related butterfly, the African monarch, *Danaus chrysippus*, which also feeds on milkweeds and may accumulate cardenolides so that vertebrate predators are caused to vomit when they eat it (Swynnerton, 1919; Reichstein *et al.*, 1968). *D. chrysippus* commonly feeds on *Asclepias curassavica* in East, Central and West Africa, but this plant is introduced and is not its normal food plant. In East Africa Owen and Chanter (1968) found that its main food in the Rift Valley is *Gomphocarpus physocarpus* which is probably slightly emetic (Brower, 1969), but in West Africa I have found that it commonly eats *Calotropis procera*, which has a higher concentration of cardenolides than *A. curassavica*, and *Pergularia daemia* and *Leptadenia hastata* both of which have a very low cardenolide content (Brower and Edmunds, in preparation). The African monarch differs from its American counterpart in that it has four well marked colour forms. In parts of East Africa all four morphs occur together in the population, but in West Africa only one form is present. The frequencies of the morphs in the population are not constant: in Uganda there is a significant difference between samples taken in 1909–12 and in 1964–6 (Fig. 4.15) (Owen and Chanter, 1968), whilst in Tanzania (Dar es Salaam) there is evidence for a seasonal change in morph frequencies. The Dar es Salaam morph frequencies given in Fig. 4.15 are based on insects collected over a two-month period in 1967. More recent work by D. A. S. Smith indicates that the frequency of the *chrysippus* morph is low in the dry seasons (down to 5 per cent of the population in February 1973) and high in the wet seasons (up to 50 per cent in June 1972). In addition there are further complications including presence of two other morphs at very low frequency, production of all female broods at certain seasons, and seasonal variation in mating preferences (Smith, 1973*a*, *b*, and personal communication).

This raises two questions not applicable to the monomorphic American monarch: first why is *Danaus chrysippus* polymorphic? and

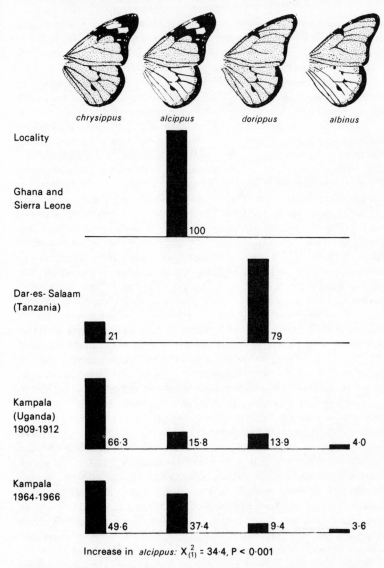

Fig. 4.15 Frequencies of the four morphs of *Danaus chrysippus* in the population in different parts of Africa. (*Data from* Owen and Chanter, 1968; and Edmunds, 1969*a*.)

second, why do all morphs occur in some areas whilst only one or two occur in others? One advantage of the polymorphism at Dar es Salaam seems to be that *chrysippus* is adapted to the wet season whilst *dorippus* is adapted to the dry season. However, it is necessary to explain why these two morphs with different physiological characteristics should also have different colour patterns, and why parts of West Africa, which have very similar climate to Dar es Salaam, have neither the *chrysippus* nor the *dorippus* form present in the population. The different colours of the four morphs must surely be adaptive in terms of other organisms in their environment which have good eyesight, that is either conspecifics or possible predators. Owen and Chanter (1968) supposed that polymorphism in an aposematic model may be a means of breaking down the precise resemblance of the mimics since they are unlikely to be able to mimic closely all the forms of the model. An alternative explanation of why the African monarch is polymorphic is that some predators find it palatable and so develop a searching image for it (they could find it palatable either because they are tolerant of cardenolides or because the caterpillars had fed on non-, or only slightly, emetic species of food plant). Under these conditions selection would favour any insects which differed from the normal pattern since predators would be likely to form a searching image for the common form and hence ignore the rarer ones (Edmunds, 1969a). Experiments using butterflies and grasshoppers which contain cardenolides have shown that there is very considerable variation in the response of predators to these insects — some predators vomit after a single trial, whereas others, such as the crowned hornbill *Tockus alboterminatus*, can eat several insects without apparent ill effects (Swynnerton, 1915b; Rothschild and Kellet, 1972). It is possible that with the greater diversity of birds in East Africa compared with West Africa, more predators occur in East Africa which occasionally or regularly prey on *Danaus*. Hence it is possible that polymorphism is of advantage in East Africa but not in West Africa, because there is a greater probability that at least one predator in East Africa will develop a searching image for it. A further complication is that when reared on the same food plant, West African *Danaus* accumulate less cardenolide than do East African *Danaus*. This suggests that West African insects are less distasteful than East African ones, but obviously a survey of cardenolide content throughout Africa is required before the full significance of this can be understood. In addition, different morphs may have different food preferences and hence different palatabilities. There is evidently much scope for research into the palatability spectrum of different populations of *Danaus chrysippus*.

Owen (1970) points out that in East Africa most populations of the butterfly *Acraea encedon* contain 60—100 per cent individuals which mimic *Danaus chrysippus*, whilst in West Africa only 0 or up to 9 per cent of the populations are normally mimetic. He argues that where there are many mimics the model suffers and so selection favours polymorphism, but where there are few mimics selection favours uniformity of colour pattern. But the situation may be much more complex than this. *Acraea encedon* is highly polymorphic with at least fifteen different colour forms in East Africa (Owen and Chanter, 1969). There are

(d) The bark-mimicking mantis *Theopompella westwoodi*. Note the outer section of the wing which is angled downwards to reduce lateral shadow. Tafo, Ghana.

(c) The praying mantis *Phyllocrania paradoxa* mimicking a dead leaf. Legon, Ghana.

(b) The leaf-mimic longhorn grasshopper *Zabilius aridus*. Legon, Ghana.

Plate 3
(a) *Argiope flavipalpis* resting in the centre of its web with a cross-shaped white webbing device. Legon, Ghana.

Plate 4

(a) Three orange dorid nudibranchs, *Discodoris tema*, resting on a rock covered with red algae and orange sponges: a brightly coloured animal that is cryptic. Tema, Ghana.

(b) The eolid nudibranch *Trinchesia coerulea*, possibly an example of aposematism. Plymouth, England.

(c) The aposematic dorid nudibranch *Phyllidia varicosa* from the coral reef. Dar es Salaam, Tanzania.

(d) Caterpillar of the African monarch, *Danaus chrysippus*, on *Calotropis procera*, possibly an example of an animal that is cryptic from a distance but aposematic from close to. Nungua, Ghana.

(e) Caterpillar of the moth *Chrysopsyche mirifica*. Note aposematic coloration and irritant hairs. Mount Atewa, Ghana.

(f) Cinnabar caterpillar (*Callimorpha jacobaeae*) on ragwort (*Senecio jacobaea*). Aposematic coloration, though possibly cryptic from a distance when on the flowers. Plymouth, England.

a	b	c
d	e	f

significant differences in behaviour and survival rate between some of these fifteen morphs, and some populations have a highly abnormal sex ratio with the frequency of males sometimes falling to less than 1 per cent. In West Africa there are fewer morphs present, but the frequencies of some of these vary during the year. It is not clear if this variation is the result of chance survival of certain egg batches during conditions when the population is low, or if it is due to certain morphs being at a selective advantage at particular times of the year (Owen and Chanter, 1971). *Acraea encedon* exudes a fluid that releases hydrogen cyanide in contact with air, so it is probably highly distasteful. Hence it is more likely to be a müllerian mimic of *Danaus chrysippus* than a batesian one: indeed, in populations where there is a high frequency of monarchs reared on non-emetic plants, *Danaus chrysippus* may actually be a batesian mimic of *Acraea encedon*.

In various parts of Africa there are other butterflies which mimic *Danaus chrysippus* and these are generally believed to be palatable species and hence batesian mimics. One of the commonest and most widespread of these is the nymphalid *Hypolimnas misippus*. The male has the black and white colour typical of the genus, but females are mimetic and occur in the same four colour morphs as *Danaus chrysippus*. There are, however, many female *Hypolimnas* with colour patterns intermediate between the four main morphs, so the genes controlling colour pattern are evidently not so tightly controlled by a master switch gene as in the African monarch or in *Papilio dardanus*. In West Africa only the *alcippus* form of the African monarch, with white hindwings, occurs (Fig. 4.15), so it is reasonable to suppose that where it is common the population of *H. misippus* should also have a high frequency of insects with white hindwings (form *alcippoides*). However, the area of white on the hindwings varies so that it is difficult to give the frequency precisely, but less than 5 per cent of female *H. misippus* have at least half of the hindwing white, whilst all of the monarchs have at least half of it white (Edmunds, 1969a). A much higher proportion of *H. misippus* have some small area of white on the hindwings, and it is possible that this slight resemblance to the model is sufficient to give it some protection from predators. In May 1965 at Legon, Ghana, 44.4 per cent of all females had some white present on the hindwings, but by July this had declined to only 12.5 per cent. This correlated with a decline in the frequency of the models in the population (Fig. 4.16). There were two reasons for the decline in the proportion of whites in the population of *H. misippus*; first a very low survival rate of white insects (see Table 4.1), presumably correlated with the scarcity and eventual disappearance of the model; and second an increasing survival rate for the orange insects during the course of the season (see Table 4.2) (the reason for this is not known). Selective values calculated by two different methods show that white was at an overall disadvantage relative to orange of 63 per cent and of 82 per cent per day (Edmunds,

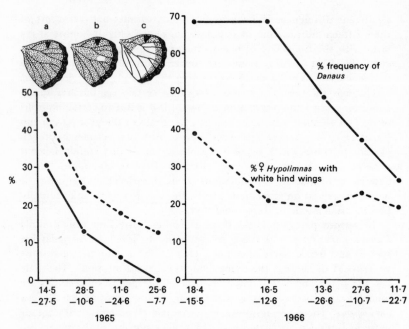

Fig. 4.16   Comparison of the relative frequency of models (*Danaus chrysippus*) in the population with the frequency of those mimics which closely resemble the models (i.e., female *Hypolimnas misippus* with white on the hindwings). Solid line: frequency of *Danaus* out of the combined population of *Danaus* and female *Hypolimnas*. Dotted line: relative frequency of female *Hypolimnas* in which there is some white present on the hindwings. Note that all *Danaus* are of the *alcippus* form (see Fig. 4.15) with white hindwings. (*a*) is an orange hind-winged *Hypolimnas*, (*b*) and (*c*) are white hindwinged forms. (*Data from* Edmunds, 1969*a*.)

1969*a*).

In 1966 *Danaus* was much more abundant than *Hypolimnas*, and although it declined in frequency it never disappeared. The orange *Hypolimnas* again showed an increase in survival rate as the season progressed (Table 4.2), so there was still a slight decline in the frequency of whites between April and July (see Fig. 4.16). But the white mimics had a high survival rate (Table 4.1), presumably because the model was abundant, so the decline in frequency of whites was not

Table 4.1   Survival rates of orange and white hindwinged female *Hypolimnas misippus* in 1965 and 1966 in Ghana. (*Data from* Edmunds, 1969*a*.) Survival rates per day with 95 per cent confidence intervals (in parentheses)

Date	Orange	White
May–July 1965	71% (56–86)	21% (0–48)
April–July 1966	64% (50–77)	70% (43–97)

Table 4.2 Numbers of marked orange hindwinged female *Hypolimnas misippus* released and recaptured in 1965 and 1966 in Ghana. (*Data from* Edmunds, 1969a)

Date	Marks released	Marks recaptured	% recaptured	$X^2_{(1)}$	p
1965					
May 21–June 24	118	13	11.0 ⎫	⎫ 2.24	⎫ 0.2 > p > 0.1
June 25–July 7	58	12	20.7 ⎭	⎭	⎭
1966					
April 19–June 15	76	4	5.3 ⎫	⎫ 11.276	⎫ <0.001
June 16–July 20	71	19	26.8 ⎭	⎭	⎭

Note: 1. Marks released includes all insects released, or retrapped and released, in the period indicated starting on the first day minus one, and omitting the last day. An insect captured and released twice will carry two marks and hence count as two releases.

2. Although the figures for 1965 are not significant, confirmation that survival does improve later in the season was obtained by comparing the proportions of worn and of fresh insects caught in June and July. Significantly more insects were fresh in early June than later on ($p < 0.01$), thus indicating that survival was better in later June and July.

so marked as in 1965 when *Danaus* was scarce. Recapture data and observations on the insects indicated that white insects fly less strongly than orange ones, and this disadvantage presumably balances the advantage which they possess when models are common. The situation is further complicated by the fact that during the long dry season when insects are scarce, the whites increase in frequency relative to the orange, and by the fact that there are also morphs with different forewing patterns which show similar changes in frequency and in survival rate during the year, but for reasons that are not clear. It would be interesting to know if there is any correlation between the survival rate of the white *Hypolimnas* and the degree of palatability of the local population of *Danaus* during the course of the year, but so far this has not been studied.

Thus the association between *Danaus chrysippus* and its mimics is very complex, but there is evidence that in Ghana those individuals of *H. misippus* which have some white present on the hindwing derive protection because of their resemblance to *Danaus*. Much more work is required on these species in Africa before we can claim to understand the situation at all adequately.

Mimetic beetles

A great many beetles are mimics, either of other beetles or of insects from some other order — at least as judged by appearance in museum cases. Very few examples have been studied in the field. The European wasp beetles *Clytus arietis* and *Strangalia* spp. are both black and yellow like the wasps (*Vespula* spp.) which they are said to mimic. *Strangalia* has a very close resemblance to a wasp in colour, but in its behaviour it is obviously a beetle. *Clytus arietis* differs very considerably from a wasp in the details of its black and yellow markings (Fig. 4.1), but when alive, its active, jerky movements give it a very close resemblance to a wasp searching for prey. Some species of Buprestidae such as *Acmaeodera* spp. also have spots or bands of white or red on the otherwise black elytra and are perhaps mimics of wasps or carpenter bees. These beetles are unusual in that they fly with the elytra lying over the abdomen, not spread horizontally as in other beetles. This means that even in flight their colour resembles that of the hymenopteran model. This is probably an example of batesian mimicry since the beetles were eaten readily by a captive bird (Silberglied and Eisner, 1969).

Lycid beetles are often coloured red and black or yellow and black, are slow moving, conspicuous, and rejected as food by lizards, birds, mice and even some invertebrate predators due to a nauseous smelling fluid secreted from the hind femora when they are attacked (Linsley, Eisner and Klots, 1961). In North America and in other parts of the world two or more species of lycid often occur together forming a group of müllerian mimics, and there may be longicorn (cerambycid) beetles and moths in the same area with similar colour patterns, some being edible batesian mimics and others unpalatable müllerian mimics. The models normally outnumber the mimics by 100:1, or even more. In one area the two lycids *Lycus loripes* and *L. fernandezi* outnumbered their longicorn mimics *Electroleptus ignitus* and *E. apicalis* by 100:1. The longicorn is edible to mice so is presumably a batesian mimic. Eisner *et al.* (1962) found that the longicorns are predators on the lycids. When attacking a lycid, the longicorn beetle climbs on to its back and attacks the dorsal surface of the thorax. The lycid makes little or no attempt to escape, perhaps because most of its predators make one attack and then leave it alone (and in any case, many predators are more likely to attack a moving prey, even if nasty, than a prey which stays still). It may also remain still because this is the normal position for copulation: males often palpate the thorax of the female lycid during mating, so that a female lycid does not attempt to throw off a beetle on her back which may be a male. The longicorn sucks blood and may occasionally kill its lycid prey. In one population 1.5 per cent of the lycids had evidence of injury caused by *Electroleptus*. Undoubtedly the longicorn derives protection from vertebrate predators due to its resemblance to *Lycus*, but obviously it cannot become too numerous or it will wipe out its model by predation. It is also possible that the *Electroleptus* may itself be genuinely unpalatable immediately after it has been feeding on a lycid due to having its gut full of partially digested lycid tissues. If this is so, then it may be either a batesian or a müllerian mimic depending on how recently it last fed on a lycid.

Ant mimics

Ants form large colonies and are particularly abundant in tropical regions. Most species can either bite or sting, or both, and few vertebrates regularly eat ants. It is not surprising therefore that ants are models for a variety of mimetic insects. The first instars of several species of praying mantids, grasshoppers and Heteroptera all closely resemble particular species of ants (Fig. 4.17). Most mantid larvae are green or brown and cryptic like the adults, but first instar larvae of *Tarachodes afzellii* are black with large heads, very like the common West African ant *Camponotus acvapimensis*, and the first instar larvae of the green *Sphodromantis lineola* are brownish red and of the same size and hue as the vicious red weaver ant, *Oecophylla longinoda*. Although the morphological resemblance between mantid and ant is not close (Fig. 4.17), the two have similar colour and behaviour. When alarmed *Oecophylla* ants normally raise the abdomen vertically, and young mantids regularly rest with the abdomen curled upwards in a similar position. Increasing size inevitably makes an ant 'disguise' impossible and some other form of protection becomes necessary. Later

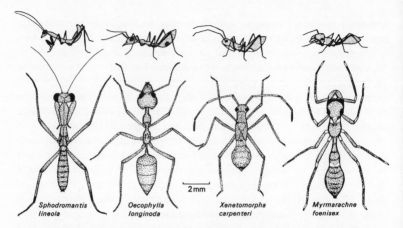

Fig. 4.17   The red weaver ant (*Oecophylla longinoda*) and some of its mimics: first instar mantis (*Sphodromantis lineola*), last instar heteropteran (*Xenetomorpha carpenteri*), and adult female salticid spider (*Myrmarachne foenisex*). All are brownish red with black eyes. Note the transparent sides near the front of the abdomen of *Xenetomorpha* and the white hairs on *Myrmarachne* behind the second legs and on the opisthosoma (shown white without stipple). These give the illusion of a narrow petiole. The live animals resemble *Oecophylla* more closely than do these anatomical drawings, and the sketches above are intended to emphasize the similarities in colour and shape of the live animals.

instars lose their resemblance to ants and become cryptic — *Tarachodes* becomes a brownish grey, flattened bark mimic, whilst *Sphodromantis* becomes green or occasionally brown in the dry season (Edmunds, 1972). Ant mimicry presumably affords some degree of protection against predation since it is known that many birds do not eat ants. However, *Sphodromantis* first instars can be found far away from any *Oecophylla* colony, so these mantids are not restricted to colonies of the particular ant they mimic. Grasshoppers and Heteroptera that mimic ants probably also gain protection from birds as a result of living amongst ants. My impression is that some species are definitely associated with particular species of ants for example the mirid *Xenetomorpha carpenteri* with *Oecophylla*, but they have not been adequately studied.

Several spiders from different families have also developed a resemblance to ants. The best known belong to the salticid genus *Myrmarachne*. The spider resembles an ant in body shape, with a narrow waist or petiole, and it holds its front pair of legs forwards like the antennae of an ant. In Africa and India the red weaver ant (*Oecophylla*) often has the red *Myrmarachne foenisex* or *M. plataleoides* in association with it. Similarly *Camponotus acvapimensis* and other black ants have a black species of *Myrmarachne* associated with them. But red *Myrmarachne* have never been reported with black ants, nor black *Myrmarachne* with red ants. Clearly the association is very specific and it is reasonable to suppose that it protects the spiders from birds or other visually hunting predators. There is only one record that I know of in which a species of *Myrmarachne* uses its resemblance to creep up on its model and prey on it (Hingston, 1927*b*). Normally *Myrmarachne* keeps well clear of its host and avoids the main runways of ants. Mathew (1954) reports that if an ant gets close and a spider fails to escape, it will be killed. In a glass tube, and in field observations on three species of *Myrmarachne* with their host ants, I have never observed spiders attack ants, and the spiders usually escaped by their keen eyesight and quick evasive responses.

The disadvantage of living with *Oecophylla* is that the ant is liable to attack anything living. Adult and juvenile spiders may be able to escape from the ants by hiding in a silken retreat or by speed of movement, but the eggs are occasionally found and eaten by the ants. Marson (1947*b*) reports that *Myrmarachne plataleoides* males chase away *Oecophylla smaragdina*, but in Africa I have found *Oecophylla longinoda* eating eggs whilst the displaced male and female *M. foenisex* rested a few centimetres away and made no attempt to defend them.

There is evidence that the mimicry of *Myrmarachne* to ants gives protection against predation by spider-hunting wasps. Many wasps prey on salticids but they very rarely take species of *Myrmarachne*. For example seven out of nine nests of the wasp *Sceliphron* in Ghana contained salticid spiders, but none of the thirty-two salticids captured was an ant mimic. Wasps such as *Pison xanthopus*, however, specialize in capturing salticids, and do take some *Myrmarachne*. But *Pison* apparently takes fewer *Myrmarachne* and more of other salticids relative to their frequencies in the environment (Fig. 4.18). *Myrmarachne* is sometimes the commonest salticid in shrubs at Legon, Ghana, especially if spiders hiding in silken retreats and so not available to a hunting wasp are ignored.

There is also evidence that *Pison* hunts by the searching image method (see p. 41) since individual wasps hunting in the same area sometimes catch quite different species of prey. It is clear from Fig. 4.19 that wasps *a* and *c* formed searching images for *Myrmarachne* whilst *b* and *d* did not.

Another wasp that may specialize in catching ant-mimicking spiders is *Trypoxylon placidum* from Malaya of which one female collected ten ant-mimicking salticids, four other salticids, and three other spiders (Richards, 1947).

It is therefore probable that ant-mimicry protects *Myrmarachne* because *Pison* is less likely to find and attack a *Myrmarachne* than it is to find and attack some other salticid, and because *Pison* is less likely to form a searching image for *Myrmarachne* than it is for other species of salticid. Furthermore, *Pison* never takes red *M. foenisex* which mimic and associate with *Oecophylla*, but it does take large black *Myrmarachne* from amongst *Camponotus* ants. Presumably the wasps do not

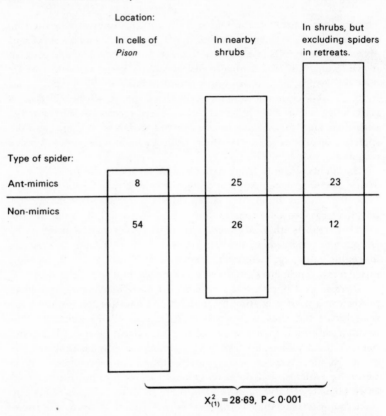

Location:

	In cells of *Pison*	In nearby shrubs	In shrubs, but excluding spiders in retreats.
Type of spider:			
Ant-mimics	8	25	23
Non-mimics	54	26	12

$$X^2_{(1)} = 28 \cdot 69, \ P < 0 \cdot 001$$

Fig. 4.18   Comparison of the frequency of ant-mimicking spiders (mostly *Myrmarachne* spp.) in cells of the wasp *Pison xanthopus* with that in nearby shrubs. The third column is derived from the second by omitting all spiders which were found in their retreats and hence were not available to a hunting wasp. All data collected at Legon, Ghana, in January and February 1973.

hunt amongst *Oecophylla* colonies where they might easily be killed, but there is little danger in searching amongst the foraging trails of the comparatively harmless but equally common *Camponotus* ants. Thus it is likely that ant mimicry by salticids is of protective value against predation by wasps, and also perhaps, by birds.

The thomisid ant-mimic *Amyciaea* from India, South-East Asia and Australia is superficially much less like *Oecophylla* than is *Myrmarachne*, and its movements are not particularly ant-like. Unlike *Myrmarachne*, *Amyciaea* is a predator on ants (Hingston, 1927*b*; Clyne, 1969; Mathew, 1954). It apparently lures ants towards it by making rather ineffectual movements like those of an ant in distress. The ant approaches, and is promptly pounced on and sucked dry by the spider. This is therefore an example of aggressive mimicry with the mimic using

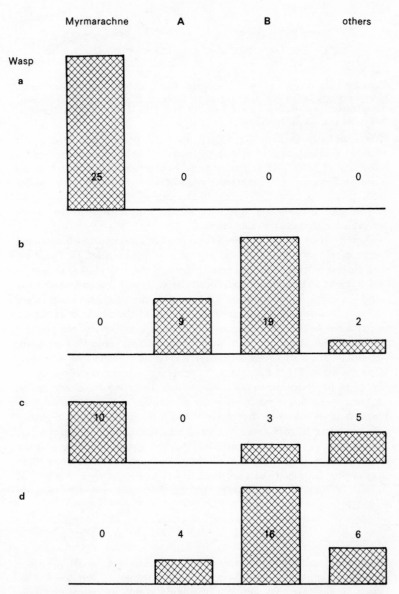

Fig. 4.19 Numbers of *Myrmarachne* spp. and of other salticid spiders in cells constructed by four wasps (*Pison xanthopus*) at Legon, Ghana. Cells made by wasps (*a*) and (*b*) were 2 m distant and collected on 6 July 1972; cells (*c*) and (*d*) were 14 m distant and collected on 15 February 1972.

Mimanomma spectrum   *Dorylus nigricans*

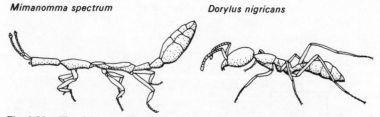

Fig. 4.20   The driver ant *Dorylus nigricans* (right) and the beetle *Mimanomma spectrum* (left) which runs with the driver columns. (*Redrawn from* Hölldobler, 'Communications between ants and their guests'. Copyright © 1971 by Scientific American, Inc. All rights reserved.)

its resemblance to approach closely towards its prey and to attack it without warning. Since the ant has poor eyesight, the resemblance is more one of behaviour and movement than of detailed structure. There is no evidence as to whether the resemblance also protects the spider from predators, but it is likely that spiders with only a superficial resemblance to ants are aggressive mimics whilst those with a close resemblance are true batesian mimics.

A number of species of beetles also regularly associate with ants, and here too mimicry appears to be involved (Hölldobler, 1971). There are two aspects to this mimicry: resemblance to the ants so that predators which avoid ants will avoid the beetle as well; and resemblance to the ant so that the ant is deceived into mistaking the beetle for another ant which it therefore does not attack. Living with the ant or in its colony protects the beetle since even an ant-predator is unlikely to prey on the few beetles amongst a multitude of ants. Some beetles that regularly run in the trails of driver ants have a very close resemblance to their hosts in shape (Fig. 4.20). Since the driver ant is blind, the resemblance can be of little benefit in inducing the ants to accept the beetle as one of themselves, but it must surely be of benefit in deceiving predatory birds. It is well known that many forest birds in the tropics regularly follow parties of driver ants and prey on the insects they disturb, but they do not eat the ants themselves. Hence a beetle travelling with the ants and resembling them is likely to be well protected. The role these beetles play in the life of their hosts is not known. (Other examples of protective associations between beetles and ants are discussed on p. 211.)

Chemical mimicry

Mimicry is usually thought to involve visually detected signals, probably because we ourselves find these the easiest to detect. But there is no reason to suppose that mimicry of sounds, scents, electrical or tactile signals do not also occur. Rothschild (1962) has pointed out that many insects have repellent odours, either as primary or secondary defence, as a warning to predators. If the predator can be deterred by a nasty smell without touching the insect, the insect will stand a better chance of surviving than if it has to be touched and tasted. It is possible that some of these odours are mimetic in that they resemble the odours of genuinely unpalatable insects, although their producers are in fact edible. No examples are, however, known. More elaborate cases of chemical mimicry occur in some social insects, but since the deception is a secondary defence rather than a primary defence, they will be discussed later on p. 259.

Mimicry of inedible objects

Typically the model of a mimetic association is an edible, but unpalatable and conspicuous animal which moves around advertising its presence. However, some animals resemble plants, animals or inanimate objects which are inedible to the predator concerned, and which may not move at all. These models are often very abundant, for example leaves, sticks or blades of grass. The mimic may be quite conspicuous, but since its predators do not normally eat sticks or grass, they ignore it. Stick mimics and grass mimics normally rest motionless amongst their models during the day and feed mainly at night. Geometrid caterpillars, stick insects and mantids such as *Hoplocorypha, Danuria* (Fig.

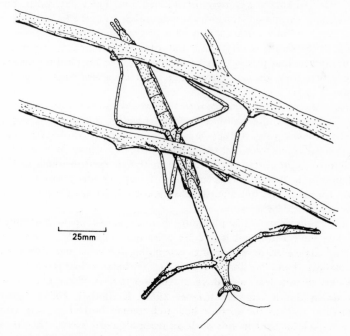

25mm

Fig. 4.21   Female *Stenovates strachani*, a stick-mimicking mantid from Ghana, in typical resting posture. (*Reproduced from* Edmunds, 1972, Fig. 5.)

S1.1) and *Stenovates* (Fig. 4.21) are all stick mimics. The caterpillars feed on leaves at night, but the mantids feed during the day. However, since mantids are predaceous they only feed occasionally when a prey insect comes within striking distance. The movements of the mantid itself are very slow with much rocking from side to side and teetering backwards and forewards — as if the insect is a twig swaying in the wind — except when actually striking at the prey (Edmunds, 1972). Although positive evidence is lacking, it is possible that the mimicry may increase the mantid's chances of approaching its prey undetected: if this is so, then the mimicry has an aggressive as well as a defensive component. Stick mimics are probably protected from birds because birds either fail to notice them or are deceived into mistaking them for the inedible sticks. De Ruiter (1952) found that caged jays (*Garrulus glandarius*) could not discriminate stick-like caterpillars (*Ennomos* spp. and *Biston* spp.) from sticks and twigs of the trees on which the caterpillars normally fed, but they could easily distinguish them from twigs of other trees. Moreover, once a bird accidentally found a caterpillar (for example by treading on it) then it quickly found most of the remaining caterpillars. Presumably it did not recognize a caterpillar initially, but once it had found one, it formed a searching image for the caterpillar and so quickly found the rest. Some birds did this by trial and error, pecking at any twig that looked a bit like a caterpillar, but others soon learnt to distinguish the caterpillars from the twigs without error. Stick mimics can only be well protected therefore if the resemblance to the model is very close, and if the models greatly outnumber the mimics so that predators have to search for so long before finding mimics that they give up and do not build up a searching image for them.

Other insects are similarly protected by a resemblance to blades of grass, for example the mantid *Pyrgomantis pallida*, and the grasshopper *Cannula linearis* (Fig. 4.22). One bird, the potoo (*Nyctibius*), rests during the day on trees and resembles a broken-off branch. It is often possible to approach to within a few metres or even to touch such a bird before it flies off. It even broods its egg in this position, fully visible for all to see, but almost indistinguishable from a piece of wood (Haverschmidt, 1964).

Other animals including phasmids, mantids, grasshoppers, butterflies and even some fish mimic leaves. Here the animal resembles a complete leaf and may even have markings resembling the veins and blemishes. The mantid *Phyllocrania paradoxa* (Plate 3c) bears a close resemblance to a dead brown leaf. and there are similar species of leaf-mimicking mantids in South America (Crane, 1952; Kettlewell, 1959). Many butterflies, such as the tortoiseshell and comma butterflies, rest with the wings closed and in this position resemble dead leaves. The green longhorn grasshoppers (bush-crickets) *Zabilius* and *Plangiopsis* have two alternative resting postures: they may rest with the wings held close

Pyrgomantis pallida   Cannula linearis

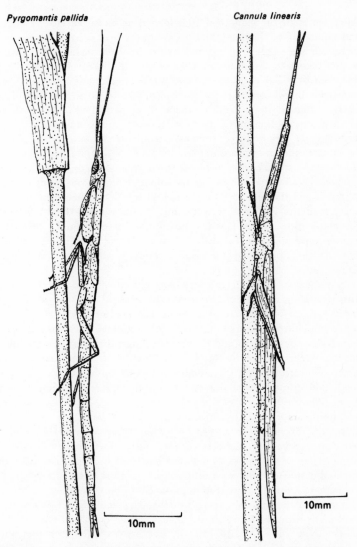

Fig. 4.22 Two grass-mimicking orthopteran insects: the mantid *Pyrgomantis pallida* (left), and the grasshopper *Cannula linearis* (right), from Ghana. (*Pyrgomantis* reproduced from Edmunds, 1972, Fig. 3; *Cannula* original.)

together over the back in a narrow angled inverted V, and so resemble a complete leaf, or they may rest on a leaf and hold the wings flat in a very shallow ⌃ so that they appear to be a part of the leaf itself (Plate 3*b*).

Lobotes surinamensis is a Caribbean fish which, as a juvenile, lives amongst mangroves (*Rhizophora mangle* and *Avicennia nitida*). The fallen leaves of the mangrove float for a time on the surface of lagoons and pools before they sink and decay on the sea bed. The fish is of the same size and shape as a leaf and also rests at the surface of the water, or just below it, with one side uppermost, and the head angled slightly downwards. It has the same yellowish brown colour as a dead leaf with numerous brown spots, and if leaves are thrown on the surface of the water it moves slowly towards them. When disturbed it moves slowly by means of its transparent pectoral fins, or it may make sudden swift darts before becoming motionless again (Breder, 1949). In Ghana *Lobotes* occurs in pools amongst leaves of *Thespesia populacea* and *Dahlbergia ecastaphyllum* both of which are yellow, blotched, or black (Fig. 4.23). *Lobotes* can change its colour to match any of these three patterns, and the resemblance is so close that local village children had never seen the fish until it was pointed out to them (Pople, personal communication). When it is touched, *Lobotes* swims quickly away, but if only mildly disturbed it drifts down in zig-zag fashion like a sinking leaf. At the bottom of the pool it suddenly rights itself and swims rapidly away.

A number of insects have a very close resemblance to bird droppings. Bird droppings are often deposited conspicuously on leaves and they are not normally eaten by other birds. Caterpillars of several species of swallowtail butterflies are mimics of bird droppings when they are small (e.g. *Papilio demodocus* and *P. memnon*), and they rest conspicuously on the upper surface of flat leaves. When these caterpillars become larger they are too big to resemble a bird dropping and they become green and cryptic, normally resting on the stems and undersides of leaves (Plate 2d).

Caterpillars of the African bombycid moth *Trilocha* also mimic bird droppings. From the second instar onwards the caterpillars are black and white, gregarious, and rest conspicuously on the upper surfaces of leaves (Plate 2e). It looks as if a bird has been roosting just above the leaf and deposited a mass of droppings. The final instar caterpillars are too large to resemble bird droppings. In *T. obliquissima* they are brown, solitary in habit, and they rest on the brown stems and petioles of their food plant instead of on top of the leaves (Carpenter and Ford, 1933). In *T. kolga* they are also brown, but my experience is that they continue to rest conspicuously on leaves. They appear to resemble the large brown faecal masses of larger birds or lizards, but further observation is required on this point (Plate 2f).

The European alder moth (*Apatele almi*) also has a caterpillar which mimics a bird dropping when it is small, but after the third moult it is banded with black and yellow, these presumably being warning colours (illustrated in colour by Rothschild, 1971).

Another unusual model for a mimetic resemblance is the Pacific gastropod *Mitrella carinata* which is mimicked by the amphipod *Pleustes platypa* (Crane, 1969). This model is certainly not unpalatable

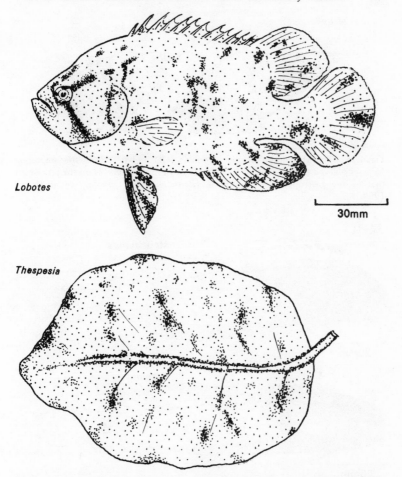

Fig. 4.23 Young *Lobotes surinamensis* and a leaf of *Thespesia populacea* as they appear floating on the surface of a rock pool in western Ghana. Both are yellow with brown blotches (*Drawn from a colour transparency taken by* Mr W. Pople.)

to some predators, but because of its protective shell and operculum comparatively few species can attack it successfully. Thus many predaceous animals will not recognize it as potential food, in the same way that insectivorous birds fail to recognize a leaf as potential food. The principle predators of *Mitrella* are likely to be slow-moving starfish and carnivorous gastropods. The amphipod probably derives protection from predators of *Mitrella* because it can escape with a sudden burst of rapid swimming, and from predators of small crustacea because they mistake it for *Mitrella* which is, to them, inedible. The colour of both model and mimic is very variable, both having bands of dark grey, yellow and brown (Fig. 4.24). This probably disrupts the outline of the

Mitrella   *Pleustes*

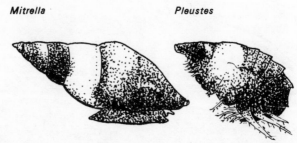

Fig. 4.24   The gastropod *Mitrella carinata* (left) and its mimic the amphipod *Pleustes platypa* (right). Notice that the head of the amphipod is on the left whilst that of the gastropod is on the right. (*Redrawn from a photograph in* Crane, 1969.)

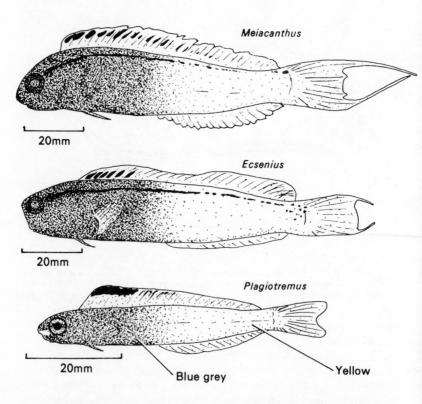

Fig. 4.25   Mimetic fish: *Meiacanthus nigrolineatus* (top), *Ecsenius gravieri* (middle), and *Plagiotremus townsendi* (bottom). All are blue-grey anteriorly, yellow posteriorly, with black stripes or blotches dorsally. For most predators *Meiacanthus* is distasteful whilst the other two are palatable and are hence mimics of it. See text for further explanation. (*Redrawn from* plates 2a, b, c of Springer and Smith-Vaniz, 1972.)

animal on the seaweed *Macrocystis pyrifera* where both species live. The shape and resting posture of *Pleustes* is similar to that of *Mitrella* except that the anterior and posterior ends of the animal are reversed.

The sea urchin *Astropyga radiata* is the model for another mimetic animal. Sea urchins are often conspicuous, move slowly, and rely for defence on their long spines. Various species of fish derive protection from living between the spines where they are inaccessible to predators (see p. 213). The fish *Siphamia argentea* commonly lives between the spines of *Astropyga* and it has the same coloration as the urchin. However, one sea urchin can only accommodate a limited number of fish. If there are too many fish, a group of *Siphamia* leave the urchins and live nearby in a dense shoal of the same size and shape as a large sea urchin. Here the grouped fish are obviously typical batesian mimics deriving protection from predators by their deceptive resemblance to an urchin. The mimetic behaviour has obviously evolved from the protective association with *Astropyga*, and from the habit of living in dense shoals (Fricke, 1970).

Another case of a group of animals together mimicking a single object is the cicada *Ityraea* which characteristically rests on plants in clusters resembling a head of flowers (Wickler, 1968).

Mimicry in fish

Several cases are known of mimicry in fishes: Wickler (1968) gives some examples including the use of lures in angler fish and of dummy eggs in *Haplochromis*, and the mimicry of leaves and of sea urchins by fish has already been described. Two further examples of mimicry in fishes will now be described.

Species of the poison-fang blenny *Meiacanthus* have canine-shaped teeth in the lower jaw which are hollow and contain a poison gland. Their bite causes inflammation to the human skin, and species of *Meiacanthus* are rejected as food by several species of predatory fish (Losey, 1972; Springer and Smith-Vaniz, 1972). *M. nigrolineatus* from the Red Sea is blue-grey anteriorly, yellow posteriorly, with black stripes on the body and on the dorsal fin (Fig. 4.25). Another blenny from the Red Sea, *Ecsenius gravieri*, has similar coloration but is acceptable as food to predatory fish. However, if they have sampled *Meiacanthus*, predators refuse to touch *Ecsenius*, so *Ecsenius* is clearly a batesian mimic of *Meiacanthus*. A third species of blenny with similar coloration, *Plagiotremus townsendi*, lives in the same area. *Plagiotremus* is palatable to most predators, so is also a batesian mimic, but it is rejected as food by the lionfish (*Pterois volitans*). Thus for the lionfish *Plagiotremus* is a müllerian mimic of *Meiacanthus*. A further complication is that whilst *Meiacanthus* and *Ecsenius* are not aggressive and permit other small fish to approach them, *Plagiotremus* is aggressive and attacks other fish. It is therefore likely that small fish, which have learned that it is safe to be close to *Meiacanthus* or *Ecsenius*, will be deceived by its coloration into ignoring the approaching *Plagiotremus* as well. Hence *Plagiotremus* may be able to approach and attack its prey with a greater chance of success than if it had a different colour pattern. This is therefore an example of 'aggressive mimicry' — of a predator using its resemblance to a harmless model to approach prey without their being alerted. Similar mimetic associations involving species of *Meiacanthus* and *Ecsenius* occur throughout the Indian and Pacific Oceans. For example, at Eniwetok atoll Losey found that *M. atrodorsalis* is very common and outnumbers the mimetic *E. bicolor* by from 2:1 to 50:1 in different parts of the reef. On this reef there is also an aggressive mimic of *Meiacanthus, Runula laudandus*. On the shallow reefs where *Meiacanthus* outnumbers *Runula* by only 9:1, Losey found that fish were very wary of *Runula* and often evaded its attacks. But on deeper reefs, where *Meiacanthus* outnumbered *Runula* by more than 100:1, fish were less wary, and many more attacks by *Runula* were successful. No doubt *Runula* also derives protection against predators from its resemblance to *Meiacanthus*, so it is both a batesian and an aggressive mimic.

The second example of mimicry in fishes is even more complex since the defence is probably not so important for the mimic as are other factors. This is the cleaner mimic (summarized by Wickler, 1968). Many species of fish and of shrimps in the sea obtain food by cleaning the bodies of other fish of fungal growths and parasites. They may have other foods as well, but much of their time is spent searching over the bodies of fish for parasites, and they even enter the mouth and gill chamber in their search. The best known cleaner is the sea swallow, *Labroides dimidiatus* which is common on coral reefs in the Indian and Pacific Oceans. Each cleaner or pair of cleaners occupies a small territory on the reef, and other fish regularly come to this place to be cleaned (Potts, 1973). The cleaner is conspicuous with black and white stripes, but it is not unpalatable, so the colours are neither cryptic nor aposematic. It swims slowly towards other fish in a characteristic way using the pectoral fins and allowing the tail to rise and fall repeatedly in the water (Fig. 4.26). Other fish respond by presenting their bodies to be cleaned and by ceasing all active swimming and feeding movements. Even normally predatory fish will allow the cleaner to enter their mouths and gill chambers without attempting to swallow it — although very occasionally cleaner fish are eaten by other fish. The relationship is obviously symbiotic: the customer fish gain from the association by removal of parasites whilst the cleaner gains by obtaining a readily available source of food. The cleaner is also protected by its coloration and behaviour which inhibits other fish from attacking it. This is analogous to predators learning to avoid warningly coloured animals, but here there is a reward instead of a punishment or unpleasant experience. In aquaria fish which come from different regions of the world quickly

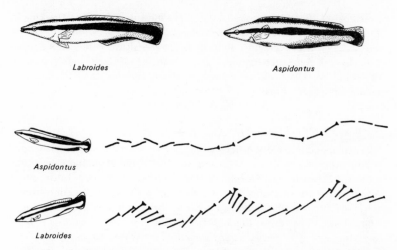

Fig. 4.26   Above: the cleaner fish *Labroides dimidiatus* (left), and its mimic the sabre-toothed blenny *Aspidontus taeniatus* (right). Below: swimming movements of *Aspidontus* and *Labroides* when soliciting a potential customer to be cleaned. Each line represents a jerky forward movement with the triangular wedge indicating vertical movements of the tail. (*Redrawn from* Wickler, 1968.)

come to recognize the cleaner and to make use of its facilities instead of attacking it, so the learning is positive conditioning, not negative conditioning as with predators of typical aposematic animals. The result as far as the cleaner is concerned is the same as for aposematic species — predators learn to recognize it as something which is not to be eaten and so it is protected by its coloration.

Another fish with very similar coloration to the cleaner *Labroides* is the sabre-toothed blenny, *Aspidontus taeniatus*. Customer fish may visit a place where they see a fish dancing just like *Labroides* does, but when the fish approaches them it bites a piece out of a fin or tail and eats it. The customer has been deceived, and may flee, or at least not return to that particular cleaning station again. Since *Aspidontus* is territorial, if the customer visits other stations it may not be attacked in this way again, and gradually its avoidance response towards cleaners wanes. But such a fish often develops the habit of keeping its head towards the cleaner so that the cleaner is forced to clean its mouth and gills first, rather than its tail and fins. *Aspidontus* always tries to go to the rear of a fish to make its attack. In this way an experienced fish gradually learns to distinguish the cleaner from its mimic, but young fish, or fish that only visit cleaners occasionally, continue to suffer from attacks by the mimic. Thus *Aspidontus* is clearly an aggressive mimic, using its resemblance to one species of fish, and the deception caused by the resemblance, to attack another animal. But it is also a batesian mimic since the attacked fish, though it may be hurt and may flee, has not been observed to turn round and attack the mimic.

Presumably the coloration of *Aspidontus* inhibits fish from attacking it. In fact the resemblance between *Aspidontus* and *Labroides* in colour and behaviour is very close indeed. *Labroides* shows some colour variation in different geographical areas, and *Aspidontus* shows parallel variation so that selection favours those mimics that deceive the greatest number of fish.

Mimicry in amphibia

The salamanders *Notophthalmus viridescens viridescens* and *Pseudotriton ruber schencki* are both red and black and resemble each other closely. Howard and Brodie (1971) found that wild birds avoid *Notophthalmus*, probably because they have already been conditioned to avoid them. Chickens will attack them initially but find them unpalatable and drop them. After from one to seven trials with anaesthetized *Notophthalmus*, four chickens learnt to avoid them on sight without pecking at them. They then avoided the similarly coloured *Pseudotriton*, even after a break of seven days (Fig. 4.27). Control birds were given *Pseudotriton* before they had experience with *Notophthalmus*, and they ate it readily. These birds were then given *Notophthalmus*, which they soon learnt to reject. Subsequently two out of the three birds also rejected *Pseudotriton*, but the third bird was able to discriminate between model and mimic and ate all the eight *Notophthalmus* it was given. Non-mimetic control salamanders were eaten throughout the series of trials. These experiments show that *Pseudotriton* is a batesian mimic of *Notophthalmus*, although, given favourable learning conditions, some predators may be able to discriminate the model from the mimic.

Another mimetic association has been described between the red-cheeked morph of the Allegheny Mountain salamander, *Desmognathus ochrophaeus*, and the red-cheeked salamander *Plethodon jordani* (Huheey, 1960; Orr, 1967, 1968). Predators such as the garter snake (*Thamnophis sirtalis*), the shrew *Blarina brevicauda* and the salamander *Desmognathus quadramaculatus* all kill and eat both species of salamander so if mimicry does occur it is of no relevance to these species of predators. However, the hawk *Falco sparverius* and the shrike *Lanius ludovicianus* both found *Plethodon* distasteful but ate *Desmognathus* readily, Hence it is possible that *Desmognathus* is a batesian mimic of *Plethodon* with relation to these predators. If this is so, then one would expect to find a higher proportion of red-cheeked morphs amongst the populations of *Desmognathus* in areas where *Plethodon* is abundant than in areas where it is scarce. This was found to be the case in a part of the Great Smoky Mountains: areas where *Plethodon* comprised more than 35 per cent of the combined total of the two species of salamander corresponded with areas in which there was a high frequency of red-cheeked morphs amongst the population of *Desmognathus* (Fig.

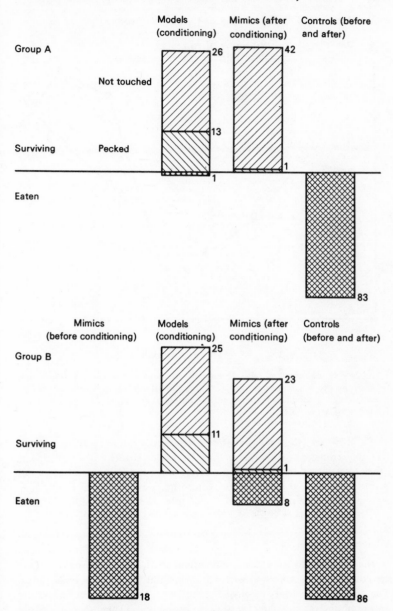

Fig. 4.27 Behaviour of chickens to model and mimic salamanders. Group A chickens were offered models first and then mimics. Group B chickens were offered mimics, then models, and then more mimics. Control salamanders were offered before, during and after the experiments and were always eaten. The model is *Notophthalmus viridescens*; the mimic is *Pseudotriton ruber*; and the control is *Desmognathus* sp. (*Data from* Howard and Brodie, 1971.)

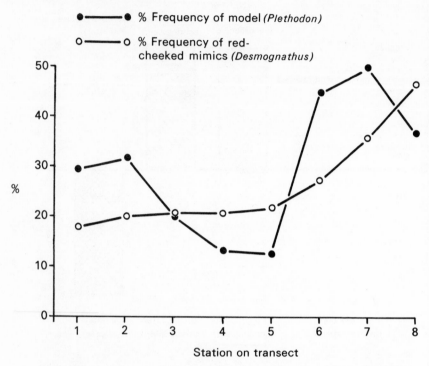

● ——● % Frequency of model *(Plethodon)*

○ ——○ % Frequency of red-
cheeked mimics *(Desmognathus)*

Station on transect

Fig. 4.28 Comparison of the frequency of model salamanders (*Plethodon jordani*) in the entire population with the frequency of those mimics which closely resemble the models (red-cheeked morphs of *Desmognathus ochrophaeus*). (*Data from* Orr, 1968.)

4.28). However, the fact that the correlation is not very close and the fact that some predators do not find *Plethodon* distasteful suggests that other factors are also involved in determining the frequencies of model and of red-cheeked mimics. This, and other supposed cases of mimicry amongst the Amphibia (see, for example, Nelson and Miller, 1971), require much further study.

Mimicry in snakes

In the New World there are a large number of species of snakes which are conspicuously coloured with red, black and yellow transverse bands. They occur in a variety of environments from tropical forest to savanna and semi-desert, and they are always conspicuous. They are collectively known as coral snakes. Some of the coral snakes are very poisonous Elapidae (*Micrurus* spp.), some are slightly poisonous though not dangerous Colubridae (e.g. *Erythrolamprus* spp.), and others are harmless Colubridae (e.g. *Lampropeltis* spp.). The colour pattern varies considerably in the arrangement of the bands and in their width, but

Micrurus fulvius   *Lampropeltis elapsoides*

Fig. 4.29   Coral snakes: the poisonous elapid *Micrurus fulvius* and the non-poisonous colubrid *Lampropeltis elapsoides*. Both are banded red (stipple), yellow (white) and black (black). (*Based on photographs in* Ditmars, 1953.)

with a sinuously moving snake it is difficult to recognize a particular species in the field with certainty as the movement blurs the colours (Fig. 4.29). In one area of Brazil coral snakes were collected by trappers over a period of four years. It was found that only 17 per cent of the snakes were one of the deadly species of *Micrurus*. If the other snakes are batesian mimics then one would expect a somewhat higher frequency of the model (see Table 4.3, second column). Further, if *Micrurus* spp. are the models it is difficult to see how a predator could learn the colour pattern since a bite from a *Micrurus* is likely to be fatal. *Micrurus*, however, is not very aggressive when attacked, but some of the less poisonous species such as *Erythrolamprus aesculapii* are very vicious and attack at the least provocation. Wickler (1968) has suggested that it is perhaps these less poisonous colubrids that should be regarded as the models of this mimetic assemblage of coral snakes. The harmless species are then typical batesian mimics, whilst the deadly elapids are mertensian mimics (named after R. Mertens, who did so much work on the coral snakes). The advantage gained by the mertensian mimics is that predators learn to avoid the coral pattern through attacking and being bitten by a moderately poisonous snake which cannot kill them. If a predator attacked a deadly snake, the predator

Table 4.3 Proportions of model and mimics amongst coral snakes from Brazil. (*Data from* Wickler, 1968)

Type of snake	Numbers captured	% of models amongst snakes caught	
		deadly snakes as models	slightly poisonous snakes as models
Deadly elapids (*Micrurus*)	214	17% models	17% mertensian mimics
Slightly poisonous colubrids	906 ⎫		74% models (müllerian mimics)
Harmless colubrids	107 ⎭ 83% mimics		9% batesian mimics

might indeed be bitten and killed, but not before it had struck and possibly fatally wounded the snake. Here the snake would be sacrificed, but so would the predator. No learning would have taken place and therefore other coral snakes would not be protected by their coloration. If the data from Brazil are reorganized, taking the slightly poisonous snakes as models and both deadly and harmless snakes as mimics, then the models are found to outnumber the mimics as is usual in cases of batesian mimicry (Table 4.3, last column). Hence a predator is most likely to encounter one of the very aggressive, slightly poisonous species of snake which may bite him and cause him to avoid all coral snakes in future encounters.

The situation becomes more complex in North America where the harmless coral snakes occur further north than some of their models. *Lampropeltis doliata* is a mimic of *Erythrolamprus aesculapii* in the southern part of North America. Further north *Erythrolamprus* is absent, but *Lampropeltis* may perhaps gain protection from its resemblance to the very poisonous *Micrurus fulvius*. Further north still, *Micrurus* does not occur, and here the pattern of the black, red and yellow on *Lampropeltis doliata* is not in rings (which are conspicuous) but in blotches and saddle-shaped marks which probably break up the outline of the snake (Hecht and Marien, 1956). Thus, as with *Limenitis arthemis* in areas where the model is absent, the survival value of the conspicuous mimetic pattern is lost, and selection has favoured the evolution of a disruptive colour pattern.

On the island of Tobago *Erythrolamprus aesculapii* is the only poisonous snake present, and here its colour pattern is ocellate, not annulate (var. *ocellatus*) (Emsley, 1966). Presumably selection favours the conspicuous annulate pattern where coral snakes are common and frequently encountered by predators, but where they are rare it is of greater protective value to be camouflaged. The ocellate pattern breaks up the outline of the snake and perhaps makes the snake as cryptic as are many vipers with similar patterns. So here too, in the absence of other coral snakes, the mimicry has broken down.

Mimicry in birds and mammals

Very few examples of mimicry are known to occur in birds or mammals but this may be simply because they have not been carefully studied. Thus the drongos (*Dicrurus* sp.) are black, conspicuous birds of the tropics which have distasteful flesh (Cott, 1947), and so it is possible that black shrikes, flycatchers and other birds in the area mimic them. But these birds have been very little studied and the relationship, if any, must be regarded as purely conjectural. There are other possible advantages in having black plumage besides the possibility that predators are deceived into mistaking the bird for a drongo. A more satisfactory example of batesian mimicry is that between the rufous flycatchers

Stizorhina fraseri in East Africa and *S. finschi* in West Africa and the ant-thrushes *Neocossyphus rufus* in East and *N. poensis* in West Africa. The similarity in colour between each of these species pairs is very close and cannot be due to chance. Further, the ant-thrush has unpleasant smelling flesh (Ziegler, 1971), possibly due to formic acid from its diet of ants, and so is probably not eaten by predators, whilst the flesh of the flycatcher is not repellent. Hence the flycatchers appear to be batesian mimics of the ant-thrushes.

Summary

Batesian mimicry is the close resemblance of a palatable animal (the mimic) to an unpalatable conspicuous organism (the model) such that some predators which avoid the model are deceived into avoiding the mimic as well. It has been shown to be of survival value in many insects with birds or amphibia as predators. Since predators normally learn through experience to avoid the models, a mimic cannot derive protection in a place where the model is very rare or absent. Sex-limited and polymorphic mimicries (with different morphs mimicking different species of models) are means whereby a species can maintain a high population relative to that of its model and still gain protection. Some predators also learn to distinguish the mimic from its model so that selection will favour any individual mimics which the predators cannot distinguish from the model. Batesian mimicry has evolved progressively through selection favouring any individuals with slightly greater similarities to the model than have the rest of the population. However, unless a model is very nasty, it may be disadvantageous for the model to have too many mimics. The result is that whilst the mimic is evolving towards the model, the model is evolving away from the mimics. However, it is clear from the examples described in this section that the details of a mimetic association vary with the palatability of the model and with the particular predators which are deceived by the mimicry. There are very few cases in which we know anything like the full story of the advantages gained by the mimic, the advantage or disadvantage to the model, and the effect of the various predators involved on the mimetic association.

Section 2
Secondary defence

Secondary defensive mechanisms are those defences of an animal which operate during an encounter with a predator. Sometimes the prey animal's defences operate as soon as it detects a predator and this may be before the predator has detected the prey. In other cases the defences only operate after the predator has initiated a prey-capture attempt. This definition of secondary or indirect defences differs slightly from those given by Robinson (1969a) and by Kruuk (1972). The function of secondary defences is to increase the chances of an individual surviving in an encounter with a predator.

Primary, passive, defence is never perfect since a predator may discover a prey by chance encounter or by systematic search for food. Once the prey has been found, the predator may develop a conditioned searching image or searching behaviour pattern and be better able to find further individuals of the same prey species in future. Hence secondary defensive adaptations have evolved in many animals. Some secondary defences are, like primary defences, *passive*. For example the heavily armoured sea urchins are immune to attacks from most predators because the predator cannot penetrate the armour of shell and spines. Some animals with chemical defensive mechanisms have the repellent chemical in the body tissues so that the animal has to be killed and eaten before the predator experiences any unpleasant effect. Thus *Danaus* butterflies have to be eaten by a bird which then vomits and becomes conditioned to avoid other individuals of the same species in future encounters. In both of these examples the secondary defence is passive and involves no energy consuming process on the part of the animal when it is attacked. In *Danaus* the insect which actually utilizes the defensive mechanism is killed and it is other individuals of the same species in the vicinity of the predator that derive protection. This is further discussed on pp. 104–8 and 199.

Active secondary defences operate only after a prey animal has detected a predator, which may be either before or after the predator has detected and attempted to capture the prey, depending on the sensitivities of the sensory equipment of the two animals. The first response of the prey when it becomes aware of a probable predator may be simply an exaggeration of the primary defence. The elongation of stick-mimics has already been described on p. 1, but other examples are considered here. If the predator continues to approach or to attack, this may elicit further active defensive responses, which will be considered under the following headings in the next six chapters: withdrawal to a prepared retreat, flight, deimatic behaviour, thanatosis, deflection of an attack, and retaliation.

Chapter 5
Withdrawal to a prepared retreat

Many anachoretic animals must emerge from their hole or burrow in order to feed, and they then respond to the presence of a predator by rapidly retreating back into their hole. Rabbits playing or feeding near the warren bolt down their burrows; spiders withdraw into crevices or into a silken retreat; and polychaetes of several different families retract into their tubes.

An animal with a more specialized retreat is the pearl-fish, *Carapus acus*, which feeds on shrimps and other small crustaceans. When disturbed it retreats tail first through the anus and thence into the body cavity of a sea cucumber (*Holothuria* spp.). Arnold (1953) has shown that this is a very complex association. The young fish are planktonic but later they enter the anus of a holothurian (head first) and become obligatory parasites, feeding principally on the gonads. As the fish grows, it leaves the host from time to time, but continues to feed parasitically. The adult eats shrimps as well as its host, and it may move occasionally to a new host. *Carapus* is thus a parasite which, as an adult, can live freely in the sea, but which regularly returns to its host for protection when disturbed by a predator.

Hermit crabs also have an unusual retreat: they live in empty gastropod shells into which they can withdraw when disturbed. The larvae of hermit crabs are free living in the plankton until they metamorphose into the benthic juvenile stage and search for a shell of suitable size in which to live. Choice of shell is based on visual and tactile cues (Reese, 1963), and as they grow hermit crabs move to progressively larger shells. Different species are adapted to living in different types of shell: for example, the West African *Clibanarius chapini* has a longer and more coiled abdomen, and usually lives in long and tightly spiralled shells (such as *Cerithium* or *Turritella*), whilst *C. senegalensis* has a shorter, less coiled abdomen, and normally lives in short spired shells (such as *Nerita* or *Thais* (Fig. 5.1)). The two species are sympatric intertidally on the West African coast, and each has been shown to prefer its own type of shell when given a choice (Ameyaw-Akumfi, 1971). The advantages of different shell preferences in the two species are not fully understood, but they may be related to the ease with which an animal can be removed from its shell. Hermit crabs often

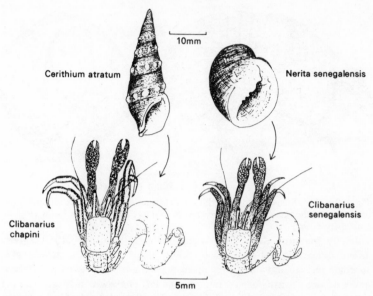

Fig. 5.1   Hermit crabs *Clibanarius chapini* and *C. senegalensis* from Ghana with their preferred shells, *Cerithium atratum* and *Nerita senegalensis*.

eject other hermit crabs from their shells which they then use themselves, and a naked hermit crab is likely to be killed if it cannot quickly find another empty shell. I have found that the starfish *Asterina gibbosa* also removes *Clibanarius* spp. from their shells and eats them. It is possible that starfish find *C. chapini*, with its coiled abdomen and long-spired shell, more difficult to remove than is *C. senegalensis*, but this hypothesis has not been tested.

Animals which are not anachoretes may also withdraw and present a predator with a hard shell or some other secondary defence. Gastropods and bivalves retract into their shells and stay tightly closed, protected by the hard calcareous shell; tortoises withdraw the head and limbs inside the body shell; the hedgehog (*Erinaceus*), echidna (*Tachyglossus*), armadillo (*Tolypeutes*), pangolin (*Manis*), woodlouse (*Armadillidium*) and millipede (*Glomeris*) can each roll up into a ball protecting the delicate head and ventral parts of the body by means of dorsal and lateral spines (hedgehog and echidna) or horny plates (the others) (Fig. 5.2).

One disadvantage of withdrawal is that the predator may be able to follow the prey into its retreat, and the prey may then be cornered and unable to escape by flight. The ferret is used commercially in many parts of the world to follow rabbits down their burrows and kill them. The polecat (*Mustela putorius*), from which the domesticated ferret is probably descended, may also do this, but its method of hunting is not known. The related weasel (*Mustela nivalis*) regularly hunts mice and

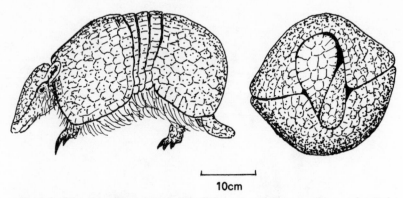

10cm

Fig. 5.2   The three-banded armadillo *Tolypeutes tricinctus* walking and rolled into a ball. (*Redrawn from photographs in* Walker, 1964. © The Johus Hopkins University Press.)

voles by pursuing them down their burrows (Southern, 1964). Many animals have evolved opercula or other means of blocking the entrance to their retreat and so of preventing a predator from entering. Hermit crabs of the genus *Diogenes* have a large left chela which is used to plug the aperture of the shell in which they live. Serpulid polychaetes have a calcareous operculum which is often armed with spines (Fig. 5.6), and the polychaete *Sabellaria* has a bristly operculum composed of closely arranged chaetae. The fairy armadillo (*Chlamyphorus*) also has an operculum to block its burrow. In this animal the posterior part of the body has horny plates which plug the burrow and (presumably) prevent predators from following it underground (Fig. 5.3). The tubicolous coelenterate *Cerianthus* sometimes pinches its tube closed after it has withdrawn (Wobber, 1970), and this prevents the predatory nudibranch *Dendronotus iris* from pursuing it into its tube and eating it (but how it constricts the tube is not known). Bagworm caterpillars (belonging to the family Psychidae) live in silken tubes to which they attach sticks, leaves or other debris. When alarmed they withdraw inside the tube and pinch the aperture closed (Fig. 5.4), but they maintain a hold on the surface to which they were attached by a silken thread. Sand wasps (*Ammophila* spp.) provision their egg cell with caterpillars and then block the entrance with pebbles and sand. They return from time to time to inspect the cell and bring more food, but each time they block and camouflage the entrance (Baerends, 1941; summarized by Tinbergen, 1958). Various mammals which nest underground also block the entrance when leaving the young in order to collect food. Both of these latter examples are cases of parental care: blocking the burrow protects the young, not the individual animal which does the blocking. However, the African ground squirrel (*Xerus erythropus*) regularly blocks up the entrance to its burrow when it retires for the night, clearly an adaptation likely to protect itself from predators (Ewer, 1966).

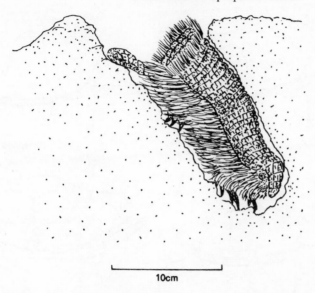

Fig. 5.3   The fairy armadillo or pichichiego, *Chlamyphorus truncatus*, digging a burrow with the posterior end of the body protected by bony plates. (*Redrawn, with modifications, from* Carrington, 1963.)

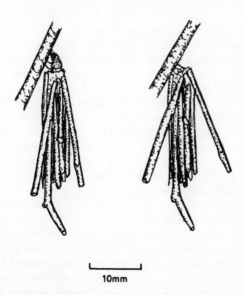

Fig. 5.4   Bagworm caterpillar (*Psychidae*) in its case. Left: clinging to a twig; right: retracted into its case with the opening pinched closed, but maintaining its hold on the twig by a silk thread.

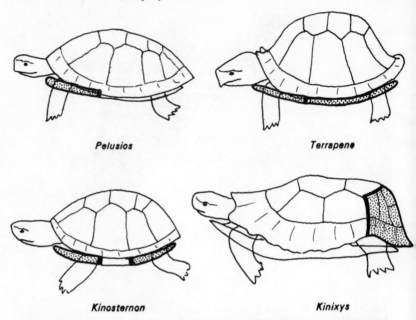

Fig. 5.5   Terrapins and tortoises with hinged parts to the shell. In each animal the hinge is shown as a thick black line, and the part that moves up or down is stippled. For further explanation see text.

Armour-plated animals may also protect the softer parts of the body by means of opercula or hinged plates. Bivalves simply close their two shells tightly by means of one or two powerful adductor muscles so that there is no opening left. (Often there is a striped part of the adductor which gives rapid closure, and a smooth part which is able to sustain a long pull to keep the shells tightly closed for a long period of time.) Many gastropods plug the entrance to the shell with a horny operculum. The box terrapins *Terrapene* and *Cuora* (family Testudinidae) have a hinge on the ventral shell (plastron) so that both anterior and posterior openings can be tightly closed when the animal withdraws (Fig. 5.5). *Staurotypus* (family Kinosternidae), *Bellemys* (Testudinidae) and the side-necked terrapin *Pelusios* (Pelomedusidae) all have an anterior hinge on the plastron and so can protect the head but not the hind legs and tail. The tortoise *Kinixys* (Testudinidae) has a hinge on the carapace posterodorsally which partially closes off the posterior opening whilst the head is protected by a pair of horns on the plastron, and the terrapin *Kinosternon* (Kinosternidae) has both an anterior and a posterior hinge on the plastron. Thus in terrapins and tortoises mechanisms for closing off one or both openings have evolved independently at least five or six times, from which we can infer that some predators may be able to kill a retracted terrapin — but only so long as the openings are not closed off.

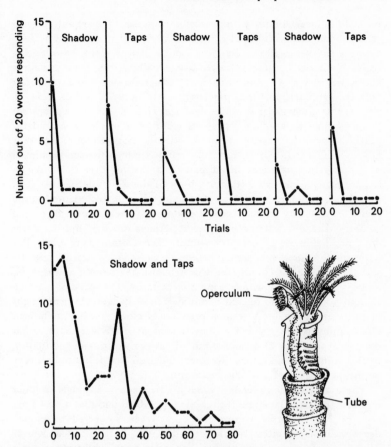

Fig. 5.6 Habituation of the serpulid polychaete *Mercierella enigmatica* to two stimuli, a moving shadow and a sharp tap on the aquarium. Above: the number of animals responding to each stimulus on its own with stimuli alternating after every twenty trials. Below: the number of animals responding to both stimuli together during eighty trials. (*Redrawn from* Wells, 1968, Fig. 8.10, and Day, 1967, Fig. 38.5.)

An alternative strategy to blocking the entrance is found in some burrowing mammals which build a second exit to their retreat. The predator that follows a rabbit down its burrow may find that it has escaped through the second exit.

A second disadvantage of withdrawal is that stimuli that cause the prey to withdraw may occur when there is no predator about. The prey may be unable to feed or to perform other essential activities whilst in its retreat, yet it must remain there for some time since it has no means of detecting whether the predator is still in the vicinity or not. For many animals predator recognition is very crude, and most marine tubicolous invertebrates withdraw as a response to a shadow passing across them. The withdrawal is very rapid since it is mediated by fast-acting giant nerve fibres. But one can imagine a worm withdrawing regularly simply because a seaweed repeatedly drifted back and forth above it. Hence we find that many invertebrates habituate: after repeated stimulation, the percentage of animals withdrawing in response to a particular stimulus declines. This implies that a persistent predator may succeed in getting its worm, but in nature a predator probably provides several different stimuli, not just one. As Wells (1968) points out, worms that show habituation to the simple stimuli of tapping the aquarium or of passing a shadow over the tentacles require many trials to habituate to both of these stimuli together (Fig. 5.6). The habituation behaviour of an animal is probably adapted to the predators normally encountered. Predators are likely to provide several stimuli — shadow, water vibration and perhaps touch — and the worm does not habituate to complex stimuli except after many trials. But with simple stimuli that are unlikely to be caused by predators (e.g. the sun going behind a cloud), the worm habituates rapidly and thus no longer wastes time by withdrawing unnecessarily. The speed of habituation to stimuli is also related to the degree of danger: decreased light intensity need not imply that a predator is close, and habituation to this stimulus is very rapid. Touching the crown of tentacles implies that the predator is very close and about to eat the worm, and habituation to this stimulus is very slow (Krasne, 1965), so that it has to be a *very* persistent predator if it is to get its worm.

There is also considerable variation between individuals in their response to particular stimuli: thus Nicol (1950) found that only 43 per cent of *Branchiomma* worms withdrew when subjected to sudden decrease in light intensity, the rest did not. Further, some worms habituated after a few trials, others only after many, whilst others again did not re-emerge from their tubes. This is perhaps an example of protean behaviour (see p. 145), by which is implied that selection favours diversity in the responses of different individuals in a population since then a predator will be unable to develop a conditioned strategy for prey capture.

Summary

The most usual secondary defence of anachoretic animals is withdrawal into a retreat. Other animals which have a protective shell or exoskeleton withdraw into this when they are attacked. Some predators may still be able to capture such prey, for example by following it into the retreat, and so further anti-predator adaptations such as opercula have evolved to block the entrance to the retreat. One disadvantage of withdrawal is that an animal may waste time by withdrawing to trivial stimuli, so many animals habituate to repeated stimuli.

Chapter 6
Flight

The most usual response of an animal when it perceives a predator (or a similar stimulus) is rapid locomotion away from the stimulus source. This takes the form of running, jumping, swimming, flying or passive dropping, according to the species. Usually the flight is fast but sometimes it is comparatively slow and feeble, yet adequate for the animal to escape its predator. Escape movements may be in a straight line, as with a blackbird flying off from a stalking cat, but more often they are erratic in direction so that it is not possible for a predator to predict the direction of its next move (Fig. 6.1). Unpredictable or 'protean defence' of this type has been reviewed recently by Humphries and Driver (1971). Birds such as snipe and ptarmigan both fly from predators in a characteristic zig-zag, whilst the European hare sometimes runs zig-zag, and sometimes runs in a straight line. If it is far from a suitable retreat, one might expect the zig-zag to be more often a response to a predator faster than itself such as an eagle, and the direct flight path to be a response to a predator slower than itself such as a felid or a canid, but there is no evidence on this point. Peregrines (*Falco peregrinus*) can easily catch flying waders in direct flight, but Kruuk (1964) found that the slower flying black-headed gulls show frequent changes of direction in the presence of a peregrine. Roeder (1965) describes how various arctiid, geometrid and noctuid moths respond to ultrasonic pulses (as from a hunting bat) by a variety of loops, dives, rolls and turns so that the path they follow is unpredictable (Fig. 6.2). If the pulse intensity is less, they usually fly in a straight line at top speed. The moth can detect the bat before the bat detects the moth, and under these conditions the most effective escape is to get out of the way as rapidly as possible. However, since the bat can overhaul the fastest moth, once the bat has detected the moth, protean movements are the more effective escape response. The responses of these moths to the ultrasonic pulses of bats are examples of escape behaviour which is only released by a particular type of predator. Many other animals have predator specific escape responses, and these are discussed further on p. 232.

The escape response of many spiders when they are attacked is dropping on a silken thread. When danger has passed the spider climbs

Fig. 6.1   Erratic escape movements: path followed by a stickleback (*Gastero-steus*) when chased by a merganser duckling. (*Redrawn from* Humphries and Driver, 1971, Fig. 1*d*.)

back up its thread. Dropping on a thread also occurs in some cater-pillars and pulmonate molluscs, and Gotwald (1972) has shown that this may occur in response to attack by driver ants (*Dorylus* spp.).

Many small insects of the aerial plankton (e.g. ceratopogonids, aphids) as well as butterflies have very erratic flight at all times, not only when a predator is in the vicinity. This may be a form of protean insurance in species liable to sudden attack from any direction such as by swifts and swallows.

Flash behaviour

Escape movements may continue until a pursuing predator gives up the chase. Provided that the prey animal has greater speed and stamina than the predator, escape by flight will be successful. Escape movements, whether direct or erratic (protean), may also be successful when the prey animal is able to flee to a hole or some other retreat where the predator cannot follow it. A third strategy for a prey animal is to flee a short distance and then to rest motionless and cryptic so that a predator may not observe it. This form of escape is used by animals such as frogs and grasshoppers which can move a short distance very quickly so that they can get away from the predator, but they then have no stamina for a long chase. Such animals often deceive predators by a flash of colour as they move which disappears when they come to rest. The predator may be caused to hesitate by the sudden movement and appearance of the bright colour (see below under deimatic behaviour), and it may then follow this colour and be deceived by its sudden disappearance into assuming the prey has vanished, whereas in reality the prey has come to rest in its normal cryptic posture with the coloured structures hidden. There is no proof that this is how flash colours work, but they often deceive man in this way, so it is likely that they deceive other vertebrates which hunt by eyesight as well. The red and yellow underwing moths (*Catocala nupta* and *Triphaena pronuba*) are both brown cryptic insects when at rest, but when they fly the bright coloured hindwings make them very conspicuous. When they

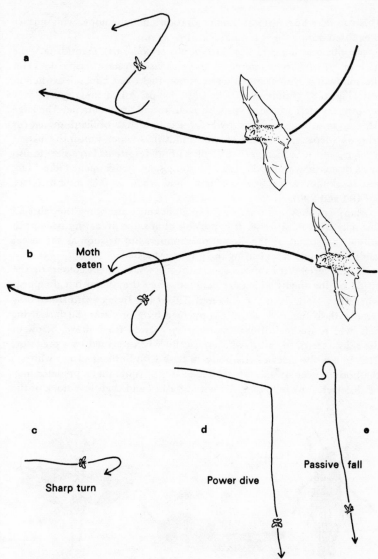

Fig. 6.2   Escape movements of various moths in response to the sonar clicks of a hunting bat. In (*a*) and (*b*) the moth flies erratically, but whilst the moth in (*a*) escapes, that in (*b*) is captured. (*c*), (*d*) and (*e*), other escape responses of moths. (*Redrawn from photographic traces of moth–bat encounters by* Roeder, 1965.)

come to rest again, the wings are folded and the bright colour disappears. Many moths including some hawkmoths (*Hippotion osiris*) have similar flash coloration and behaviour. Butterflies are often brightly coloured on the upper surface but cryptic on the lower surface.

This too may be a form of flash coloration since the upper wing surface is exposed during flight but hidden when at rest.

Various grasshoppers (e.g. *Trilophidia tenuicornis*), mantids (e.g. the African *Pseudoharpax virescens*) and cicadas also have flash colours on the hindwings which are exposed during flight but concealed when at rest (Fig. 6.3) (Edmunds, 1972; Cott, 1940). All of these species rely for primary defence on crypsis, though not all are palatable. The large African grasshopper *Phymateus* for example has brilliant orange or crimson (depending on the species) hindwings which vanish the instant it lands, and it is then very difficult to find despite its large size. It also has a dramatic warning display with these same wings opened like a fan, and it exudes a repellent smelling fluid when further molested (see pp. 154 and 200).

Many grasshoppers buzz or make a clicking noise when they fly and this may be a flash noise. It is possible that some predators attempt to follow the sound of the flying grasshopper, but as soon as the insect lands the sound ceases leaving the predator baffled.

In the vertebrates various species of frogs have bright colours on the inside of the thighs which are exposed when they jump but disappear when they land. Even the coloured frill of the flying lizard *Draco* may serve a flash function when it is pursued by a predator. Surface living fish, which are beautifully cryptic when viewed from above, below or the sides (see p. 9), may swim erratically when pursued by a predator. This causes the sides of the body to flash with brilliant silvery white as the body is seen at an atypical angle. It is possible that a predator may be deceived into following the white flashes and then lose track of the

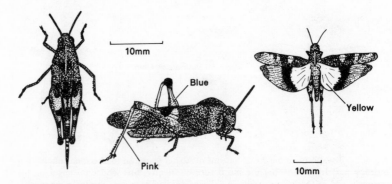

Fig. 6.3   The acridid grasshopper *Trilophidia tenuicornis*. Left: dorsal view; and centre: lateral view showing cryptic coloration (dark brown) with disruptive markings on the hindlegs and hindwings. Right: insect flying directly away from an observer showing the conspicuous black and yellow flash colours on the hindwings. Note also the pink hind tibia and blue on the femur: possibly these colours attract a predator to the most dangerous part of the insect, the spines on the tibia, which can give a powerful kick.

fish when it reverts to normal swimming. Many birds too show sudden flashes of colour when they fly (white rump or tail feathers) which are hidden when they land, and it is possible that this too is a flash colour. These flashes may in addition serve the function of warning other birds (of the same or of different species) that there is a predator nearby, and this would explain why many species in the tropics which form mixed flocks share the same colours on the rump or under tail coverts (Moynihan, 1968). This has been called 'social mimicry' and it may apply to mammals as well as birds, for example the white rumps of many ungulates and rabbits which vanish when the animals stop running and turn to face their pursuer or bolt down a hole.

Summary

The most usual secondary defence of active animals is movement or flight away from the predator. Such escape movements may be in a straight line or erratic, the advantage of the latter being that it is not easy for a pursuing predator to predict the direction of the prey animal's next movement.

Flight may be accompanied by exposure of bright colours (flash behaviour) which are hidden when the animal comes to rest. It is believed that predators may be induced to follow this bright colour instead of including other features of the prey in their searching image, and so when the bright colour vanishes the predator gives up the search. Noises made by fleeing prey may also deceive predators in a similar way.

Chapter 7
Deimatic behaviour

When discovered by a predator many animals respond by adopting a characteristic posture which appears to be designed to intimidate the predator. Such displays have been called startle displays (Crane, 1952; Edmunds, 1972), dymantic displays and dymantic coloration (Young, 1950, 1959; Packard, 1972), and deimatic responses (Maldonado, 1970). Deimatic, from the Greek $\delta\epsilon\mu\alpha\tau\acute{o}\omega$ — I frighten — appears to be a more natural derivation than 'dymantic', so I propose to use the term *deimatic behaviour* or *frightening behaviour* to include all such displays, postures and frightening noises. Deimatic behaviour produces mutually incompatible tendencies in a predator: it stimulates an attacking predator to withdraw and move away. This results in a period of indecision on the part of the predator (even though it may eventually attack), and this gives the displaying animal an increased chance of escaping (Humphries and Driver, 1967). A deimatic display may be a genuine warning that the animal can attack and harm the predator, or it may be a bluff, but there is no clear distinction between the two: a prey animal may be unpalatable to one predator but not to another, so that its display is a genuine warning to the first but a bluff to the second. This is discussed further below.

Deimatic behaviour may occur in both cryptic and aposematic animals. In any encounter between predator and prey it will be of advantage to the prey if the predator can be intimidated and driven off before it has actually seized the prey. If the prey is grabbed, this may be fatal even if the predator subsequently releases it. Predators normally reject aposematic or well armed prey only after having discovered its disagreeable properties during an initial attack. A deimatic display may improve the chances of an aposematic animal's surviving in an encounter with a naive predator by intimidation, and with an experienced predator by providing a set of unambiguous signals which it has learned to avoid.

In the case of palatable animals which cannot harm their predators, the conditions eliciting release of deimatic (bluff) behaviour, and the details of the display itself, are rather variable. Thus it is very hard for a predator to learn the characteristics of the display and hence to learn how best to attack a displaying animal.

Survival value of deimatic behaviour

There is little quantitative evidence of the survival value of deimatic behaviour, but there is a wealth of anecdotal evidence to the effect that some frightening displays successfully intimidated certain predators. To demonstrate the survival value of a genuine warning display it is necessary to show that a displaying animal is not attacked and also that other non-displaying animals are sometimes attacked. This is clearly demonstrated in the behaviour of moose when encountering wolves (Mech, 1970). All moose which stood firm and faced a pack of wolves survived, usually without actually having to strike out at the wolves, whilst some at least of those that fled were caught and killed (this is further discussed on p. 272 and in Fig. 12.28). Musk oxen form a defensive ring round the cows and calves when they are attacked by wolves, and often this threatening circle is sufficient to intimidate the wolves and cause them to move away without attacking (Fig. 7.1). Eland also ward off predators by a group defensive formation, sometimes without the predators (hyaenas) actually attacking (Kruuk, 1972, and see also p. 280 and Fig. 12.35).

There is also evidence that some of the bluff deimatic displays are of survival value, though not necessarily to all predators that an animal may encounter. Crane (1952) found that neotropical mantids that gave a startle display were not attacked further by a variety of monkeys and lizards. But African mammalian predators (civets, kusimanse and bushbaby) are not intimidated by startle displays: on the contrary, they may actually be attracted by a displaying mantid which they had not previously noticed, and proceed to attack it (Edmunds, 1972). An exception to this generalization is described on p. 156. Eye-witness accounts of encounters between mantids and reptiles (geckos, chameleons and agamid lizards) suggest that insectivorous reptiles are also not deterred by a display from attacking a mantis. The display is, however, of protective value when directed towards birds. Maldonado (1970)

Fig. 7.1   Defensive ring of musk oxen (*Ovibos moschatus*) when threatened by a pack of wolves. The cows and calves are in the centre of the ring. (*Based on photographs in* Walker, 1964, and Mech, 1970.)

found that two species of insectivorous birds from South America were intimidated by the displays of the large mantid *Stagmatoptera biocellata*. Although the mantid can strike with its clawed forelegs, it is doubtful if it could do any serious damage to a bird, so the display is largely a bluff. In some cases Maldonado observed that the bird eventually killed the mantid, but this was in a cage where there was no possibility of escape. In a natural situation the initial display and lunge of the mantid might well cause the bird to fly away altogether. It is also known that many tropical insectivorous birds regularly eat mantids (Edmunds, 1972) so obviously any behaviour which inhibits an attack, and so reduces predation will be of survival value.

Thus there is evidence that some genuine warning displays and some bluff displays are of protective value with relation to certain species of predators.

Evolution of deimatic behaviour

Deimatic displays often involve complex sets of movements and colour patterns (see below), and it is at first difficult to see how such complex behaviour patterns could have evolved.

In the deimatic threat or warning displays of mammals it is usual for the prey animal to display its weapons (teeth or horns), and it is obvious how this behaviour has evolved by selection favouring a hesitation before an overt attack on the predator. This may give the predator time to hesitate in its turn, or even to move away, whereas overt attack would compel the predator to attempt to kill the animal. However, bluff displays probably succeed because some predators have an innate avoidance response to any unusual stimulus such as a bright colour (Coppinger, 1969). Blest (1957a) offered normal tortoiseshell butterflies (*Aglais urticae*) and butterflies with the orange scales rubbed off (so that the wings were dull grey) to laboratory reared yellow buntings which had had no previous experience of butterflies. He found that the normal orange insects elicited more escape responses than the grey experimental ones (Fig. 7.2), indicating that the birds have an innate avoidance response to the bright orange colour. Domestic chicks hesitate before pecking at their first mealworm, and they are very reluctant to test a new type of food. Birds that have been fed on a variety of different foods are more likely to attack a novel food than are birds fed on a monotonous diet (Coppinger, 1969). There are in fact two extreme types of behaviour in predators: some are inquisitive and will investigate and sample any new object they encounter, whilst others are very conservative in their feeding habits and preferences and may ignore a common edible food so long as a familiar food is still available (Klopfer, 1961; Alcock, 1971). These conservative predators are the more likely to be intimidated by a rather simple deimatic response, but some of the more elaborate displays may be directed at

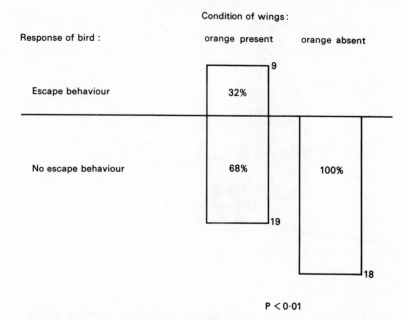

Fig. 7.2   Responses of inexperienced yellow buntings (*Emberiza citrinella*) to tortoiseshell butterflies (*Aglais urticae*) with and without orange scales on the wings. (*Data from* Blest, 1957*a*.)

inquisitive predators. Thus it seems likely that deimatic displays have evolved initially from some simple, sudden movement or exposure of a coloured surface. This would be of some protective value against certain predators, but selection would favour those prey individuals in which the behaviour or the colour markings were even more striking in their appearance, since this would be more likely to cause the predator to withdraw or to hesitate in its attack. Thus we find that some deimatic behaviour patterns involve a variety of different movements and include bright colours, stridulation and sometimes glandular secretions as well. These are described further below.

Some examples of deimatic behaviour

Deimatic behaviour of arthropods

Many orthopterous insects have dramatic anti-predator displays and in some of them the display is clearly a bluff rather than a genuine warning of noxious qualities. Stick insects often display by erecting the wings thus giving an apparent increase in size and revealing bright colours or eyespots on the hindwings. This has been described for *Metriotes diocles* and the male of *Pterinoxylus spinulosus* by Robinson

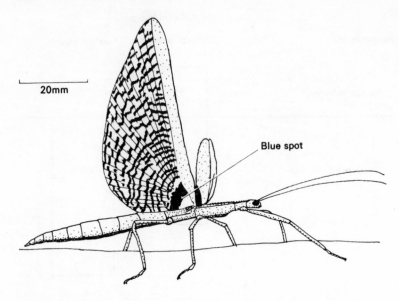

Fig. 7.3 Deimatic (startle) display of the stick insect *Metriotes diocles*. (*Redrawn from* Robinson, 1968*b*, Fig. 5.)

(1968*b*, *c*), species in which the display is apparently entirely bluff since they have no active defence which might harm a predator (Figs. 7.3, 7.4). Similar displays occur in female *Pterinoxylus spinulosus*, as well as in *Eurycnema goliath* and *Tropidoderus childrenii*, but in all these the display is accompanied by stridulation (Robinson, 1968*c*; Bedford and Chinnick, 1966). Possibly the stridulation resembles the hissing of a snake or the stridulation of a threatened scorpion. *Orxines macklotti* has a similar display but in addition it has glands which may produce a repellent secretion (Robinson, 1965). The wingless phasmid *Oncotophasma martini* has a dramatic display involving the flexion of the abdomen dorsally similar to the stinging movement of a scorpion, and this is accompanied by stridulation (Fig. 7.5). This animal also has a genuine warning element to the display since when handled it can jab the spines on the hind femora into an aggressor (Robinson, 1968*a*). The grasshopper *Phymateus cinctus* has a dramatic display involving expansion of the coloured hingwing membranes (Plate 6*a*), but here there is evidence that the display is genuine warning. There are strong spines on the hindlegs, as in other species of grasshoppers (though they are not often used in this species when it is handled); there are also repellent glandular secretions from the thorax; and there is evidence that the body tissues are themselves unpalatable to some predators (see p. 200). Hence there is not always a clear-cut distinction between a warning display and a bluff display.

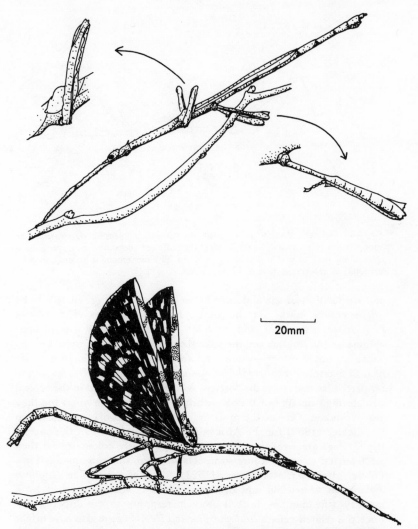

20mm

Fig. 7.4 Defensive behaviour of the male stick insect *Pterinoxylus spinulosus*. Above: resting posture with two insets showing the positions adopted by the middle and hindlegs. Below: deimatic display with erection of coloured hindwings. (*Redrawn from* Robinson, 1968c, Figs. 1 and 2.)

The deimatic displays of mantids are probably also bluff since mantids cannot harm a large predator, but the spines on the forelegs may nevertheless be able to do considerable damage to small lizards, so there is an element of genuine threat in the display as well. The details of mantid displays vary in different species (Edmunds, 1972), but there is often erection of the fore- and hindwings giving apparent increase in size, erection and abduction of the flexed forelegs, opening of the jaws,

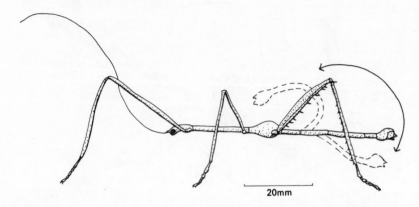

20mm

Fig. 7.5 Deimatic display of the male stick insect *Oncotophasma martini*. The abdomen is repeatedly lowered and then flexed dorsally in a movement resembling that of a scorpion about to sting. The movement is accompanied by stridulation. (*Redrawn from* Robinson, 1968a, Fig. 1.)

and stridulation of the abdomen between the hindwings. There may be bright colour marks on the wings, forelegs and jaws. The African *Polyspilota aeruginosa* and *Statilia apicalis* have very conspicuous colours on the forelegs and the ventral surface of the pronotum, but the wings are not very conspicuous and are not used for stridulation (Plate 6*b*). Conversely *Pseudocreobotra ocellata* and *Stenovates strachani* have no particular marks on the forelegs which are not flexed in the typical way during display, but they have bright colours on the wings and they also stridulate. One of the most dramatic startle displays is illustrated by Wickler (1968) for the African *Idolium diabolicum*. In this species the forelegs are swollen in size and conspicuously coloured, and they are reported to act as a lure to insects which mistake the mantis for a flower. However, Carpenter (1921) describes how an *Idolium diabolicum* reared up and displayed when it was approached by a monkey (the monkey backed away in response to the display), and since the related genera *Idolomorpha* and *Hemiempusa* also have bright colours on the forelegs which are only exposed when they are attacked, it is clear that in *Idolium* too the posture is an anti-predator display. In *Sphodromantis lineola* I have found that it is virtually impossible to induce a displaying mantis to strike at a prey insect, and this further indicates that it is unlikely that the startle display of *Idolium* could evolve into a lure to attract prey.

Crane (1952) considers that the startle display of mantids is caused by a conflict between the drive to escape by flying and the drive to escape by remaining motionless and cryptic. Such a conflict would naturally cause elevation of the wings, and this on its own may initially have had some protective value in frightening a potential predator.

Later selection perfected the display by favouring individuals which had brighter coloured wings or forelegs, or which stridulated. The colours on the forelegs may possibly attract a predator to the most dangerous feature of the mantid – the tibial claw and the spines on both tibia and femur – and it is certainly true that species which have brightly coloured forelegs also have the power to draw blood if carelessly handled. But some species without bright markings on the forelegs are also capable of drawing blood, so this is not the only factor involved.

It is interesting to compare the three groups of arthropods that have evolved similar raptorial forelegs. Large mantids such as *Sphodromantis* display them against predators; *Tarachodes* displays both against predators and against conspecifics; whilst the smaller mantids *Oxypilus* and *Catasigerpes* usually display only against conspecifics. Mantis shrimps (Stomatopoda) have only been reported to display against conspecifics (Dingle and Caldwell, 1969), but possible display against predators has not been investigated. In the third group, the mantispids, I have found no conspicuous marks on the forelegs which are purely raptorial in function. Mantispids are mostly small and are protected by their mimicry to wasps and by the ability to fly quickly and effectively. It is probably no coincidence that small mantids, which can also fly well, have no deimatic behaviour.

A rather different type of deimatic behaviour occurs in the chrysalis of the death's head hawkmoth (*Acherontia atropos*) and in a few other pupae belonging to the lepidopteran families Hesperidae, Lycaenidae and Sphingidae. These all produce squeaks or other sounds when handled (Hinton, 1948). It is possible that these noises intimidate some potential predators and cause them to refrain from eating the pupa.

Many moths which characteristically rest on vegetation have deimatic displays when they are disturbed during the day-time. In the case of hawkmoths the displays are usually bluff since the insects are themselves palatable. They cannot fly immediately they are disturbed but have to 'warm up' first by rapid wing vibration. Hence it is important to deter a predator from attack by some form of startle display, at least until the moth has warmed up so that it can escape by flight. Such displays, usually involving exposure of bright colours on the hindwings, are more sustained in larger species and in cooler climates than in smaller species and in warmer climates (Blest, 1963a).

Arctiid and ctenuchid moths also have deimatic displays which expose bright colours on the hindwings or on the abdomen, but these are usually genuine warning displays since most of the insects are distasteful (Blest, 1964). Some arctiids emit a repellent smelling froth from the thorax when they are disturbed (Carpenter, 1938), and this may further warn predators of their unpalatability (Fig. 7.6). Some arctiids and ctenuchids produce high frequency sounds (stridulation) from tymbal organs in the metathorax as part of their deimatic behaviour (Blest, Collett and Pye, 1963; Blest, 1964). In *Melese*

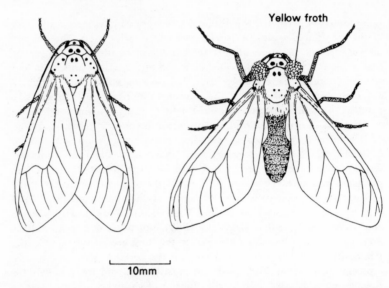

Yellow froth

|← 10mm →|

Fig. 7.6 The arctiid moth *Rhodogastria leucoptera* from West Africa. Left: normal resting position. Right: deimatic display as a result of being sharply poked on the thorax. During display the wings are raised and sometimes opened slightly to expose the red abdomen, and a smelly yellow froth is exuded from glands on the thorax. Stipple indicates red colour, the rest of the insect is white with black spots. Other species of *Rhodogastria* have brown wings with red or yellow on the abdomen and legs.

laodamia the sound is only produced when the moth is handled, but species of *Haploa* and *Halysidota* click in response to sonar pulses of bats (Dunning, 1968). These responses are evidently deimatic in function and are only produced when the moth detects a predator, so they are clearly secondary defences. The buzzing of bees and some flies is continuous and so is a primary defence, not a deimatic behaviour. Some reduviid and pentatomid bugs (e.g. the African *Bathycoelia*) also buzz when attacked, presumably as a deimatic response.

Deimatic noises are also produced by some scorpions. When it is attacked a scorpion adopts a warning display posture, either with the pedipalpal chelae raised and protracted, as in *Heterometrus* (attack threat), or with the pedipalps retracted and held close to the body, as in *Pandinus* (defensive threat). Correlated with these differences in position of the pedipalps during display, these two genera have evolved different stridulatory mechanisms on the coxae of the pedipalps and first legs (Alexander, 1959, 1960). *Opisthophthalmus* can adopt either an attack threat or a defence threat posture (Fig. 7.7). In this scorpion the stridulatory surfaces are on the chelicerae and the carapace so that the position of the pedipalps does not affect its ability to stridulate. Alexander (1958*a*) found that stridulation successfully intimidated inquisitive hedgehogs (*Erinaceus*) but did not prevent shrews (*Suncus*)

Fig. 7.7   The South African scorpion *Opisthophthalmus nitidiceps* in attack
threat (left) and defensive threat posture (right). Notice the difference in the
position of the pedipalpal chelae. (*Redrawn from photographs in* Alexander,
1959, plates 1*a* and *b*.)

or meercats (*Suricata*) from attacking and eating scorpions. Hedgehogs
did attack and eat silent scorpions, and thereafter they would also eat
ones which stridulated. Hence stridulation is presumably only a genuine
warning to some predators which are likely to be harmed by a
scorpion's sting, but it may be a bluff towards other predators which
are not affected by the sting.

A different type of deimatic response is shown by spiders such as
Argiope and *Pholcus* which rapidly vibrate themselves and their webs
when they are disturbed. The animal becomes blurred in appearance,
possibly with an apparent increase in size, and it must be very difficult
for a predator to launch a successful attack on such a target. The
behaviour is probably intimidating and hence deimatic in function.

Some forms of deimatic behaviour can be regarded as examples of
mimicry, but whilst batesian mimicry is typically a primary defence in
which one animal resembles another (usually) aposematic animal, in the
examples I shall describe here the mimicry is of a secondary defence
and is a bluff. The darkling beetles, *Eleodes* spp., are typically black,
and when disturbed they adopt a posture with the head down and the
tip of the abdomen directed towards the predator. If further molested
they spray a fluid from the tip of the abdomen on the attacker (Eisner
and Meinwald, 1966). Beetles of the genera *Megasida* and *Moneilema*
are similar in appearance and adopt a similar posture to *Eleodes* when
they are disturbed, but their display is pure bluff since there is no
defensive fluid to be sprayed (Raske, 1967).

Deimatic behaviour in vertebrates

Vertebrates such as rodents, carnivores and some of the larger ungulates
will turn and attack a pursuing predator when they are cornered, using
teeth, claws, or whatever weapons they possess. But before overt attack
they may display their weapons since even at this stage if they can

prevent an attack by the predator they may survive. Cats and rats display their teeth and erect the hair on the back giving an illusion of increased size. The larger ungulates display their horns or make preliminary kicks with their hooves. Some tortoises and young mammals make explosive snorts or hisses when encountered by a predator, and this noise is obviously a bluff. Cats make explosive spitting noises when cornered and such behaviour by a kitten can sometimes intimidate quite a large dog. The aardwolf *Proteles* raises its hair in the usual way when attacked but keeps its mouth closed, unlike typical carnivores (Ewer, 1968). This is evidently because it has rudimentary teeth and to expose them would demonstrate to a predator that it had no weapons.

Erecting hair to give an apparent increase in size may help to intimidate a predator, but toads (e.g. *Bufo regularis*) inflate their lungs with air and hence swell up when cornered by a snake (e.g. *Philothamnus* spp.). It is probable that sometimes the enlarged toad may be too big for the snake to swallow and so it may survive. Similarly puffer and porcupine fish inflate themselves with air or water when they are attacked.

House mice rattle or lash the tail in conflict situations such as when cornered by a predator (Crowcroft, 1966; Ewer, 1968), but it is not clear if this is a genuine warning that they may attack if further provoked or if it is really a bluff. Probably any sudden movement or explosive noise can be intimidating and cause a predator to hesitate before attacking. Partridges and francolins normally remain close to the ground, cryptic, until a predator is close when they fly up with a sudden noise of beating wings and often a clacking call as well. Rabbits also bolt with considerable noise as they brush vegetation aside, the noise probably causing a momentary hesitation on the part of a nearby predator.

A few species of mammals which are very effectively protected against predation have more elaborate deimatic displays. When a skunk is threatened by a predator it erects its tail and stamps its forefeet on the ground (Bourlière, 1955). *Spilogale* may even rear up on its forepaws and advance towards the predator presenting the full length of its black and white body to view, thus giving an apparent increase in size (Fig. 7.8). If the predator persists in its attack both genera of skunk (*Mephitis* and *Spilogale*) squirt a nauseous smelling fluid from the anal sacs. The zorilla and African polecat have similar behaviour and are also black and white, but they have not been so intensively studied as the skunks (Alexander and Ewer, 1959).

African porcupines make a warning sound by rattling their quills before charging backwards and impaling would-be predators on their sharp quills. The American porcupine hisses, erects its spines, and presents its rear to predators (Fig. 7.8). Since the quills are barbed and easily detached, they can cause severe injury to a predator. The streaked tenrec from Madagascar is another spiny animal with similar

Fig. 7.8 Deimatic behaviour of three mammals: (*a*) the tenrec *Hemicentetes semispinosus* in normal posture with spines lowered; (*b*) tenrec when facing a predator with spines raised and repeatedly bucking the head up and down; (*c*) the American porcupine *Erethizon* sp. in normal posture; (*d*) American porcupine with quills erected when encountering a possible predator; (*e*) spotted skunk *Spilogale putorius* displaying towards a possible predator. Not drawn to same scale (*Tenrec based on photographs in* Eisenberg and Gould, 1970; *porcupine redrawn from* Carrington, 1963; *skunk redrawn from* Bourlière, 1955, Fig. 55.)

defensive behaviour. It is black and yellow (possibly warning coloration) and has detachable spines. When attacked it faces the predator,
erects the spines on the head and back, and repeatedly bucks its head
up and down, stridulating as it does so (Fig. 7.8). The bucking may
drive spines into the predator if it comes too close (Eisenberg and
Gould, 1970). The American rattlesnake (*Crotalus* spp.) and the
African brush-tailed porcupine (*Atherurus*) also rattle scales or quills on
the tail when they are disturbed, apparently as a warning to predators
(Cott, 1940; Ewer, 1968) (Plate 6*c*). *Atherurus* is protected by sharp
quills, and the rattlesnake has a venomous bite. All of these anti-
predator displays are deimatic in function.

A number of other deimatic behaviour patterns occur which appear

to mimic the behaviour of snakes. I have already referred to the stridulation of various mantids and stick insects which may resemble the
hissing of a snake and hence cause a predator to hesitate from attacking
further. Some hole-nesting birds such as great tits, blue tits and the
wryneck (*Jynx torquilla*), also mimic a snake if a predator appears at
the nest entrance when the bird is inside. They hiss like a snake and bang
their wings against the wall of the hollow. Since it is dark in the nest a
carnivorous mammal may perhaps be intimidated by the display
although the bird itself has very little anatomical resemblance to a
snake (Hinde, 1952). It is possible that many mammalian predators
have an instinctive avoidance response towards rustling or hissing
noises.

Some snakes may themselves mimic the pattern and warning displays
of other poisonous snakes. In the New World many poisonous snakes
are aposematically coloured (coral snakes) and are mimicked by various
harmless snakes (see p. 132). In the Old World most of the poisonous
snakes are cryptic. Probably this is because although they may be
poisonous and can kill a predator with a single bite, they are themselves
vulnerable to attack. A well placed bite or peck from a strong predator
could kill a viper, even though the predator might also be killed. Hence
it may be of greater protective value to be cryptic than to be conspicuous unless a species is so common that it can withstand the losses
caused by inexperienced predators attacking it. Once it has been discovered it will still be of value to a poisonous snake if the predator can
be frightened away before it has actually launched an attack. Hence we
find that normally cryptic vipers hiss and swell up when disturbed and
before they actually attack. The African carpet viper (*Echis carinata*)
has a particularly dramatic display in which the body is thrown into a
series of C-shaped turns and constantly swirled by throwing waves of
movement back down the body from the neck. The coils slide over one
another and produce a rasping or hissing noise, and the head remains
motionless except when the animal makes a periodic strike at the
aggressor. The swirling also displays the light and dark marks on the
body, and such a display may intimidate a predator and inhibit it from
further attack. Eventually the swirling movement carries the snake
away from the predator until it is able to turn and flee without being
attacked. This display is a genuine secondary defensive warning, but it
is difficult to see how any predator could learn that it is dangerous
since one bite from a carpet viper is likely to be fatal. It is possible,
however, that such occasional predators as baboons may see one of
their number attack and be killed by a snake and then themselves learn
to avoid such snakes in future (observational learning — see also
p. 65). They might further discipline juveniles into keeping away from
similar snakes (Gans, 1964). Thus the swirling display of the carpet
viper may be a genuine warning that social animals can learn through
experience to avoid, or it may be a frightening display which inhibits an

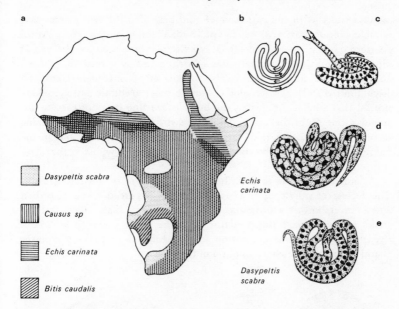

Fig. 7.9 (a) map of Africa showing the distribution of the egg-eating snake *Dasypeltis scabra* and of the three vipers whose deimatic coloration and swirling display it mimics; (b) movements of the coils of *Dasypeltis* during the swirling display indicated diagrammatically; (c) *Dasypeltis* striking at a predator whilst swirling; (d) swirling display of carpet viper *Echis carinata*; and (e) swirling display of *Dasypeltis scabra*. (*Map based on* Gans, 1961, *modified by recent information provided by* Mr B. Hughes; (b) *from* Gans and Richmond, 1957; (c), (d) *and* (e) *based on colour slides, film, and specimens in the possession of* Mr B. Hughes.)

attack even in a naive predator.

Egg-eating snakes of the genus *Dasypeltis* are all harmless, but some of them have a very similar coloration to that of various vipers. The most widespread is *Dasypeltis scabra* which occurs over much of Africa and is very variable in its colour pattern in different areas (Fig. 7.9). Gans (1961) has shown that over much of Africa it has the same colour pattern as the night adder (*Causus* spp.). In some areas both species are brown with conspicuous dorsal blotches or diamonds, and with lateral bars of dark brown or black outlined with cream or white. In other areas both species are more uniformly brown. When attacked, both species inflate the front part of the body, hiss or wheeze, and strike at the stimulus source. The adder can inflict a poisonous bite, but the strike of the egg-eating snake is harmless. In parts of South-West Africa, night adders do not occur and here *Dasypeltis scabra* resembles the Cape horned viper (*Bitis caudalis*). This mimetic assemblage is complicated by apparent müllerian or mertensian resemblances as well. *Bitis caudalis* and *Causus rhombeatus* occur together in Zambia, and here they have similar colour patterns. In West Africa the night adder occurs sympatrically with the carpet viper, and here also the two species have similar colour patterns. In Egypt, Sudan and parts of West Africa (eastern and northern Ghana) *Dasypeltis scabra* resembles the carpet viper in coloration. *Dasypeltis* has a swirling display very similar indeed to that of its model, and it also strikes at an intruder (Gans and Richmond, 1957). The detailed colour patterns of mimic and model are not all that similar, but when the body is swirling it is not easy to distinguish one from the other. Thus the swirling display is a frightening display in both the carpet viper and the egg-eating snakes, but in the former it is backed up by the threat of a deadly bite whilst in the latter it is purely bluff.

A similar situation is described from Australia by Bustard (1968). The legless lizard *Pygopus nigriceps* flattens its head when disturbed and even strikes at a potential predator just as a snake does. But the strike is aimed past the predator, and the lizard's mouth is closed. *Pygopus* has the same coloration and head markings as the venomous elapid *Denisonia* which is common in the area.

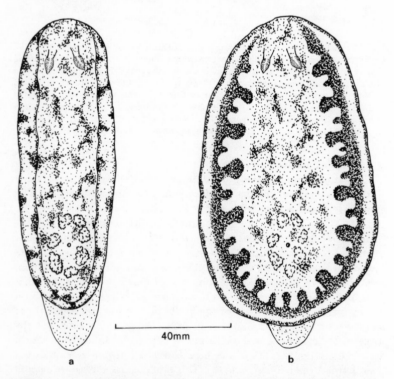

a   b

Fig. 7.10   (a) nudibranch mollusc *Hexabranchus marginatus* in normal crawling posture with mantle margin inrolled; (b) *Hexabranchus* during deimatic display with mantle fully exposed. (*Reproduced from* Edmunds, 1968b, Fig. 2.)

Deimatic behaviour in molluscs

A very impressive startle display occurs in the nudibranch mollusc *Hexabranchus marginatus*. This animal rests or crawls with the mantle margin inrolled and it is then mottled cream and pink and so is cryptic on the stones or corals of the sea-bed. When disturbed it extends the mantle margin fully on each side and exposes brilliant red and white markings (Fig. 7.10) (Edmunds, 1968*b*). Unfortunately it is not known how potential predators react to this behaviour nor is it known if *Hexabranchus* is palatable or unpalatable; but the circumstances which elicit the response suggest that it is designed to frighten predators.

The cuttlefish *Sepia* also has well developed deimatic behaviour (Holmes, 1940). When touched, two black spots suddenly appear (by expansion of black chromatophores) on the back (Fig. 7.11*a*). With further stimulation the rest of the animal pales so that the two spots are very conspicuous. Then the animal may dart rapidly away, darkening as it moves, so that a pursuing predator is left attempting to follow a white object that no longer exists (flash behaviour). With further irritation the cuttlefish changes colour repeatedly as it swims, sometimes adopting a zebra pattern (cryptic), sometimes with two black spots

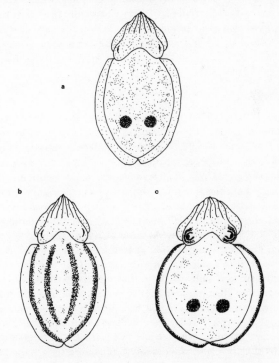

Fig. 7.11   Deimatic displays of the cuttlefish *Sepia officinalis*. For further explanation see text. (*Redrawn from* Holmes, 1940.)

(deimatic), and sometimes becoming pale with four black lines (deimatic) (Fig. 7.11*b*). Finally it may eject a cloud of ink and come to rest motionless beneath the cloud where it cannot be seen. Sometimes instead of behaving in this way a disturbed cuttlefish flattens itself, thus giving the illusion of increase in size, pales, and adopts the colour pattern shown in Fig. 7.11*c*, with two black spots, a black rim to the fin, and black rings round the eyes. This is probably also a deimatic response, and the eyelike marks on the head as well as on the back suggest that it may release the same responses in predators that large eyespots do (see below). The wide variety of deimatic responses of the cuttlefish may reflect the ability of predators to learn that they are bluff and to persist with an attack. Displays or other behaviour in response to predators are variable (protean) and hence may confuse predators simply because they are unpredictable and hence difficult to learn (Chance and Russell, 1959; Humphries and Driver, 1971). Similar deimatic behaviour involving striking colour displays occur in other cephalopods such as *Octopus vulgaris* (Packard and Sanders, 1971) and *Argonauta* (Young, 1959). *Heteroteuthis* can eject either a luminous secretion or a black ink when it is disturbed. The luminous secretion breaks up into a shower of sparkling droplets which may perhaps frighten an attacker (Nicol, 1971). Searciid fish and the shrimp *Acanthephyra* also eject luminous clouds, but it is not clear if these are deimatic responses or if they protect the animals in some other way (see p. 175).

Eyespots

Coppinger (1969, 1970) and Blest (1957*a*) showed that the sudden appearance of bright colours releases escape responses in some birds, and hence selection would be expected to favour those prey individuals which have the colours or displays that are most effective in releasing escape in their predators. This leads to dramatic frightening displays as already described, but it can lead instead to the development of circular eye patterns with associated displays. Blest placed dead mealworms on a box and allowed birds (yellow buntings, chaffinches and great tits) to approach and feed on them. As soon as a bird alighted on the box it completed an electric circuit that caused a bulb below the mealworm to light up and reveal a pattern on either side of the bait. By using different patterns he was able to show that circular patterns released more escape responses than parallel lines or crosses, and that the more similar the circle was to a vertebrate eye, the more effective it was in eliciting escape responses (Fig. 7.12). He concluded that the escape response to sudden presentation of a vertebrate eye is innate. Presumably in nature this means that if a bird is suddenly confronted by a large predator it will instinctively try to escape, and this behaviour has obvious survival value (see p. 66). But many insects mimic the eyespots of vertebrates

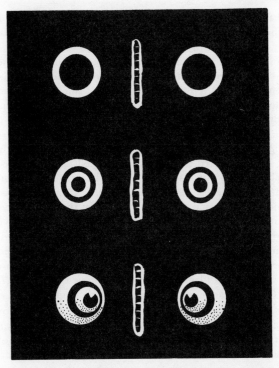

Fig. 7.12   Three of the models used by Blest in his experiments on the responses of birds to the sudden appearance of different shapes. As soon as the bird approached the mealworm (centre) a circle or eyespot was lit up on either side of it, and the response of the bird was recorded. Of the three shapes shown here the single circle elicited fewest escape responses in the birds whilst the eye with excentric pupil elicited most escape responses. (*Based on drawings in* Blest, 1957*a*, and Wickler, 1968.)

and hence gain protection when they are themselves attacked by birds or mammals. When the peacock butterfly (*Nymphalis io*) is at rest or feeding it often keeps the wings closed exposing the blackish under-surface which resembles a dead leaf. When disturbed the wings are opened (abducted) and protracted, and a hissing may be produced by rubbing the edges of the wings together. The wings are then lowered and raised several times in succession exposing and concealing four large eyespots (Plate 7*c*). When normal butterflies and butterflies in which the eyespot scales have been rubbed off are presented to yellow buntings in equal numbers, there are significantly more escape responses to the normal insects (137) than to the eyeless ones (37). However, the birds rapidly habituate and learn to ignore the eyespots, although Blest found that some individual birds became conditioned to avoiding eyespots permanently. Hence although the avoidance response to eyespots is innate a predator can learn to ignore it if the stimulus is presented too frequently. Eyespot displays then can only be of survival value if they are not too common and not exposed to predators too frequently.

Most insects that have large eyespots normally keep these hidden and only expose them when they are disturbed. Butterflies have them on the upper side of the fore- and hindwings, since they normally rest with the lower surfaces of the wings exposed, whilst moths have them on the upper surface of the hindwings since they rest with the hindwings concealed under the forewings. The best known moth with eyespots is the eyed hawkmoth (*Smerinthus ocellatus*). When disturbed the forewings are protracted exposing the big ocelli, and the insect moves rhythmically up and down so that there is the impression of movement associated with the eyes. Since it rests on tree trunks the resemblance to the sudden appearance of an owl at a hole in the tree is very striking.

The way in which such perfect eyespots as those of the eyed hawkmoth have evolved is indicated by comparison with related species. Many sphingids have a deimatic response when poked: in the convolvulus hawkmoth the body is raised and the wings depressed thus exposing crimson and black bars on the abdomen, whilst in *Hippotion osiris* the forewings are raised exposing crimson hindwings. In species of *Platysphinx* the forewings are protracted exposing bright yellow on the hindwings. At the proximal anterior corner of the hindwings there is a black spot which has a slight resemblance to a vertebrate eye, especially in species such as *P. constrigilis* in which the black spot has a pale excentric spot in it (Plate 7d). One can see how the perfect 'eye' of the eyed hawkmoth could have evolved from a simple startle display with each modification being more effective than the last at intimidating predators.

The eyespot displays of saturniid moths, many of which have very perfect 'eyes', are described by Blest (1957b). Some of the displays of mantids (*Pseudocreobotra* and *Stagmatoptera*) and stick insects (*Metriotes diocles* and *Eurycnema goliath*) also include eyespots in the colour patterns displayed (Edmunds, 1972; Maldonado, 1970; Robinson, 1968b; Bedford and Chinnick, 1966).

Some very dramatic eyespot displays occur in sphingid caterpillars. Most sphingid larvae are cryptic, but some have deimatic displays when they are molested. The deaths head caterpillar hunches its thorax up and literally gnashes its jaws in a rapid clicking. The striped hawkmoths *Hippotion osiris* and *H. eson* inflate the region just behind the head by retracting the head, and they may lash the head from side to side. There are two large spots on the thorax which are thus displayed. These eyespots, however, cannot be concealed, so they must not be too conspicuous or the resting animal would be easily seen and attacked by birds which might not have noticed a purely cryptic animal. In *H. eson* the eyespots are dark green with a yellow rim in green caterpillars, and dark brown with a red and yellow rim in brown caterpillars (Plate 8c,

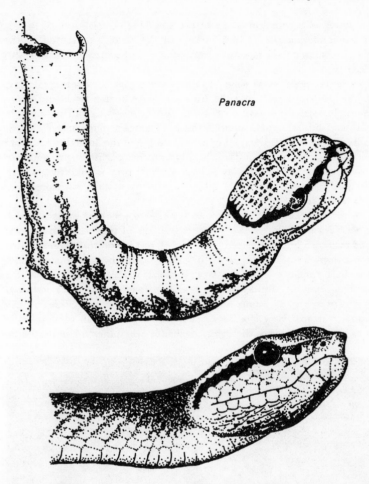

Panacra

Trimeresurus

Fig. 7.13 Deimatic display of the sphingid caterpillar of *Panacra mydon* compared with the head pattern of the young ·Wagler's pit viper (*Trimeresurus wagleri*), both from Malaya. (*Redrawn from photographs in* Morrell, 1969.)

d). In *H. osiris* the brown caterpillar has blue eyespots with a black rim. Thus the eyespots are cryptic or not noticeable from a distance, but conspicuous when seen from nearby. Other sphingids normally have the eyespots hidden under a fold of skin and they are only exposed when the thorax is inflated (e.g. the oleander hawk, *Deilephila nerii*, Plate 8*a*, *b*). Even more dramatic displays occur in *Panacra* and *Leucorampha*. Both of these caterpillars rest upside-down beneath a leaf or branch. When disturbed the head is raised from the substrate and inflated. In *Panacra* the dorsal pattern of the head with false eyes resembles that of the young of a common snake of the area (Fig. 7.13) (Morrell, 1969). In *Leucorampha* the ventral surface of the head is presented to the stimulus source and it has conspicuous eye-like marks (Moss, 1920) (Plate 7*b*).

Some animals emphasize genuine eyes during an eyespot display. Unlike the adult owl, juvenile owls are quite unable to defend themselves. Instead they have a dramatic display involving fluffing out the wing and head feathers so that the bird appears much larger than it really is, the beak is snapped open and shut, and the animal sways from side to side. With its huge eyes the appearance is frightening to a human observer although the display is largely bluff: even when it lunges with the beak it does not actually bite at a stimulus source, but stops short of it (Plate 7*a*).

When attacked the octopus also takes on a colour pattern which emphasizes the eyes. In the deimatic displays of the cuttlefish the true eyes may also be emphasized (Fig. 7.11*c*), but in addition there are two very conspicuous false eyes in the centre of the back. Other cephalopods have similar eyespot displays (Packard, 1972). A vertebrate with a dramatic eyespot display is the Brazilian toad *Physalaemus nattereri* which presents its rump towards predators. On each side of the abdomen in rear view is a large eye mark (Fig. 7.14). In some populations the toads display the eyespots readily when they are molested, in

Fig. 7.14   Deimatic behaviour of the toad *Physalaemus nattereri* in which two large eyespots are displayed. (I. Sazima, Brazil.)

others they fail to direct their rump towards the stimulus source, and in yet others they do not display at all (I. Sazima, personal communication). Presumably this variation in behaviour between different populations reflects different predators to which they are exposed.

Thus eyespot displays can be regarded as a special type of deimatic behaviour in which an animal with no retaliatory defence capitalizes on the innate avoidance responses of predators to large eyes. The eyespots themselves may be genuine eyes displayed so as to make them conspicuous, or they may be false eyes which are concealed except during display.

Summary

When discovered by a predator many animals attempt to frighten the predator — this is called deimatic behaviour. Deimatic behaviour can be a genuine warning to a predator that an animal is unpalatable or well armed, or it can be a bluff. Predators have an innate avoidance response to bright colours which appear suddenly, and the elaborate deimatic displays of some insects may have evolved from quite simple movements of coloured surfaces. Predators may also have an innate avoidance response to suddenly appearing eyes, and many deimatic displays involve the exposure of large eyespots.

Chapter 8
Thanatosis

Deimatic behaviour inhibits an attack from a predator by stimulating it to withdraw. A different method of inhibiting an attack is by feigning death — thanatosis — so that an animal fails to release a killing response in the predator. Some beetles, bugs (e.g. *Ranatra*), grasshoppers, spiders, stick insects (see Robinson, 1969*b*) and mantids (e.g. *Phyllocrania*, see Edmunds, 1972) become inert when they are attacked, either with the legs extended or (in beetles) withdrawn close to the body so that they cannot be bitten off. The position may be maintained only briefly or for several minutes according to the species. There are some vertebrates which also exhibit thanatosis such as the American opossums (*Didelphis* spp., hence the expression 'playing possum') and the African ground squirrel which can be picked up and handled for several minutes without their moving (Ewer, 1966). The protective value of giving up the attempt to escape is that the predator may relax its attention, thus allowing the animal to recover and escape actively. For example, some small birds rest motionless after being captured and held loosely in the hand, but then they suddenly revive and fly away without warning. A further advantage of thanatosis is that many predators, including cats, lizards and mantids, strike to kill only on prey that shows active movement. Absence of such a stimulus to strike may therefore protect the prey since although it may be examined by the predator, it does not necessarily receive a lethal blow or bite. When cats play with a captured but injured mouse it is easy to see how the mouse would often do better to remain motionless as if dead instead of persistently trying to hobble away, only to be repeatedly recaptured by the cat. Some snakes roll into a tight, motionless ball when they are attacked (Bustard, 1969). The advantage of 'balling' compared with typical thanatosis may be that the vulnerable head is protected from injury in the centre of the ball.

Summary

Some animals respond to an attack by a predator by feigning death — this is thanatosis. Thanatosis may be of survival value because it causes a predator to relax its attention, thus giving the animal a chance to escape actively, and also because a motionless animal fails to elicit a killing strike by the predator which may eventually lose interest and move away.

Chapter 9

Deflection of an attack

Another way in which a prey animal can escape when attacked by a predator is to cause the predator to attack in the wrong place. It may deflect the attack away from itself or from its young onto some other less vulnerable object (diversion behaviour), or it may deflect the attack on to some part of its own body where an attack is unlikely to be fatal. The part of the body attacked may be edible so that the predator gets some slight reward even though the prey animal escapes, or it may be particularly nasty so that the predator is inhibited from further attack on the body of the prey animal.

Diversion behaviour

Diversion behaviour occurs when an animal diverts an attack away from itself or from its young onto some other object. Such 'distraction displays' are well known in birds, especially waders and ducks, and are typically given in response to the presence of a large ground predator such as a dog or man. The parent bird moves quietly away from its eggs or young, then gives an elaborate, conspicuous display which gradually takes it still further away from the young. Typically one wing is flapped conspicuously as if the bird is injured and unable to fly. This occurs on the ground in the case of waders, on water in the case of ducks, and it is likely to attract the predator away from the young to pursue the parent. When closely approached by the predator the bird quickly 'recovers' and flies off. When far away from the chicks the bird may give a static display, as if it is so badly injured that it cannot move (Fig. 9.1) (Simmons, 1952). Some birds such as the dunlin (*Calidris alpina*) creep along the ground uttering squeaky notes rather like a small mammal. Possibly predators mistake it for a small mammal and give chase, whereupon it flies off whilst the chicks lie hidden and ignored until the predator has moved away (Simmons, 1955).

Some other animals also give distraction displays to divert the attention of a predator away from the young. The freshwater holostean fish *Amia* thrashes about when large predatory fish are near its young, apparently resembling an injured fish (Lagler *et al.*, 1962); and the

Fig. 9.1   Distraction display of the kentish plover *Charadrius alexandrinus* in response to a potential predator approaching its young. (*Redrawn from* Robert Gillmor's *drawing in* Simmons, 1952, Fig. 1.)

mother tiger moves conspicuously away from her cubs when man is nearby (Schaller, 1967).

All of the examples of diversion behaviour given so far involve an attempt by a parent to divert a predator away from the eggs or young. There are also cases in which the predator is diverted away from the animal itself. Many theridiid and argiopid (orb-web) spiders wrap up insect remains or egg cocoons and place these in their webs. Since they are of similar size and shape to the spider itself, it is likely that predatory wasps may be deceived into attacking one of the false spiders and hence overlook the true spider, or at least give her the chance of escaping. This has already been described on p. 37 since it is a primary defence which operates whether or not a predator is in the vicinity.

Cephalopods also employ a form of diversion behaviour in defence. When attacked, the cuttlefish may eject a cloud of black ink (Holmes, 1940; Boycott, 1958). The cloud is rather viscous and remains as a discrete unit for some time instead of diffusing rapidly. At the same time as it ejects the cloud the cuttlefish pales in colour and swims rapidly away leaving the predator to attack the black cloud in mistake for itself. Some deep-sea squids (e.g. *Heteroteuthis*) discharge a luminous cloud which may function in the same way, or possibly it blinds pursuing predators and has a deimatic function. When the cuttlefish is further provoked it can eject quantities of a rapidly diffusing black ink which forms a dense fog. Under cover of this fog it comes to rest and takes on one of its cryptic colour patterns or partly buries itself in sand. Searciid fish and the shrimp *Acanthephyra* can also emit luminous secretions which may perhaps function in similar ways (Nicol, 1971). The sea hare *Aplysia* ejects a cloud of purple fluid when attacked, but although this has been observed many times it is not known if it diverts an attack away from the animal, or if it is deimatic, or simply has a nasty taste.

Deflection marks

Deflection marks direct an attack either at a non-essential part of the body or at a positively distasteful part of the body. Many lycaenid butterflies rest on vertical surfaces with the head downwards, contrary to most butterflies which usually rest head up. Normally when a predator attacks a resting insect it must be prepared for the insect to take evasive action, and usually this will mean grabbing just above where the butterfly is resting. But with lycaenids such a manoeuvre will fail since the insect will fly off downwards. Some lycaenids, particularly in the tropics, deceive predators even more effectively. *Thecla togarna* for example quickly turns through 180° when it lands so that its hind end points in the direction of previous flight. It has thin antenna-like projections on the tips of the hindwings, and these are moved up and down for a few seconds after landing whilst the true antennae are kept motionless (Fig. 9.2). The base of each false antenna is further held open and is marked so as to resemble an eye. Presumably if a bird has been watching the insect settle it may quickly pounce on the insect, but will be deceived into mistaking the direction in which it will fly away. After having waggled the false antennae the insect then rests motionless (Wickler, 1968). In this case the false antennae and the false eyes can be regarded as deflection marks diverting an attack to the wrong end of the insect. Even if the hindwings are grabbed by a bird the insect may still escape, albeit with its wings slightly torn.

Some snakes, such as the harmless ground python *Calabaria reinhardtii*, also waggle the tip of the tail when disturbed so that a predator may mistake the tail for the head and attack it (Bustard, 1969). If the tail is injured but the snake escapes with its life, the deception will have been of survival value. Wickler (1968) discusses several other species of snake in which the tail mimics the head in this way.

Small eyespots on the wings of various butterflies probably deflect attacks by predators in a similar way, as was suggested long ago by Poulton. Evidence that this is so is summarized by Cott (1940) and Blest (1957*a*). Carpenter (1941) found that the wings of many butterflies in Africa carry triangular beak-marks implying that they have been seized by a bird but have escaped. He observed that the beak marks are concentrated around the small eyespots, and since these are normally near the edges of the wings rather than near the body, the attack is not always fatal to the insect. Swynnerton painted conspicuous eyespots on the undersides of the wings of fifty-one *Charaxes* butterflies which were then released. He subsequently recaptured forty-seven insects that had beak marks or torn wings. Thirty-six of these injuries were close to eye-

Fig. 9.2 Resting posture of the butterfly *Thecla togarna*. Note the false antennae and eyes at the tip of the hindwing, and the disruptive markings which concentrate attention in this region and away from the true head. (*Redrawn from a painting in* Wickler, 1968, Fig. 15.)

spots, and the remaining eleven could be regarded as attempts by birds to aim at the eyespot. Thus the small eyespots appear to direct attacks at a part of the insect that is not essential. Blest (1957*a*) marked mealworms with small blobs of paint to resemble small eyespots. He found that small birds (yellow buntings) pecked more frequently at the head than at the tail, and more frequently at the mealworms with painted eyespots than at mealworms with a painted end but no eyespot (Fig. 9.3). Thus he obtained evidence that small eyespots induce more pecks than do other areas.

One disadvantage of small eyespots is that they are conspicuous and may actually attract the attention of a predator that had not previously noticed an insect. This disadvantage is overcome in the grayling butterfly (*Eumenes semele*) by a behavioural action similar to that of *Thecla togarna*. When a grayling alights it closes its wings above the body but keeps the eyespots visible (Fig. 9.4). It may make a few movements, and if a predator is in the vicinity it is likely to attack at this time. But if the insect is not attacked or disturbed, after a few seconds it orients its body in line with the sun, so reducing the shadow, and lowers the forewings between the hindwings so that the eyespots are hidden and the insect is entirely cryptic (Ford, 1945; Tinbergen, 1958). The insect even leans over to one side in line with the sun, which is never directly overhead in Europe, so that there is minimal shadow (Fig. 9.4). Whilst resting, if disturbed it may raise the forewings briefly and hence elicit

% of all pecks which were at painted end

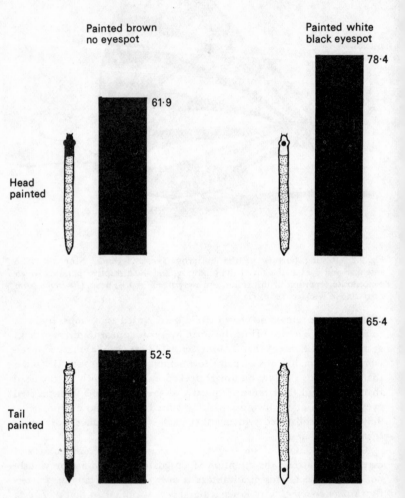

Fig. 9.3   Responses of yellow buntings (*Emberiza citrinella*) to mealworms with a painted mark at one end. (*Data from* Blest, 1957a.)

an attack at the eyespot rather than elsewhere. But a more drastic disturbance causes it to fly away.

Some species of fish (e.g. *Chaetodon capistratus*) have an eyespot on a fin or near the tail, and it has been suggested that these attract predatory fish to attack the wrong end of the fish which is able to escape by a quick dart in the opposite direction (Cott, 1940). It would be interesting to have confirmation of this from skin divers since

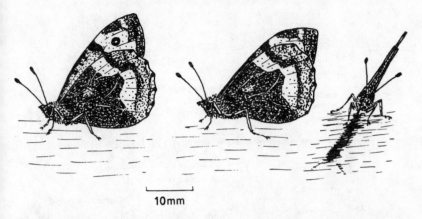

10mm

Fig. 9.4   Resting behaviour of the grayling butterfly *Eumenes semele*. Left: insect having come to rest with the eyespots exposed. Centre: normal resting posture with the forewings lowered hiding the small eyespots. Right: resting position adopted in sunlight with body in line with the sun and tilted to minimize shadow. (*Based on* Ford, 1945, Tinbergen, 1958, and others.)

dummy eyes or other small circular spots may have a variety of different signal functions in fish (Wickler, 1968).

Other colour marks may direct the attack of a predator at a positively nasty part of the prey animal. Many eolid molluscs have brightly coloured papillae on the dorsal surface (Fig. 9.5) and these are waved to and fro when the animal is molested (Edmunds, 1966*a*). A predator that decides to attack may nibble off a few papillae, which contain nematocysts and sometimes glandular secretions, rather than sample the entire animal. Brightly coloured tufts of urticating hairs on many caterpillars possibly serve a similar function with the added signal to the experienced predator not to touch them again. The bodies of both the eolid and the caterpillar may actually be palatable.

Autotomy of a non-essential part of the body

A variety of animals have the ability to constrict and break off a part of the body when they are attacked (autotomy). If a predator is left with the broken limb whilst the animal escapes, the defence will have succeeded in protecting the animal.

Eolid molluscs and some dorid and sacoglossan molluscs autotomize their papillae when they are attacked. The papillae continue to writhe for some minutes after detachment, and presumably they may keep a predatory fish occupied whilst the mollusc crawls away. Such papillae can be regenerated later. They also contain defensive glands and (in eolids) nematocysts, so they appear to be the most unpalatable part of the animal (Edmunds, 1966*a*, *b*). *Discodoris fragilis* autotomizes part of

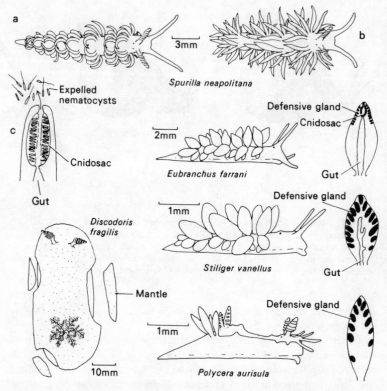

Fig. 9.5   Defensive adaptations of some sea slugs: (*a*) an undisturbed eolid crawl-ing; (*b*) the same animal when poked shortens the body and extends and waves the cerata papillae to and fro; (*c*) tip of a papilla with some nematocysts *in situ* and some ejected and discharged. The eolid *Eubranchus* has nematocysts and defensive glands in its papillae whilst the sacoglossan *Stiliger* and the dorid *Polycera* only have defensive glands. The papillae of all three can be autotomized and later regenerated. The dorid *Discodoris fragilis* (bottom left) can autotomize parts of its mantle.

its mantle when it is attacked, and this too can be regenerated (Alder and Hancock, 1864; Edmunds, 1971). The sacoglossan mollusc *Oxynoe* can autotomize and regenerate its tail under similar circumstances (Stamm, 1968; Warmke and Almodovar, 1972). Some bivalves which live buried in sand can autotomize and regenerate the projecting siphon if this is nibbled by a fish. (Other examples of autotomy in molluscs are summarized by Stasek, 1967.) Tubicolous polychaetes may also have powers of autotomy and regeneration. When the lug-worm *Arenicola* projects its posterior part of the body above the surface of the sand to defaecate, it immediately becomes vulnerable to predation. The hind region of the worm contains no vital organs and can be autotomized and regenerated.

Tail autotomy is well known in geckos and various other lizards, and in species such as *Ptychozoon* the tail is more conspicuously coloured than the rest of the body, presumably so as to divert an attack towards the tail rather than at the body (see Fig. 2.9). After detachment the tail continues to writhe for some time, and the animal can regenerate it in a few weeks (Cott, 1940).

Arthropods can often autotomize limbs and these are regenerated at the next moult, although it may take several moults before they attain their normal size. Defence by the use of chelae which are then autotomized (attack autotomy) is described in the crab *Potamocarcinus* on p. 189.

Some social hymenopterans (e.g. *Apis, Polybia* and certain ants) can autotomize the sting, but there is no regeneration and the loss results in death of the insect. Nevertheless for a social insect sting autotomy can be of advantage to the colony since the effect of a sting left behind in the skin of a vertebrate is much more effective at deterring it from further attack than is a sting which is withdrawn (Hermann, 1971). Autotomy of the sting, papillae or chela all leave the predator with the most noxious part of its prey. With other examples of autotomy the predator does get some small reward even though the prey animal survives the encounter.

Summary

Some animals avoid being killed when a predator attacks by causing the attack to be directed at the wrong place. Diversion behaviour directs the attack towards a non-vulnerable animal or towards an inanimate object. Deflection marks induce a predator to attack at a non-vulnerable part of an animal where no damage can be done. Sometimes the part attacked is particularly nasty so that the predator may hesitate before attacking such an animal again. Finally, a non-vital part of an animal can sometimes be autotomized (and later regenerated) leaving the predator with a small reward whilst the body of the prey animal escapes.

Chapter 10
Retaliation (aggressive defence)

If a predator is very close to a prey animal, or if it has actually seized it, then the prey animal may retaliate by attacking the predator with whatever weapons it possesses. Mammals bite, scratch or use horns or hooves; birds peck, stab or scratch; and insects bite, scratch or sting. A cornered rabbit will kick out vigorously at a stoat, and a rat's incisors can cause deep wounds to a dog's nose, so attack as a final form of defence can sometimes inflict injury on the predator.

Weapons used in defence against predators may be used for other purposes as well, or they may originally have been used for some other function but are now purely defensive, or they may have evolved purely for defence. It is convenient to consider defence by retaliation according to the nature of the weapons used.

Weapons derived from the mechanism of food capture

Defensive weapons are often structures which are used primarily for capture of food. Mantids bite with the jaws and slash with the forelegs when picked up; both of these weapons are also used for food capture or for eating. Many mammals attack predators with their teeth, and birds peck or stab with the bill. Reduviid bugs such as *Platymeris* (Plate 5*b*) pierce and suck their prey and they also 'bite' a predator when they are attacked. To be more precise, they pierce the attacker with the mouthparts and inject salivary fluid which can be very painful. When molested *Platymeris* can eject saliva up to 30 cm, the tubular sucking mouthparts being naturally preadapted to squirting (Edwards, 1960). This fluid is harmless to insects (unless injected into the haemocoel as occurs when they feed). Probably insects are not serious predators of *Platymeris* because of its large size, although the large argiopid *Nephilengys* does occasionally catch it. The fluid is very painful to the eyes and nasal membranes of vertebrates (Roth and Eisner, 1962), and probably it is against insectivorous vertebrates that this defence has evolved.

The scorpion uses its sting to inject poison into an attacker, just as it does into its own prey. Similarly wasps, which normally use their sting to paralyse prey, also use it in their own defence. Bees have retained the sting as a purely defensive weapon but have drastically changed their feeding habits so that the sting is never used for food collection. If a *Halictus* bee is caught by the orb-web spider *Araneus diadematus* it can occasionally sting the spider before it is wrapped up and bitten. The sting paralyses the spider for long enough for the bee to bite its way free from the web (Bristowe, 1958). Snakes are another group that use teeth and sometimes venom for capture of prey, and the venom is also used for defence.

Weapons evolved primarily for intraspecific behavioural reasons

Other weapons have evolved primarily for use in intraspecific behavioural situations but may also be used interspecifically. The antlers of many deer are of use in encounters between rival males (Geist, 1966). In most species the females have no antlers, yet if these were primarily for defence against predators it is the females that one would expect to require the more efficient protection. Even the males shed their antlers in the winter when predator pressure due to shortage of food is likely to be most critical. However, in many antelopes and bovids both sexes have horns, and it is not easy to decide if the primary function of the horns is related to intraspecific social organization or to defence against predators. In the moose the horns are used principally in intraspecific encounters whilst hooves are used in warding off wolves by kicking and trampling (Mech, 1970). Giraffe also use their horns for fighting one another and their hooves for kicking at predators. Elk and musk oxen, however, do use their horns against predators, and even the cow wildebeest, which does not attempt to retaliate in its own defence, does use its horns in attacking hyaenas or jackals when these predators are attempting to kill its calf (see Chapter 12). The oryx has very long stiletto-like horns, beautifully adapted to ritualized intraspecific combat, but it has been known for oryx to stab lion with them as well. Thus whilst ungulate horns have probably evolved as a result of intraspecific social selection pressures, in a few species they are also important in defence.

In most carnivorous mammals the secretion of the anal sacs is of social importance in marking territories or home ranges, or in sexual behaviour, but in the skunk and a few other carnivores this secretion is its most potent defensive weapon. Here again the defensive function is obviously secondary. However, despite the nauseous smelling fluid, skunks are eaten by vertebrate predators occasionally, probably when the predator is close to starvation and when other prey is not available (Verts, 1967).

The noise (song, or stridulation) of cicadas is normally made by the male to attract females. In the American seventeen-year cicada

(*Magicicada* spp.), the males aggregate in large numbers. Birds gorge themselves on the cicadas, but they do not remain in the vicinity of the aggregation after feeding. It has been suggested that the noise produced by large numbers of cicadas is so loud that it acts as a repellent to birds. Possibly it is painful to the birds' hearing apparatus, or it may interfere with their normal means of communication (Simmons, Wever and Pylka, 1971). At all events, the birds fly into the swarm, feed quickly, and then fly out and away as quickly as possible. Grackles certainly appear to avoid areas of cicada song. Probably the sound prevents the birds from hearing anything else. The song of *Magicicada cassini* has a sound frequency of 4 500–6 000 Hz, so presumably birds in which sounds of this frequency range are of importance will be disoriented in a swarm of this species. The song of *M. septendecim*, however, is in the range 900–1 600 Hz. This difference between the two species no doubt enables females to come to males of the appropriate species, but it may also jam a much wider range of frequency of hearing for the predatory birds in areas where both species of cicada occur than in areas where only one species occurs.

The serrations on the leading edge of the forewing of African *Charaxes* butterflies are used to beat at and drive off other conspecifics and other species of *Charaxes* from a supply of food (tree sap or rotten fruit). These butterflies also struggle vigorously when seized by mantids or spiders and the damage sustained by the serrations may occasionally help them to escape. It is not clear, however, if the serrations evolved primarily to drive off other butterflies from a restricted supply of food or if they evolved for defence against predators.

Physical weapons of purely defensive function

Other retaliatory defensive structures have apparently evolved purely for interspecific defence (as anti-predator devices), for example the spines of hedgehogs, porcupines, tenrecs and sea urchins. The spines of some fish such as sticklebacks and trigger fish have a locking mechanism so that when erected they remain fixed in that position. Perch and pike have been found to have difficulty in swallowing spined fish (Hoogland *et al.*, 1957). The erected spines jam in the mouth of a predator which may be forced to spit out the prey, then grab it again in a more satisfactory position, or even to reject it altogether (Fig. 10.1). Pike which have experienced one or several sticklebacks reject them in future encounters. As soon as they see a stickleback, they follow it with their eye, but the hunting sequence then stops without the fish turning its body to follow the stickleback. If given equal numbers of minnows and sticklebacks, the minnows are usually eaten first in preference to sticklebacks. With minnows, three-spined and ten-spined sticklebacks, the minnows are eaten first, and the three-spined sticklebacks survive the longest (Fig. 10.1). Hence the ten-spined stickleback is less well

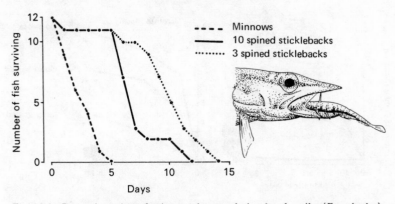

Fig. 10.1 Protective value of spines against predation by the pike (*Esox lucius*). Twelve minnows (*Phoxinus phoxinus*), twelve three-spined sticklebacks (*Gasterosteus aculeatus*), and twelve ten-spined sticklebacks (*Pygosteus pungitius*) were placed in a large aquarium with a single pike, and the numbers surviving each day are shown. The minnows were all eaten first because they lack spines, whilst the three-spined sticklebacks survived longest because they have the longest spines which jam in the mouth of the predator (shown on the right). (*Redrawn from* Hoogland *et al.*, 1957, Figs. 10 and 12.)

protected by its spines than is the three-spined, and this accounts for it being more timid, cryptic in colour, and its habit of living close to vegetation. The three-spined stickleback, by contrast, often breeds in exposed places where its bright colours (of the male when courting) render it conspicuous. Predators probably learn to avoid the three-spined stickleback after a number of trials, so its coloration may have an aposematic as well as a sexual function; but the ten-spined, with shorter and less effective spines, is better protected by remaining hidden. If the ten-spined is chased by a predator it swims directly away from its pursuer, always presenting its tail to the predator's mouth. In this way the predator is forced to seize it tail first when the spines jam in its mouth, rather than head first when they would not jam so effectively and when it would be easier to swallow (Morris, 1958).

The long spines of anomuran and brachyuran zoea larvae, of stomatopod larvae, and of the swimming crabs (*Callinectes*) probably also jam in the gullet of fish and are therefore of defensive importance.

Long spines may also be a deterrent against other predators besides fish. The rotifer *Brachionus calyciflorus* may have short or long posterolateral spines, or be spineless. Gilbert (1967) and Halbach (1971) have shown that individuals with long spines are rarely eaten when contacted head-on by the predaceous rotifer *Asplanchna*, whilst spineless individuals are often eaten (Fig. 10.2). Further, the time taken for *Asplanchna* to catch and swallow the first *Brachionus* is significantly longer in a colony of spined animals than it is in a colony of spineless ones. When the *Brachionus* is attacked by *Asplanchna*, it withdraws its corona of cilia and the increased body pressure causes the posterolateral

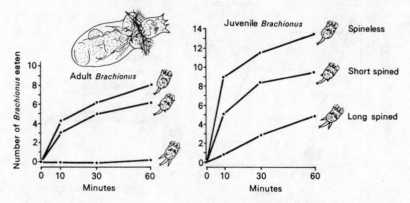

Fig. 10.2   Predation by adult *Asplanchna* on *Brachionus calyciflorus* with long spines, with short spines, or spineless. Left: predation on adult *Brachionus* with (above) drawing of the spines of a *Brachionus* jamming in the mouth of the rotifer *Asplanchna*. Right: predation on juvenile *Brachionus*. (*Data from* Halbach, 1971; *drawing from a photograph in* Halbach, 1971.)

spines to be protracted. They then jam in the mouth of *Asplanchna* and make swallowing difficult or impossible (depending on the sizes of the two rotifers and on the length of the spines). Even short spines have some protective value according to Halbach, but they are not nearly as effective as long spines. When alternative prey is available (spineless *Brachionus rubens*) both long-spined and short-spined forms are less heavily preyed upon than are the spineless ones (Fig. 10.3).

The factors producing short spines are not clear but may be environmental (e.g. shortage of food and low temperature), but the long spines are only produced in the presence of *Asplanchna*. *Asplanchna* produces a substance in the water, either by secretion or excretion, which is probably proteinaceous, and which induces spine formation in developing eggs of *Brachionus*. Consequently the proportion of *Brachionus* with spines should be correlated with the abundance of the predatory *Asplanchna*. Thus this defence is not only specific to a few closely related species of predators, but it is only developed if these predators are present. Halbach found no evidence that long spines impair the swimming or feeding ability of *Brachionus*, but presumably there must be some advantage in having no spines when *Asplanchna* is absent or the spined condition would have come under genetic rather than environmental chemical control, and would have spread throughout the population.

Some fish have poison glands attached to spines. The poison of a few species is toxic to man, but there are few detailed studies of the use of spines on natural predators. *Scorpaena plumieri*, the scorpion fish, attacks with raised dorsal spines a stick or other object placed near it (Hinton, 1962). This species is normally cryptic on the sea-bed. When poked or disturbed it expands the pectoral fins and displays their

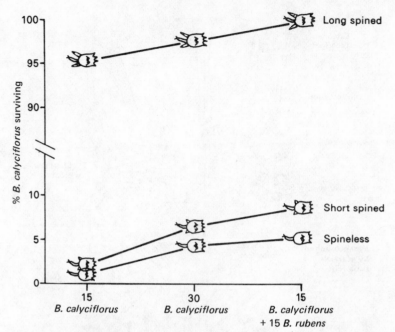

Fig. 10.3 Survival of adult *Brachionus calyciflorus* when exposed to predation by *Asplanchna* with and without spineless alternative prey (the rotifer *Brachionus rubens*). (*Data from* Halbach, 1971.)

ventral surface which is yellow and black with blue spots. Possibly this is a warning display, but it has not been demonstrated that a predator learns to avoid a fish that displays in this way. If it is further molested it rams the attacker with its spines (Breder, 1963). *Dactylopterus* and *Pterois* also attack intruders with their spines. *Scorpaena guttata* similarly raises its spines if approached by another fish or by man, but it flees from a hunting octopus (*Octopus bimaculatus*). Resting or fighting octopus are ignored (Taylor and Chen, 1969). The octopus eats *Scorpaena* in aquaria, and there is one record of this in the sea, so the spines are not effective against all predators.

The spines of sea urchins are also not effective against all predators, and hence many sea urchins have other defences. Pedicellariae are small jawed organs which snap closed on animals that succeed in getting between the spines of a sea urchin. Most pedicellariae probably perform the function of holding and killing larvae of encrusting animals (e.g. bryozoans and serpulids) which might otherwise foul the skeleton of the urchin. But some of the pedicellariae of *Strongylocentrotus lividus* and *Psammechinus miliaris* contain poison and appear to be a defence against attacks by the starfish *Marthasterias glacialis*. As the starfish crawls over the urchin the pedicellariae bite and immobilize the tube feet, and usually this induces the starfish to retreat (Fig. 10.4). A

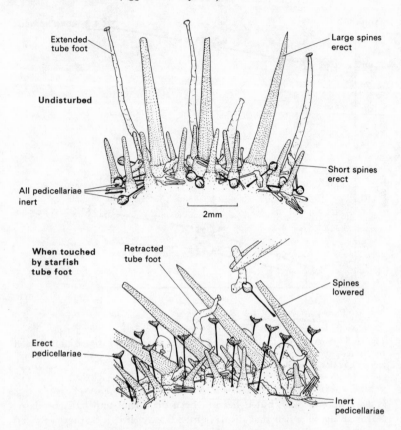

Fig. 10.4   Response of the sea urchin *Psammechinus miliaris* to contact by the starfish *Marthasterias glacialis*. Above: surface of an undisturbed sea urchin with all pedicellariae inert. Below: surface of sea urchin when contacted by a tube foot from the starfish (held in forceps). Note that the spines are lowered, tube feet retracted, and one type of pedicellaria (shown with black stalks) are erected with their jaws open — one has even closed onto the tube foot and been broken off at its base. (*Redrawn with modifications from* Jensen, 1966.)

persistent starfish, however, may get its meal since there is only a limited number of these pedicellariae, and once all of these are broken off and attached to tube feet, there is no further defence (Prouho, 1890; Jensen, 1966).

Earwigs (Dermaptera) flex their abdomen over the head when seized and pinch a predator with their pincer-like cerci. I have found that they are very adept at biting their way out of a spider's web, even after being wrapped up in silk by *Araneus diadematus*. If the pincers make contact with the spider they sometimes break a leg or even break open the abdomen and kill it.

Crabs and other chelate decapod crustacea use their chelae in defence. When attacked many crabs display with their chelae held open and raised towards the predator. They pinch if further provoked. Robinson *et al.* (1970) studied the behaviour of several species of crabs when attacked with a toy teddy bear (to represent a mammalian predator). Most species run away displaying their chelae, but *Xanthodius* becomes motionless and cryptic. If further provoked most species attack and sometimes a chela is autotomized (i.e. broken off by the crab) whilst still pinching the predator. In more than 50 per cent of trials with *Gecarcinus* and *Potamocarcinus richmondi* attack was accompanied by autotomy of a chela. When *Potamocarcinus* attack-autotomizes a chela on the otter *Lutra annectens*, the otter retreats in pain until it has managed to remove the chela. In the meantime the crab makes good its escape. Thus crabs use chelae in defence, and attack-autotomy can be an effective defence against mammalian predators. Autotomy can only occur twice of course after which the crab has no chelae left. However, land crabs (*Gecarcinus*) which have lost a chela moult much sooner than intact crabs (Skinner and Graham, 1970), and regeneration of the lost chela, although of small size, occurs at this moult. (Autotomy may be of defensive value in other situations as well and this is discussed in Chapter 9 on p. 179.)

Pupae of many beetles have also evolved elaborate defensive devices which probably protect them from attacks by predatory mites, beetles and ants. These devices have been called 'gin traps' by analogy with the gin traps used by man to catch various birds and mammals (Hinton, 1946). Gin traps occur in many families of beetles including the Dermestidae, Dryopidae, Cerambycidae, Scarabaeidae and Tenebrionidae, and they have evidently evolved several times independently. *Dryops* has five median gin traps, each with a toothed anterior jaw which closes onto a ledge beneath the posterior jaw (Fig. 10.5). When beetles and mites crawled over *Dermestes* pupae, the gin traps repeatedly opened and closed. Several times tarsi were pinched in a trap and then released, but the experience was usually enough to cause the beetle or mite to crawl elsewhere. Tenebrionids have paired lateral gin traps (Fig. 10.5), and those of *Tenebrio molitor* show no response when a beetle simply crawls over it. However, when the pupa is bitten the gin traps are closed repeatedly and hence in this species they are probably also of defensive importance (Hinton, 1946; Wilson, 1971).

Some lepidopteran pupae which are not protected by a silken cocoon also have gin traps, for example the privet hawkmoth (*Sphinx ligustri*) (Bate, 1973).

Chemical defence

The use of noxious chemicals in defence has already been described for certain animals such as *Platymeris* and *Scorpaena* in which the chemical is associated with a physical weapon such as a spine. Many other

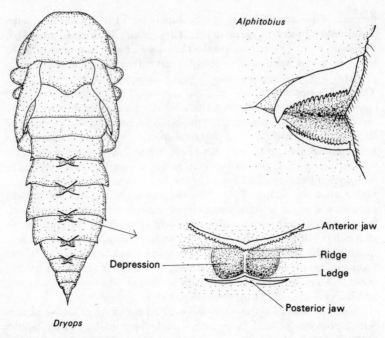

Fig. 10.5   Gin traps on beetle pupae. Left: dorsal view of pupa of *Dryops luridus* with five gin traps one of which is shown enlarged on the right. Top right: lateral gin trap of the tenebrionid *Alphitobius diaperinus*. The anterior and posterior jaws of the gin traps (which are not stippled) lie on adjacent segments of the body and can be closed together by contraction of the appropriate segmental muscles. (*Redrawn from* Hinton, 1946, Figs. 21, 22, and 27.)

animals, however, have chemical defences which are not associated with a conspicuous weapon. Some animals have the chemicals concentrated into glands so that a predator may experience the effects of the chemical before it has bitten or swallowed the animal. In other cases the chemical is inside the body so that the animal has to be injured before a predator experiences any unpleasant effect.

Defensive glands

Glands which are purely defensive in function have evolved in many animals. In gastropod molluscs there are a variety of defensive secretions including white (probably proteinaceous) materials, sulphuric acid, and mucopolysaccharides. Typically, these are only exuded when the animal is molested and there is evidence that they are of a deterrent nature. For example the secretions of *Acteon* and *Haminoea* are toxic to small planktonic animals (Fretter and Graham, 1962); acids are known to be distasteful to fish (Bateson, 1890); and many molluscs with glandular secretions are rejected as food by fish (Thompson,

1960*b*). Gastropod defensive glands are often concentrated in the part of the body most likely to be encountered first by a predator, and in some eolids, dorids and sacoglossans they are located on autotomizable papillae (Fig. 9.5). Defensive secretion of sulphuric acid of pH 1 or 2 has evolved at least five times in gastropods: it occurs in cowries, pleurobranchids, philinids, and also in species of *Discodoris* (Plate 4*a*) and *Onchidoris* (Thompson, 1960*a*, 1969; Edmunds, 1968*a*). Most of these defensive glands are probably derived from mucous glands which may originally have had a lubricative function, but this is not known for certain.

Glands which are defensive in function are also found in a variety of terrestrial arthropods. Many of these are described by Roth and Eisner (1962) and by Eisner and Meinwald (1966), and the chemical nature of the secretions is summarized by Weatherston and Percy (1970). Some defensive glands merely ooze a drop of fluid when they are stimulated (as by a predator), for example, those of millipedes. The scarlet tiger moth *Panaxia dominula* exudes fluid from the thoracic cervical glands when it is handled, and this fluid is noxious to a variety of terrestrial vertebrate predators (Ford, 1964). In the garden tiger (*Arctia caja*, Plate 5*c*) the cervical defensive glands contain an active choline ester, and there are also toxic substances in the abdomen and in the eggs (Bisset *et al.*, 1960; Frazer and Rothschild, 1962). Many other arctiid moths exude a nauseous smelling yellow fluid from the thorax when they are poked (see for example the colour plate in Rothschild, 1971), and they may also raise the wings in a static display exposing bright red or yellow on the abdomen (Fig. 7.6). The larva of the lygaeid hemipteran *Carpocoris purpureipennis* secretes a defensive fluid from the dorsal abdominal gland, then moistens its tarsi in the secretion, and brushes this on an aggressor (Remold, 1963). Since the secretion contains a paralysing aldehyde contact poison, predatory arthropods are probably repelled. Blowflies (*Calliphora*), though not predators, are paralysed if the poison penetrates through the tracheae or cuticle. The bug itself is paralysed if its cuticle is abraded, and there are mushroom-shaped structures which prevent the fluid from entering the insect's own tracheae. There is evidence also that the scent of the secretion repels ants.

A similar method of utilizing a defensive fluid occurs in the harvestman *Vonones sayi* (Eisner, Kluge *et al.*, 1971*a*). When picked up with forceps this animal regurgitates fluid from the mouth and this accumulates at the edge of the body. A quinonoid secretion is injected into this fluid from glands at the edge of the carapace. The legs then pick up this fluid and brush it against forceps or other object attacking the animal. This has been shown to be an effective defence against the ant *Formica exsectoides*. The regurgitated fluid is apparently harmless but serves to dilute the potent secretion from the carapace glands. The tenebrionid beetle *Argopsis alutacea* also exudes a defensive secretion (in this case ventrally near the tip of the abdomen) and then wipes it both onto its own body as well as onto an attacking object. This secretion contains naphthoquinones and benzoquinones (Tschinkel, 1972).

Papilionid caterpillars secrete a nauseous smelling fluid from the osmeterium situated dorsally on the thorax. In many species the principal chemicals in the secretion are aliphatic acids (Lopez and Quesnel, 1970; Eisner, Kluge *et al.*, 1971*b*), but in *Battus polydamas* the active substance is a sesquiterpene. When a papilionid caterpillar is molested the osmeterium is evaginated as a Y-shaped organ so that the glandular surface is exposed to air. The smell from the osmeterium can be very potent and may perhaps deter some predators whilst its bright red or orange colour may also startle or bluff them. Since caterpillars of *Papilio demodocus* are eaten by some birds it is possible that the smell and colour are genuine warnings of unpalatability towards some predators but are bluff displays towards others.

Other arthropods have powerfully muscular defensive glands and may eject the fluid some distance from themselves as a jet or a spray. Examples occur in cockroaches, earwigs, stick insects, bugs, caterpillars, beetles, grasshoppers, millipedes and whip scorpions (Eisner and Meinwald, 1966). The whip scorpion *Mastigoproctus* moves the tip of the abdomen and squirts a fluid containing acetic acid at a predator

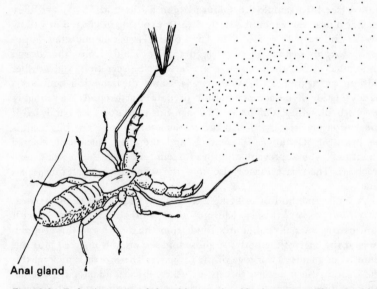

Anal gland

Fig. 10.6   Defensive spray of the whip scorpion *Mastigoproctus*. The animal has been seized by the left first leg and responded by rotating the anal gland and ejecting a spray at the aggressor (fine stipple). (*Redrawn from a photograph in* Eisner and Meinwald, 1966. Copyright 1966 by the American Association for the Advancement of Science.)

(Fig. 10.6). The stick insect *Anisomorpha biprestoides* even squirts a defensive fluid at birds before they have actually launched an attack. The tenebrionid beetle *Eleodes longicollis* practically stands on its head in order to aim the tip of the abdomen at a predator and to spray it with a quinonoid secretion. A persistent predator may overcome these defences. Grasshopper mice for example were able to capture and eat *Mastigoproctus* and *Eleodes*. By grabbing *Eleodes* by the head so that the spray was directed into the ground the mice avoided being sprayed and were able to eat it. Toads are sufficiently quick to be able to grab *Eleodes* before it has time to discharge its secretion, but even toads find the bombadier beetle *Brachinus crepitans* too quick in its response, and they reject it. *Brachinus* is aposematically coloured, blue and orange, and its abdomen contains a reservoir filled with hydroquinone and hydrogen peroxide (Fig. 10.7). These are ejected into an explosion chamber which secretes peroxidase from glands in its wall. This enzyme allows the hydrogen peroxide to oxidize the hydroquinone to quinone with release of free oxygen, and the pressure of this gas explosively ejects the fluid as a spray (summarized by Schildknecht, 1971). Such chemical defences do not prevent an animal from being killed by some predators even though the predator may not eat it. For example the carabid beetle *Calosoma* attacks julid millipedes and severs the body before the defensive fluid of the now dying millipede (Fig. 10.8) drives the beetle away. This beetle never learns to avoid a millipede, even after many trials (Roth and Eisner, 1962).

The milky secretion from the prothoracic glands of water beetles is

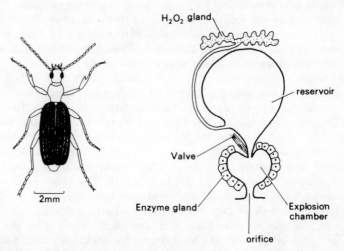

Fig. 10.7 Bombadier beetle *Brachinus crepitans*. Left: adult animal with orange head, thorax and legs, and dark blue elytra. Right: diagram of the abdominal defensive apparatus which opens close to the anus. For further explanation see text. (*Gland based on* Eisner and Meinwald, 1966, copyright 1966 by the American Association for the Advancement of Science, *and* Schildknecht, 1971.)

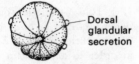

Secretion from lateral glands

Dorsal glandular secretion

Narceus gordanus   *Glomeris marginata*

Fig. 10.8 Defensive glandular secretions of millipedes. Left: *Narceus gordanus* with lateral glands. Right: *Glomeris marginata* with dorsal glands. (*Redrawn from photographs in* Eisner and Meinwald, 1966. Copyright 1966 by the American Association for the Advancement of Science.)

also toxic. Fish are partially narcotized and caused to vomit as a result of eating *Dytiscus marginalis*, and small mammals and amphibians also find this water beetle emetic (Schildknecht, 1971). The surface-swimming gyrinid beetle *Dineutes discolor* secretes a viscous fluid from the pygidial gland when it is attacked. This secretion causes fish and newts to reject it as food (Benfield, 1972).

Nasute termites have elaborate flask-shaped glands on the head and they eject at intruders a fluid which hardens in the air into sticky threads. The spray not only deters predators but also acts as an alarm substance pheromone which causes other individuals to encircle the intruder and to spray as well. When it is attacked, *Peripatus* can also eject a slimy fluid from its oral papillae which solidifies on contact with air. It can be ejected a distance of more than 40 cm, but is not poisonous. Instead it entangles predaceous arthropods and hence enables *Peripatus* to escape (Alexander, 1958*b*).

Various pentatomid bugs have defensive glands in the metathorax of the adult and on the abdomen of the larva (Remold, 1963). *Dolycoris* can eject this fluid to left or right, or in both directions, according to the direction of the attack. *Captosoma* can eject either a diffuse, unaimed long-range spray, or a precisely aimed short-range jet.

Amongst the vertebrates, frogs and toads often have defensive glandular secretions, sometimes associated with aposematic coloration. The chemical nature of the toxins varies considerably even in different populations of a single species. Daly and Myers (1967) explain such local variation of toxins in the Panamanian poison frog *Dendrobates pumilis* in terms of isolation into small populations, presumably as a result of either genetic drift or of selection for some other factors which happen to be correlated with toxicity. The toxicity of the population is not related to its coloration since some apparently cryptic races are more toxic to mice than are red and presumably aposematic races. It seems to me more likely that these local variations are evolutionary responses to predation by different species in each area. Some predators may be better able to tolerate the toxins than others, and some may quickly learn to associate toxicity with colour whilst others fail to do so. So far, however, no detailed study has been made of the relationships between anurans and their predators.

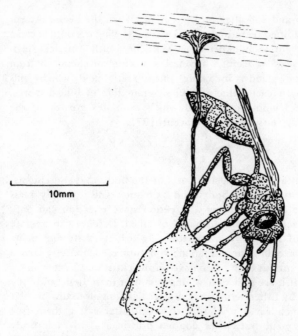

10mm

Fig. 10.9   Female wasp *Mischocyttarus drewseni* rubbing an ant-repellent secretion from its abdomen onto its nest stalk. (*Redrawn from* Jeanne, 1972, Plate IV, Fig. 7.)

Chemical defence may also be used by a parent insect to protect its young. Many social wasps suspend their comb-nests by slender stalks. *Polistes*, *Mischocyttarus* and other genera have a tuft of hairs ventrally on the tip of the abdomen. Jeanne (1970) has shown that *Mischocyttarus drewseni* periodically rubs these glandular hairs onto the nest stalk (Fig. 10.9). If the secretion is rubbed onto a glass rod with bait at the top end, ants will no longer run up it. As soon as an ant touches with its antennae a place where this secretion has been smeared, it stops and vigorously wipes its antennae. So this secretion repels ants from the nest and prevents them from attacking the wasp larvae. Ants run freely over old wasp nests where wasps no longer smear the stalk regularly. This anti-ant defence is probably not necessary for wasps in which the colony numbers several hundred or thousand individuals. But *Belanogaster*, *Polistes* and *Ropalidia* wasps in Africa often have less than twenty individuals in the colony, and such colonies could easily be overrun by ants. I have found that all of these wasps rub an ant-repellent secretion on the narrow stalk to their nest.

Other chemical defences give protection against bacteria, fungi or other micro-organisms. Dytiscid water beetles periodically leave the water and brush a secretion from the pygidial gland over the body with their hind legs. The secretion contains a phenolic compound which kills micro-organisms and solidifies on the cuticle. When the beetle returns to water the solidified secretion crumbles away leaving the cuticle clean and water repellent (Schildknecht, 1971). The South American leaf-cutter ant *Atta sexdens* secretes a complex mixture of chemicals from the metathoracic glands, including phenylacetic acid which kills bacteria, and myrmicacin which inhibits germination of fungal spores. They also secrete indolylacetic acid which promotes growth of the fungus which they cultivate (Schildknecht, 1971).

Poisonous flesh

Some animals do not have specific parts of the body that are repugnatorial, but the entire flesh or body fluid is noxious. This is usually a less efficient defence than producing a secretion since rejection can only occur after the animal has been injured or killed. Defence can operate only in terms of the predator experiencing a nasty taste and being conditioned to avoiding similar animals in future encounters. Hence animals with poisonous flesh often have warning attributes as a primary defence. Puffer fish are said to have unpleasant or toxic flesh, and they are rarely eaten by fishermen. Danaid butterflies contain toxins in their body tissues which cause predators to vomit after eating them (see pp. 104—8). Acraeid butterflies contain prussic acid in their body tissues (Owen, 1971), and so do the eggs, larvae, pupae and adults of burnet moths (*Zygaena filipendulae* and *Z. lonicerae*) (Jones *et al.*, 1962). These moths exude a nauseous smelling fluid from near the maxillary palps when they are molested, but this does not contain prussic acid. Possibly it acts as a warning smell which may deter a predator from further attack. The more potent defence, prussic acid, can be released only by damaging the moth. The prussic acid could also protect burnet moths against parasites which may be unable to tolerate the high concentration of prussic acid in their bodies. However, both *Apanteles zygaenarum* (Hymenoptera) and *Zenillia* spp. (Diptera) parasitize burnets and break down hydrogen cyanide by means of the enzyme rhodanase.

Cott (1947) points out that the flesh of some birds is much less palatable than that of others, and he used cats, hornets (*Vespa orientalis*) and man to assess palatability of the fresh carcases. Whilst it is perhaps rash on the available evidence to conclude that any bird is inedible due to toxins in the flesh, it is nevertheless true that many conspicuous birds such as drongos (*Dicrurus* spp.) and the pied kingfisher are apparently easy to approach, and their flesh is comparatively unpalatable. The flesh of the ant-thrush is also apparently unpleasant though further investigation of this is required (see p. 135). Possibly these species with distasteful flesh are used only as a reserve prey if other food is not available, but whether predators avoid them innately or because they have already killed one such bird is not known.

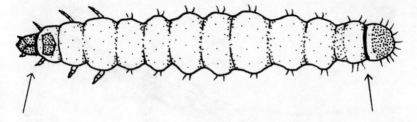

Fig. 10.10   Larva of the beetle *Diabrotica* showing the sites where reflex bleeding occurs (arrowed). (*Redrawn from* Wallace and Blum, 1971.)

Chemical defence where the body tissues are noxious can usually operate only by killing or at least injuring the animal. An exception to this is found in some insects which have toxic haemolymph and which are able to force this out through the thin intersegmental membranes on to the surface of the body. A highly specialized example is the larva of the chrysomelid beetle *Diabrotica*. When it is disturbed it shows reflex bleeding at the intersegmental membrane between the head and thorax and also between the last two abdominal segments (Fig. 10.10). The blood of the larva clots very quickly in contrast to the adults which cannot reflex bleed and whose blood clots only slowly. Wallace and Blum (1971) found that ants stick to the viscid blood and may be trapped by it until they die, but mice were unaffected by it. Since the larvae tunnel in plant roots they are only likely to meet with an ant or other predator at the front or rear of the body, so it is only in these two regions that reflex bleeding has evolved.

Defence carried over from another stage of the life cycle

Many hairy caterpillars protect the vulnerable pupal stage by spinning a silken cocoon round themselves and implanting their body hairs into the wall of the cocoon. The caterpillar of the ctenuchid *Euchromia lethe* places hairs on the branch surrounding the place in which it is going to build its cocoon, and ants which encounter these hairs usually turn and move away (Plate 2*a*, *b*). Caterpillars of another ctenuchid, *Aethria carnicauda*, erect a series of hairy barriers on each side of their site of pupation, and these prevent ants from wandering over the pupa (Beebe, 1953) (Fig. 10.11). The hairs of the gold-tail caterpillar (*Euproctis chrysorrhoea*) protect all stages of the life cycle of this species (Eltringham, 1913; Ford, 1955). The caterpillar is aposematically coloured with tufts of yellow irritant hairs. When it pupates, the male places the hairs all round itself in the wall of the cocoon, but the female concentrates them towards one end of the cocoon. When she emerges she wipes her anal tuft of hairs onto this region and many of the irritant bristles stick to her abdomen as a conspicuous yellow tuft —

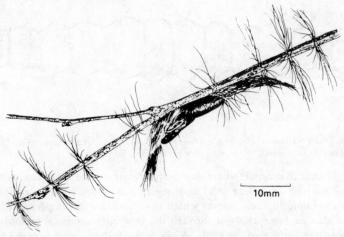

10mm

Fig. 10.11   Chrysalis of the moth *Aethria carnicauda* with protective barriers of larval hairs on either side of it. (*Redrawn from a photograph in* Beebe, 1953, Plate 1, Fig. 2.)

presumably warning coloration. When she lays eggs the female presses her tail onto the sticky heaps of eggs so that some irritant hairs attach to them and hence protect them as well. The male does not wipe the tip of his abdomen on the hairs, so he is not protected in the same way as the female, but he does nevertheless have a tuft of yellow hairs on the abdomen. Hence it is possible that these yellow hairs mimic the irritant hairs of the female so that the male is a batesian mimic of the female.

The pupal stages of other insects are also sometimes protected in a similar way by means of the larval defences. The larva of some ceratopogonid flies secretes a fluid from setae on the head, thorax and abdomen which causes the ant *Lasius niger* to drop it. When the larva pupates, the larval skin is retained on the tip of the abdomen. Groups of larvae pupate together with their abdomens facing outwards so that all are protected. The beetle *Chrysomela tremula* also retains the larval skin on the terminal abdominal segments of the pupa. When poked, the pupa responds with movements which force fluid out of the larval cuticle. The fluid is a repellent to some predators, and when stimulation ceases, the fluid is sucked back into the old larval skin (Hinton, 1951).

Defences carried over from one stage of the life cycle to the next are not confined to insects, but I shall give just one other example. The coelenterate *Bougainvillea multitentaculata*, like other animals of its group, captures prey and is protected by nematocysts. The eggs produced by this species are covered with nematocysts of the parent, and it is likely (but not proved) that these protect the egg from some predators (Szollosi, 1969). There are many further examples of parent animals protecting their eggs or young, but these fall beyond the scope of this book.

Defences derived from an animal's food

Some animals utilize the defences of their prey or plant food for their own defence. The best known case is that of eolid molluscs (Plate 4b) which feed on coelenterates and use the nematocysts of their food for their own defence. Wright (1858) first demonstrated the close correlation between the types of nematocyst found in individual eolids and in the species of hydroid on which they had been feeding. Some nematocysts pass into the digestive gland of the eolid without exploding, and these then pass up branches of the digestive gland to the sacs at the tips of the ceratal papillae. Here they are taken up by cells and stored. When the animal is suitably stimulated (as by prodding it with forceps), the muscle in the cnidosac wall contracts expelling bunches of nematocysts from the pore at the tip of the ceras (Fig. 9.5). There is considerable variation in the proportion of ejected nematocysts which actually explode, but in many species, such as *Spurilla neapolitana*, more than 70 per cent of those ejected explode and are therefore likely to be of protective value (Edmunds, 1966a). It has not so far proved possible to estimate the percentage of nematocysts which explode when an eolid is attacked by a fish, but it is known that many fish reject eolids as food, and it is likely that this is because of the nematocysts ejected by the eolid when it is sampled.

Brower and Brower (1964) have recently drawn attention to several insects, particularly butterflies, which derive their defensive chemicals from the plants on which they feed. Many insects including grasshoppers, mantids and caterpillars frequently regurgitate a drop of fluid from the gut through the mouth when they are handled. It is likely that in some cases this fluid is repellent to predators and may cause rejection before the insect is killed. The mantid *Sphodromantis lineola* often eats the flesh and cuticle of a grasshopper but leaves the gut, presumably because this part is unpalatable. The Browers also draw attention to Eltringham's observation that the caterpillar of the moth *Gonodontis bidentata* is eaten by lizards (*Lacerta viridis*) if it has recently been fed on apple, but it is tasted and rejected if it has been fed on ivy (*Hedera*). This can be regarded as the first stage in the evolution of an efficient defence system utilizing plant poisons. In the second stage, the plant poisons are assimilated by the animal and accumulate in the body tissues. This occurs in several families of butterflies. Papilionids such as *Pachlioptera aristolochiae* feed on plants of the family Aristolochiaceae and accumulate aristolochic acid-I in their bodies (Euw *et al.*, 1968). Species of *Danaus* commonly feed on milkweed plants of the family Asclepiadaceae and they accumulate cardenolides from the plants in their bodies which render them emetic when eaten by birds (see pp. 104 and 196). Caterpillars of *Danaus* fed on plants from other families or on species of milkweed which do not contain cardenolides do not accumulate them, and are not emetic to birds. Similarly the oleander aphid

(*Aphis nerii*) accumulates plant poisons when fed on *Asclepias curassavica* or *Nerium oleander* (Rothschild *et al.*, 1970). However, not all insects fed on asclepiads or on oleander accumulate toxins. Caterpillars of the oleander hawkmoth *Deilephila nerii*, for example, do not accumulate cardenolides from oleander. The plant may nevertheless be of some protective value against accidental ingestion of caterpillars by herbivorous mammals, since most mammals avoid eating oleander.

This second stage in the evolution of defences utilizing plant poisons is far from perfect since an animal may have to be tasted or swallowed by a predator before the poisons cause it to be rejected. Consequently although the predator may learn to avoid similar animals in future, some individuals may nevertheless be killed.

The final stage in the evolution of defences utilizing plant poisons is the accumulation of poisons in the defensive secretions such that an animal need not be killed when a predator samples and rejects it. No really satisfactory examples of this are known since all of the animals concerned have other defensive secretions as well. The grasshopper *Poekilocerus bufonius* normally feeds on milkweeds and accumulates cardenolide heart poisons in its body tissues, just as occurs in *Danaus* butterflies. However, some cardenolides also enter the defensive glands and are ejected together with other defensive secretions when the animal is molested (Euw *et al.*, 1967). The defensive secretions of *Poekilocerus* render both larvae and adults distasteful to a variety of vertebrate and invertebrate predators, irrespective of the larval food plant (Fishelson, 1960). Other aposematic African grasshoppers such as *Phymateus* spp. also sequester cardenolides when fed on appropriate plants (Reichstein *et al.*, 1968). Rowell (1967) found that *Phymateus purpurascens* larvae fed on *Lantana* were edible to baboons and a crane whilst adults were rejected, and he suggested that the gregarious larvae might be protected by mimicking driver ants. It seems to me more likely that the larvae as well as the adults of *Phymateus* are normally distasteful since they are brightly coloured and gregarious, and they also have defensive glands (Ewer, 1957). Possibly they normally feed on milkweeds, as does *Poekilocerus*, so that they are emetic to predators, but the glandular secretion is of protective value because it gives warning to a predator before the prey has been swallowed and hence killed. This would imply that the secretion is distasteful but harmless to a predator — except when the insect has been feeding on milkweeds. Hence a population of aposematic grasshoppers may contain genuinely emetic models and distasteful but non-emetic automimics. It is possible that Rowell's edible grasshoppers were automimics feeding on the introduced plant *Lantana* and mimicking natural populations of the same species in the area which feed on milkweeds. Clearly much further work is required on these insects.

The flightless American grasshopper *Romalea microptera* secretes a defensive froth from its thorax just as do *Phymateus* and *Poekilocerus*, and this secretion has been shown to repel ants (Eisner, Hendry *et al.,* 1971). It contains a mixture of phenols, terpenes and benzoquinone, and in one population it was found to contain 2,5 dichlorophenol as well, which is a commonly used herbicide. Evidently this population had accumulated the herbicide into their bodies and incorporated it into their defensive secretion — if they can do this with a man-made chemical it is likely that they do this also for naturally occurring plant chemicals.

Amongst the Heteroptera, *Oncopeltus fasciatus* has also been shown to accumulate cardenolides when it has been feeding on milkweeds, but it has other chemical defences as well (Feir and Suen, 1971). The coreid stink bug *Amorbus rubiginosus* has been shown to contain 2-hexenal in its defensive glands a chemical that also occurs in the eucalyptus on which it feeds (see Brower and Brower, 1964). The aposematic tiger and cinnabar moths (*Arctia caja* and *Callimorpha jacobaeae*) both feed on species of *Senecio* and accumulate alkaloids from their food in their bodies. But both of these moths have other chemical defences, and it has not been shown that these alkaloids contribute to the unpalatable properties of the insects (Aplin *et al.,* 1968; Rothschild, 1972). One reason why so many of these insects have endogenous chemical defences in addition to exogenous ones (derived from their food) may be because some predators have evolved immunity to the plant poisons. It is known that hyraxes and gazelles are immune to the cardenolides of oleander and will eat this plant freely. Hence an insect living on such a plant may be eaten by herbivores unless it has some other defence to which they have no immunity.

Summary

The final defence of a prey animal when encountered by a predator is very often to retaliate with physical or chemical weapons. If the predator can be injured at this stage it may stop attacking so that the prey animal survives. Weapons used in defence may originally have had some other function such as capture of food, or they may have evolved purely for defence against predators. Physical weapons are usually spines, teeth or claws, whilst chemical weapons are normally in the form of glandular secretions which ooze onto the surface of an animal or can be ejected some distance. Sometimes physical and chemical weapons are combined as a poisonous spine or sting. A special case of chemical defence is when a noxious chemical is in the body tissues of an animal, not in a specific gland, so that the animal has to be killed before a predator experiences any unpleasant experience. Such defences can only benefit other individuals in the population, but not the individual attacked. Many animals sequester defences from their food and use them in their own defence, but such animals very often have several defences, some of endogenous and some of exogenous origin.

Chapter 11

Defensive groups and associations

Some animals derive protection by living in close proximity to one or more other animals. It is convenient to distinguish between groups of animals comprising a single species, groups comprising two or more species, and intimate associations between two or more animals of different species. An animal living in a group or in an association may use other animals for both primary and secondary defence, and it is convenient to consider primary and secondary defence in these animals together rather than to treat them in the appropriate divisions of Sections 1 and 2.

Single species groups of animals

Many species of animals live in groups, for example shoals of fish, herds of deer and antelope, flocks of various species of birds, and social insects. For any one individual in a large group there is 'safety in numbers' − if a predator does attack the chances are that it will kill some other individual in the group. Nevertheless, a group is much more conspicuous than a single animal, and animals that rely on the group for defence always have some other defence as well: thus shoaling fish are usually cryptic and have powers of active escape (swimming); groups of herbivorous mammals and birds can usually escape by running or flying: social hymenoptera can bite or sting; and aposematic animals that live in groups are distasteful (the advantages of group living for aposematic animals have already been considered on pp. 63 and 75 so will not be discussed further here).

One of the advantages of living in a group is that it reduces the frequency of encounters with predators, as compared with dispersed animals, but at the same time it makes detection of the entire group easier because of its large size (Brock and Riffenburgh, 1960). If a large group such as a shoal of fish is attacked, a predator can only eat a certain number of prey animals no matter how large the shoal may be. Hence above the limit of the predator's appetite, increase in size of the group reduces the frequency of predator—prey encounters without affecting the consumption of prey at each encounter. In addition, when the group is attacked the individuals often scatter (Fig. 11.1). Scatter-

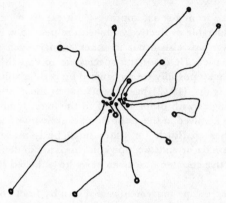

Fig. 11.1 Diagram to show scattering movements of twelve individuals from their initial clumped positions. (*Redrawn from* Humphries and Driver, 1971, Fig. 2.)

ing occurs in nestlings at the time of fledging, in various rodents, domestic kittens (R. F. Ewer, personal communication), fish, insects and young spiders (Humphries and Driver, 1971). In most of these examples the group reforms when the danger is past. Scattering reduces the chances of more than one prey being killed since the predator is unlikely to be able to follow and catch more than a single individual.

A group can be of defensive value in another way. Many fish live in huge shoals which can be described as 'anonymous' groups since other fish are recognized simply as conspecifics, not as particular individuals. In these shoals the density of the prey may actually reduce the success of the predator, for example goldfish in a dense swarm of *Daphnia* (Welty, 1934), or the cod which Radakov found catches more coalfish if these are alone than if they are in a shoal (in Nikolsky, 1963). Probably the scattering of the shoal simultaneously arouses responses in the predator to move in several directions at once. The result is hesitation, vacillation, or a longer reaction time, which gives an increased chance to the prey of escaping (Humphries and Driver, 1967). The pike has a series of fixed action patterns in its hunting behaviour, each action being triggered by the preceding one (Hoogland *et al.*, 1957). Thus first of all it sees prey and this releases eye movements. Next the fish slowly turns towards the prey and stalks it slowly until close enough to leap and seize it. The prey is then shaken to position it head first before swallowing. If at any stage of this fixed sequence the pike sees another prey, the sequence is broken and started again at the beginning with further eye movements. So in a dense swarm of prey it is possible that a predator such as a pike may catch fewer prey than in a sparse shoal. Marshall (1965), however, reports that pike can snap up individual minnows from a shoal, and it is probable that predators which regularly encounter shoaling prey have this ability. But it is not known if this is acquired by experience or if it is innate. If the ability to pick out one fish from a shoal improves with experience then shoaling will be of considerable protective value with respect to naive predators. Anonymous groups also occur in other animals such as the wildebeest of East Africa. Herds of this species are so vast that although a few individuals may possibly be recognized by certain animals, most of them are not. Kruuk (1972) has shown that hyaenas often lose their quarry as it mingles with other animals in the herd. Similar principles may perhaps be involved in the mobbing of raptors by small birds. First one then another chaffinch (or other small bird) make brief mock attacks on the predator and the individual birds are in constant motion so that an effective counter attack can never be launched (Hinde, 1954; Marlar, 1956).

More complex escape movements are shown by flocks of birds. Starlings bunch tightly together when attacked by a falcon and zig-zag in close formation (Tinbergen, 1951). Since the falcon attacks by a high speed power dive, the tight formation makes orientation on a single individual difficult and increases the chances of crashing into a bird other than the intended victim. Waders often show similar behaviour, bunching together, changing their relative positions in the flock, zig-zagging, and even splitting into several subflocks (Fig. 11.2) (Humphries and Driver, 1971). Such manoeuvres may be effective in warding off a predator even though a single individual from the group would quickly be caught.

In social insects a single ant or wasp can often do little to deter a marauding predator, but the effect of dozens or hundreds of the insects may be very much more effective at driving it away. Group defence in which the individuals of the group are known to one another can be even more effective in warding off predators, for example the defensive ring of musk ox and the defensive behaviour of eland and zebra when facing groups of predators (see pp. 151 and 280).

In Section 1 I pointed out that it is advantageous for a cryptic species if the individuals are spaced out rather than grouped together since then a predator is less likely to build up a searching image for the species. It is of further advantage to such a dispersed, cryptic species to be polymorphic. For groups of animals, although they may be initially

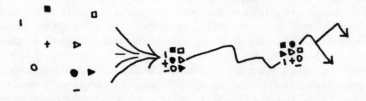

Fig. 11.2   Flock manoeuvres of waders showing how they bunch together, change their positions relative to one another in the flock, fly erratically, and split into sub-flocks. (*Redrawn from* Humphries and Driver, 1971, Fig. 4.)

less easy to find than are the same number of individuals dispersed, predators are likely to form a searching image for the species very quickly. However, if the group moves off after a kill has been made, it may be some time before the predator encounters the group again, and during this period its searching image may wane. It may actually be of advantage in a group if there is no polymorphism but if all individuals are uniform in appearance. In this way the predator cannot concentrate on any one individual and may be diverted from the quarry originally selected as prey to some other individual. The hesitation which this change may cause can increase the chances of both prey individuals escaping. If one of the prey group is different in appearance from the others this may make it easier for a predator to concentrate on this particular animal. Mueller (1971) showed that when a predatory bird is given a choice of many small rodents as prey it is likely to take any individual of different colour to the rest, even if that individual is the more cryptic. Allen (1972) exposed green and brown lard baits as prey for wild blackbirds. He found that when they were clumped together with a ratio of nine green to one brown (or nine brown to one green), the birds took more of the rarer coloured prey than one would expect if they were selecting them at random. With the prey dispersed they took more of the commoner prey than one would expect (but the results are complicated by the fact that brown were always more conspicuous and hence taken more often under similar conditions than green). Hence in the dispersed population selection is likely to be apostatic, favouring variability in appearance of the individuals in the population (polymorphism). In a dense population selection is likely to be stabilizing, favouring uniformity of appearance, probably because predators are confused by many identical prey and so concentrate on any one that appears different. Experiments such as those of Mueller and Allen can be criticized on the grounds that they bear little relation to situations in nature. But groups of certain species can be very abundant: herds of wildebeest, shoals of fish and flocks of starlings can number tens or hundreds of thousands of individuals. On a hillside of the Storr in Scotland (Skye) I have seen several hundred rabbits in quite a small area one of which was albino, and hence very conspicuous. Probably an eagle or buzzard could also see the large number of rabbits but would be more likely to attack the white one because the bird is unlikely to be distracted by movement of neighbouring animals from this different looking individual.

Another advantage of group living is that an individual need not constantly be on the alert for possible predators: in a large herd the chances are that one member of the group will detect a predator and give the alarm to the rest, so that all animals can feed or carry out other activities and only occasionally pause and be alert for predators. In some species of social animal certain individuals act as sentinels whilst the rest get on with their other activities. Baboons usually have one or

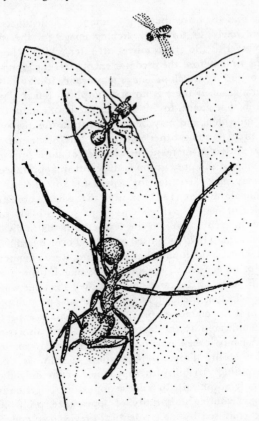

Fig. 11.3   Defence of the leaf-cutter ant *Atta cephalotes* against the phorid fly *Apocephalus*. Whilst the media worker is occupied with cutting a piece of leaf the minima worker prevents the fly from ovipositing on either of the two ants. (*Redrawn from* I. and E. Eibl-Eibesfeldt, 1968.)

more males posted at strategic vantage points such as in a tree or on a rock, and these give the alarm if a predator is nearby. In the leaf-cutter ant, *Atta cephalotes*, one important predator (strictly speaking a parasite) is the phorid fly *Apocephalus* which lays its eggs on the necks of foraging workers. The fly larva eats into the ant's head and destroys the brain. Normally an ant can fend off an ovipositing fly by using its legs and jaws, but when it is cutting or carrying a leaf it cannot do so. However, colonies of *Atta* contain many very small (minima) workers which are too small to actually cut and carry leaves. One of their functions is to accompany the larger (media) workers and to ward off phorids whilst these are collecting leaves (Fig. 11.3) (I. and E. Eibl-Eibesfeldt, 1968).

For animals living in groups the breeding season may be a particularly vulnerable time since the young may be less well able to escape

Plate 5
(a) Aposematic reduviid bug *Phonoctonus lutescens*, Legon, Ghana.
(b) Aposematic reduviid bug *Platymeris rufipes*, Legon, Ghana.
(c) Aposematic arctiid moth *Arctia caja*. Oxford, England.
(d) Six-spot burnet (*Zygaena filipendulae*) on greater knapweed (*Centaurea scabiosa*). Aposematic coloration, but possibly cryptic from a distance. Plymouth, England.

$$\frac{a \mid b}{c \mid d}$$

Plate 6
(a) Deimatic behaviour of the grass-hopper *Phymateus cinctus*. In addition to displaying the hindwings, a noxious yellow fluid is secreted on either side of the thorax. Shai Hills, Ghana.
(b) Deimatic behaviour of the mantid *Statilia apicalis* when poked. Legon, Ghana.
(c) Brush-tailed porcupine (*Atherurus africanus*) showing body spines and tail quills which can be rattled. Legon, Ghana.
(d) Orb-web spider *Caerostris albescens* resting on a branch: view of abdomen and hindlegs. Kwabenya, Ghana.

from predators than are the parents. In wildebeest and colonial nesting birds selection has favoured short, synchronized breeding seasons so that this vulnerable period is brief, and even though the predators nearby may catch and eat their fill of the young, they cannot possibly kill all during this time. Any individual in the group that breeds outside the peak season is less likely to leave surviving offspring than individuals that breed at the peak time.

Mixed species groups

Some animals live in groups containing several different species, and such groups have similar defensive advantages to single species groups. Herds of zebra, ostrich and wildebeest in East Africa probably benefit from each others presence since they each have sense organs of different sensitivities for detection of predators. As soon as one individual detects a predator, the entire mixed herd can be alerted. Similarly tick birds on zebra, buffalo or other herbivores often detect a predator before their host does, and they then transmit the alarm to the host by their agitated movements. Such mixed species associations may have evolved under the selection pressure of more efficient exploitation of food rather than that of improved defence, but they nevertheless have considerable defensive value. In mixed species flocks of tits in temperate woodlands during winter, each species exploits a different part of the habitat for food, but all may utilize a locally abundant food source, and all respond when any one individual detects a predator and gives an alarm call (Lack, 1971). Similar mixed species flocks occur in weaver birds during the non-breeding season in the tropics. Cattle egrets often live closely associated with cattle or buffalo but it is not certain if they actually give warning of danger to the cattle. This is primarily a feeding association: the birds feed on grasshoppers and other insects disturbed by the trampling of the cows, and it is known that cattle egrets associated with cows obtain more food than do egrets without cows (Heatwole, 1965; Dinsmore, 1973).

Associations between different species

With the examples of cattle egrets and tick birds in mind it is easy to see how highly specific associations between two species of animal may have evolved. Such associations may be primarily feeding associations (as with the cattle egrets and tick birds) or they may be primarily defensive (see below), or they may have several advantages, as with the commensals living in the burrow of the echiuroid *Urechis* described on p. 4. In typical defensive associations one species of animal derives protection by living in close association with another species that is protected by spines, sting, or some other defence. Predators normally avoid this second species of animal, and in so doing they avoid the first one as well.

Associations with ants

Ants form large colonies which forage over a wide area around the nest. They bite or sting both prey and predators, and a large number of attacking ants can often drive away quite large animals. In addition not many predators feed on ants so that it is of advantage for other species of arthropod to mimic them (see p. 115). An animal is likely to be protected from insect or arachnid predators if it can live amongst ants since the ants will drive these predators away. But in order to survive amongst the ants such an insect must either evade attack by the ants, or be immune to such attack, or be able to appease an attack. Spiders and other arthropods that mimic ants usually evade attack by detecting the ants before the ants detect them and by quick escape movements. Some beetles that associate with ants have such a thick exoskeleton that they are immune to attack, but most ant associates appease attack by secreting fluids which stimulate non-predatory feeding behaviour in the ants. However, the details of the relationship between insect and ant vary in different species.

Aphids and other Homoptera are well known for their association with ants (Way, 1963). In most cases the ant obtains food from the association whilst the homopteran derives protection from other predaceous insects which avoid places where there are many ants, so the association is a true mutualism (i.e. both participants benefiting from the association). When an ant encounters an aphid it caresses and palpates the aphid with its antennae in a characteristic way. This causes the aphid to exude a drop of sugary material from the anus (honeydew) which the ant eats. If the ant fails to take the droplet, it may be resorbed. In the absence of ants the honeydew is forcibly ejected from the rectum or kicked away with the hind legs. *Aphis fabae* has been shown to feed more if ants (*Lasius niger*) are attending it than if they are not, but the association is not obligatory for either the aphid or the ant. The aphid *Protrama* and the coccid *Saissetia zanzibarensis*, however, cannot eject honeydew efficiently in the absence of ants, and for these homopterans the association is obligatory. Without ants, moulds grow round the anus of *Saissetia* and kill it.

Hence it is possible that Homoptera benefit from this association by getting rid of surplus carbohydrates, but undoubtedly the most important benefit is the protection they receive from the ants. Most species of ant will attack any insect or other animal they encounter, so the aphids which are tended by ants are likely to be less often attacked than are aphids without ants. Evidence for this is rather inconclusive. *Lasius niger* is reported to attack syrphid larvae and ladybird larvae and adults; *Iridomyrmex* attacks adult ladybirds and both adult and larval lacewings and syrphids; *Formica rufa* apparently tolerates ladybirds but attacks large syrphid larvae; and *Oecophylla* tolerates ladybirds and lycaenid caterpillars. Parasites are not deliberately attacked by ants but are disturbed. It has been shown that 0—0.4 per cent of coccids inside the nests of *Oecophylla longinoda* are parasitized whilst from 5—17 per cent of coccids outside the nests are parasitized, but unfortunately the percentage parasitized away from *Oecophylla* is not known (Way, 1963). Probably the degree of parasitization and of predation depends on the distance of the homopteran from the ants' nest and hence on the frequency with which ants pass near it and disturb potential parasites or predators.

Strickland found that *Crematogaster striatulus* from tropical West Africa builds earthen shelters over scale insects (*Pseudococcus njalensis*) which it tends on cocoa. The entrances to the shelters are too small to admit ladybirds or other predators, so the scales are protected in this way by the ants (summarized by Sudd, 1967). In addition, the shelters also protect the scales and the ants from heavy rain, and ants do not build shelters in the dry season. Some ants (e.g. *Lasius flavus*) carry aphid eggs to their nests and protect them during the winter. *Oecophylla longinoda* carries coccids (*Saissetia*) to twigs close to or inside its nest but it also kills coccids if they are very numerous, so the coccid only receives protection if it is not too common. The aphid *Paracletus cimiciformis* is carried by its host ant (*Tetramorium caespitosum*) to its nest where it is protected from predators, but here the relationship differs from that of other homopterans and ants. The ants tend the aphids and regurgitate food for them so that the aphid derives all its food from the ants and is a parasite.

Kloft has suggested that ants are attracted to aphids because the abdomen of aphids resembles, and may be mistaken for, the head of a fellow worker ant (Fig. 11.4). Ants greet fellow workers and stimulate them to share food in the same way that they palpate aphids (see Sudd, 1967). However, this supposed mimicry cannot apply to coccids since their abdomen is totally different in shape to the head of an ant, and

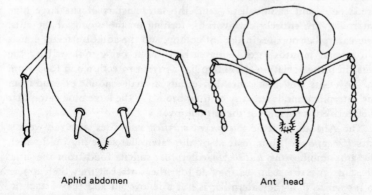

Aphid abdomen   Ant head

Fig. 11.4 Diagram to show the supposed resemblance of the abdomen of an aphid to the head of an ant. (*Redrawn from* Sudd, 1967, Fig. 5.7, *after* Kloft.)

other insects which live with ants do not resemble them in shape (see below). Hence the similarity in shape between an ant's head and an aphid's abdomen is probably coincidental and not a true example of mimicry.

Many species of lycaenid caterpillars also have protective associations with ants. The caterpillar has a gland opening on the seventh abdominal segment which exudes a fluid when appropriately stimulated by ants. Ants frequently kill caterpillars and so obviously any adaptation of the caterpillars that reduces this predation is of advantage. Probably the initial stage in the evolution of this association was the chance happening that the secretion of this gland (perhaps originally a defensive gland) proved to be palatable to the ant and hence inhibited its attack. Ants feed on the sugary nectar of plants without inflicting damaging bites. Perhaps this lycaenid secretion was sufficiently similar to nectar to switch the ants from predatory feeding to feeding without biting. This is the extent of the association in lycaenids such as the green hairstreak (*Callophrys rubi*). Ants deter other possibly predatory insects in the vicinity and so the caterpillar gains protection.

In the chalk-hill blue (*Lysandra coridon*), ants carry the caterpillars to food plants close to their colony and 'milk' them, just as they do with aphids. Here the caterpillars are obviously better protected because of the more regular traffic of ants than they would be further away from the colony. In other lycaenids, however, the relationship between caterpillar and ant is much more complex (see Ford, 1945, and Owen, 1971, for summaries). The large blue butterfly (*Maculinea arion*) of Europe has a caterpillar which is of normal behaviour until the third moult. At this stage if an ant finds a caterpillar it milks it, and this causes the caterpillar to adopt a peculiar hunched posture. The ant then carries it to its nest underground where there is no green food for the caterpillar to eat (Fig. 11.5). Possibly the first time such an incident occurred the caterpillar starved, but lycaenid caterpillars are often predaceous and cannibalistic, and the later instars of the large blue caterpillar are entirely carnivorous, feeding on the young of the ants. The caterpillar pupates in the ant colony, and the adult butterfly crawls out several months later, apparently without being molested by the ants. In this association the caterpillar derives protection and food from the host ants whilst the ants derive only a small amount of food from the caterpillar and lose many of their brood, so the large blue caterpillar is effectively a parasite of the ant colony.

The African lycaenid *Euliphyra mirifica* associates with red weaver ants (*Oecophylla*), but instead of the caterpillar secreting fluid which the ants imbibe, the *Euliphyra* caterpillar solicits food from the ants. The ants give it a drop, as they do for other ants, but they then appear to recognize that the caterpillar is not a fellow ant and they attack it. The *Euliphyra* withdraws its head under a horny cuticle, and since its body is tough and leathery it is immune to further attack, so the ants

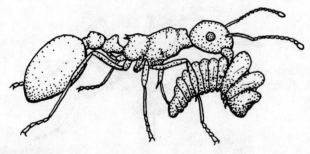

Fig. 11.5   Worker ant (*Myrmica* sp.) carrying the fourth instar larva of the large blue butterfly (*Maculinea arion*) to its nest. (*Redrawn from* Ford, 1945, Plate xvii, 5, *after* Frohawk.)

soon lose interest (Lamborn, 1913). *Aslauga* and *Megalopalpus* are two other lycaenids which feed on the homopterans tended by ants and are not themselves attacked by the ants. All of these caterpillars probably derive protection from other predators simply by living amongst ants. There are yet other lycaenids such as *Teratoneura* whose caterpillars feed on lichens but normally live amongst ants. There is no evidence that *Teratoneura* secretes a fluid that appeases the ants, so it is not known how they avoid being attacked by their hosts (Farquharson, 1921).

The nitidulid beetle *Amphotis marginata* has a rather similar relationship with the ant *Lasius fuliginosus* as has *Euliphyra* with *Oecophylla*. It normally rests on the trails of the ants and when a worker comes near it taps its labium and causes the ant to regurgitate food for it to eat. The ant then apparently recognizes the beetle as not being an ant and attacks it — but the beetle retracts its legs, flattens its body to the ground, and is then impervious to the ant's bites. This beetle neither resembles the ant in shape such that predatory birds might be deceived into mistaking it for an ant, nor does its behaviour protect it from attack by ants. Nevertheless it may encounter fewer arthropod predators by living on ant trails than if it lived elsewhere.

The staphylinid beetle *Dinarda* also associates with ants but in this case it lives in the peripheral chambers of the nest of *Formica sanguinea* where it is well protected from all predators except ants. It feeds on dead ants and other organic debris, and it also stimulates passing ants to regurgitate food so that, like *Amphotis*, it is a food parasite. Again like *Amphotis* the ant then attacks the beetle, but *Dinarda* responds to the attack by secreting a fluid which the ant imbibes (Fig. 11.6), and this enables the beetle to move away before the ant starts to attack again. *Dinarda* could never survive in the main part of the ant's nest, but it is able to live in the peripheral chambers where there are few ants to appease (Hölldobler, 1971).

In the beetle *Atemeles pubicollis* the association with ants is much

Fig. 11.6   The staphylinid beetle *Dinarda* eliciting regurgitation of food from the ant *Formica sanguinea*. (*Redrawn from* Hölldobler, 1971. Copyright © 1966 by Scientific American Inc. All rights reserved.)

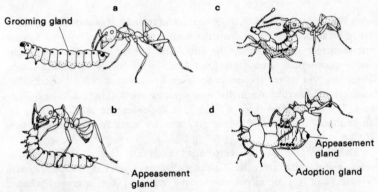

Fig. 11.7   Relationship between ants and the staphylinid beetle *Atemeles pubicollis*: (*a*) *Formica polyctena* ant about to lick a drop of fluid from the appeasement gland at the tip of the abdomen of a larval beetle; (*b*) *Formica* regurgitating food for the *Atemeles* larva; (*c*) *Myrmica* ant taking a drop of fluid from the appeasement gland of an adult *Atemeles*; (*d*) *Myrmica* taking fluid from the adoption glands of adult beetle preparatory to carrying it into its nest. (*Redrawn from* Hölldobler, 1971. Copyright © 1966 by Scientific American Inc. All rights reserved.)

more complex. Adult beetles enter the nests of *Formica polyctena* in the spring when the ants are becoming active after the winter. The beetles appease the ants by presenting the tip of the abdomen to the ant which licks a droplet of fluid from it, hence the gland is called an appeasement gland. The ant then licks glands on the sides of the abdomen (adoption glands) and this secretion stimulates the ant to pick up the beetle and carry it to the nest. Once in the nest the beetles lay eggs and the larvae hatch out and grow in the ant colony. *Atemeles* larvae tap the labium of worker ants with their mouthparts and induce them to regurgitate food in exactly the same way that ant larvae stimulate regurgitation by workers (Fig. 11.7). This is obviously aggressive mimicry since the ant does not benefit from the deception. The beetle larvae are actually more successful than ant larvae at eliciting

regurgitation of food, so presumably their mimetic behaviour is something in the nature of a super-stimulus to the ants: beetle larvae
accumulate more than their share of food compared with ant larvae
when workers are fed with radioactive labelled food. If found away
from the nest both larvae and adults can induce an ant to carry them
back to the colony. In the autumn adult *Atemeles* move out of *Formica*
nests into nests of *Myrmica* spp. which are active throughout the winter
— *Formica* is quiescent in the winter months.

Atemeles larvae are straightforward parasites on the food intended
for ant larvae, but they are also cannibalistic and eat both other beetle
larvae and ant larvae from time to time. The ants appear to gain no
benefit at all from the association whilst the beetle gains protection in
the nest as well as food.

Thus the details of the many associations between ants and other
insects are very complex and variable, in some both partners benefit
whilst in others the ant receives no benefit but only harm from the
association. However, probably in all of these associations the aphid (or
beetle or caterpillar) gains protection against other predators in the area
which either avoid places where there is much ant traffic or which do
not enter ant's nests.

Associations with echinoderms

Echinoderms are protected against many predators by their calcareous
skeleton and spicules, for only predators which can bite through the
skeleton or swallow them whole can readily eat an echinoderm. Hence
some animals live closely associated with echinoderms and thereby
obtain protection against predators. Several different species of fish
have evolved the habit of swimming between the spines of a sea urchin
when a predator is nearby and hence they are protected by the long
sharp spines of the urchin. Details of the association vary, but all fish
which do this presumably obtain protection in this way. The clingfish
(*Dellichthys morelandi*) finds its host *Evechinus chloroticus* by sight,
not by smell, and thereafter it always lives close to its host. The association in this case is detrimental to the urchin since the clingfish nibbles
off tube feet and pedicellariae (Dix, 1969). The urchin *Astropyga
radiata* has very long spines and many individuals of the fish *Siphamia
argentea* can retreat between the spines when danger threatens (Fricke,
1970). *Siphamia* is of typical fish shape and can swim fast if attacked,
but other fish associated with urchins are highly modified to living
amongst the spines in both shape and colour so that the association is
obligatory, and they are more or less defenceless away from their host.
For example *Diademichthys* and the shrimpfish *Aeoliscus* both have
elongated bodies with small, non-protruding fins, so that they can live
amongst the spines of *Diadema*, and they normally rest head down
amongst the spines where their coloration makes them difficult to see

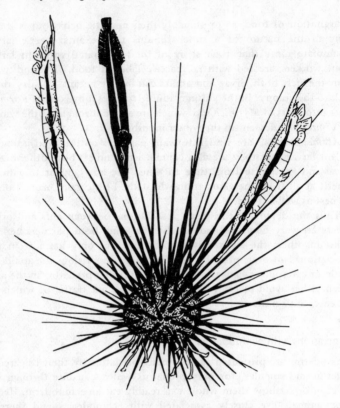

Fig. 11.8   Associations between fish and sea urchins. Two *Aeoliscus strigatus* and one *Diademichthys deversor* are shown amongst the spines of *Diadema*. (*Based on drawings in* Davenport, 1966a, and Gotto, 1969.)

(Fig. 11.8) (Davenport, 1966a; Gotto, 1969). There is also a shrimp *Tuleariocaris zanzibarica* which lives amongst the spines of *Diadema setosum*, resting head downwards between the spines. It has been shown that this shrimp finds its host by visual cues: it prefers to rest on models with dark rather than pale spines, with vertical rather than horizontal or oblique spines, and with many rather than few spines (Fricke and Hentschel, 1971). This behaviour would naturally lead it to associate with large, healthy *Diadema* rather than with small or unhealthy *Diadema* (with fewer or non-vertical spines), or with other species of urchin (with different coloured spines).

There are also some associations between organisms and other echinoderms such as starfish and sea cucumbers, but since these lack the defensive spines of sea urchins it is not usually possible for large animals such as fish to obtain protection in this way. One exception is the pearl-fish *Carapus* which retreats into the cloaca of sea cucumbers when attacked (see p. 138). Polynoid polychaete worms live amongst the tube feet of echinoderms and are here protected from all predators except those which feed on their host. In Europe the polynoid *Acholoë astericola* normally lives amongst the tube feet of the starfish *Astropecten irregularis* where it is cryptic. In addition to deriving protection from the association *Acholoë* also feeds on the food of the starfish and even enters its host's stomach to obtain food (Davenport, 1966a). If removed from its host, *Acholoë* cannot detect a new host from a distance, but as soon as contact is made, it crawls onto the starfish. The association is very specific — *Acholoë* will only crawl onto *Astropecten irregularis*, not onto other starfish from the same habitat. However, *Acholoë* associates with different species of asteroid in regions where *Astropecten irregularis* is absent, for example off West Africa it lives with *Astropecten michaelseni, Luidia atlantida* and *L. heterozonata*.

The polynoid *Arctonoë fragilis* is found with several species of starfish. If removed from its host it orients towards it by smell and shows a distinct preference for the smell of its normal host (*Evasterias troschelii*) to that of other species. Similarly, *Arctonoë pulchra* from *Dermasterias imbricata* choose water that has passed over *Dermasterias* in preference to water from other echinoderms, and worms that have been living with the sea cucumber *Stichopus parvimensis* prefer *Stichopus* water to *Dermasterias* water. However, if removed to a new host for three weeks, *A. pulchra* may become conditioned to its new host and subsequently prefer this to its old host (Dimock and Davenport, 1971). Full-grown *Arctonoë* are very aggressive, so that only one individual can normally live on a single host animal. Young worms are not aggressive and several may live on a single host, so as they mature the weaker individuals are driven away to find new hosts. If they cannot find a host of the preferred species they may be forced to live with a different one, and this may be why *Arctonoë pulchra* is sometimes found living with the gastropod *Megathura crenulata*. In this way associations with new species of hosts can evolve, and the olfactory conditioning could eventually lead to adaptation of some populations to particular hosts, and thence to speciation.

Associations with coelenterates

There are many protective associations in the sea involving coelenterates and fish or other animals. In probably every case the nematocysts of the coelenterate are the principal protective mechanism for both the coelenterate itself and for its commensal. It is convenient to consider associations with pelagic coelenterates (jellyfish, chondrophores and siphonophores) first, and then to discuss associations with benthic anemones and corals.

Stromateoid fish commonly associate with jellyfish when they are young (Mansueti, 1963; Horn, 1970). Young of the harvestfish (*Peprilus alepidotus*) and the butterfish (*Poronotus triacanthus*) in the

North Atlantic feed on plankton close to the jellyfish *Chrysaora quin-quecirrha*, or on its ectoparasites. As they grow larger they also nibble the tentacles and manubrium of the jellyfish, and when very large they may seriously damage it. Occasionally a fish may be stung by nemato-cysts or even killed and eaten, but this appears to be unusual. One reason why such fish avoid being stung by their host is that they have a swim bladder which enables them to maintain neutral buoyancy, and they are adept at manoeuvring between the tentacles of their host. Adult stromateoid fish do not normally associate with jellyfish and most have no swim bladder.

Rees (1966) noted that whiting (*Gadus merlangus*) associates with *Cyanea lamarcki*, and that it normally swims above the umbrella of the jellyfish. It only swims beneath it to take refuge when alerted by a possible predator. Off the West African coast, young of the horse mackerel *Caranx hippos* live under *Rhizostoma* and *Cyanea*, but after they have grown to about 10 mm in length they move into lagoons where there is probably less predation than in the open sea (Pople, personal communication; Kwei, 1970). All of these fish—jellyfish associations are probably protective for the fish. Some are largely fortuitous since small fish will congregate under any large object such as a boat or a floating log. Some fish probably live under jellyfish when these are available, but can also live on their own; others may be obligatorily associated with jellyfish and unable to survive without them. Most such associations have not been adequately studied, but it appears that the nature of the association varies from purely protective (to the fish) to parasitism and even predation. The only benefit the jellyfish receives is that some fish remove ectoparasites from them.

One possibly obligatory association is that between the man-of-war fish *Nomeus gronowi* and the Portuguese man-of-war *Physalia pelagica* (summarized by Gotto, 1969). The fish has vertical stripes of blue and silvery grey so it is well camouflaged amongst the long blue tentacles of *Physalia*. Away from *Physalia* it is very conspicuous and hence, one imagines, liable to be eaten, though it is known to occur occasionally with a few other species of jellyfish or siphonophore. *Nomeus* lives its entire life associated with *Physalia* and is one of the few stromateoid fish to retain the swim bladder into the adult stage (Horn, 1970). This may be because the presence of the swim bladder increases its ability to swim amongst its host's tentacles without touching them. However, *Nomeus* is partially immune to the nematocysts of *Physalia* since it is not necessarily killed when it is thrown against the tentacles of its host, though it is certainly stung. *Nomeus* also eats the tentacles of *Physalia*, so it appears to be parasitic. It may perhaps benefit its host as well by luring other predatory fish into the vicinity of the tentacles so that they are stung and captured by *Physalia*. It is a moot point whether this association should be regarded as one of a fish utilizing the aposematic properties of its host, or of one of simple crypsis or mimicry, since *Nomeus* is the same colour as the tentacles of *Physalia*.

The eolid molluscs *Glaucus atlanticus* and *Glaucilla marginata* normally live in association with the chondrophores *Velella* and *Porpita*, or with the siphonophore *Physalia* (Thompson and Bennett, 1970), though they can also swim freely on the surface of the sea buoyed up with air. They feed entirely on their hosts which they also resemble in colour. When feeding they are protected by the nematocysts of the host, and even when alone they can eject stored nematocysts in their own defence, just as can other eolids (see p. 199). *Fiona* is another eolid that feeds on *Velella*, resembles it in colour, and is presumably protected by its host's nematocysts (Bayer, 1963). Eolids appear to be immune to the nematocysts of their coelenterate food. Although the nature of this immunity is not understood, there is a correlation between the presence of characteristic vesicles in the epidermis and a coelenterate diet (summarized by Edmunds, 1966a).

A more complex association is that between the medusa *Zanclea costata* and the nudibranch *Phyllirrhoë bucephala*. The larval nudibranch attaches to the medusa and feeds parasitically on it. As it grows larger it eventually devours the entire medusa. This is a very specialized case of predation involving an obligatory association, but the nudibranch probably derives protection from the nematocysts of its host during the early stages of the association (Martin and Brinckmann, 1963).

Another group of associations involves sea anemones and pomacentrid fish (Mariscal, 1972). The clownfish or damselfish (*Amphiprion percula*) of the Pacific is often found living with the sea anemone *Stoichactis kenti*. It swims amongst the tentacles of the anemone without being harmed, but it can be stung by anemones under certain circumstances (Fig. 11.9). If a damselfish which has been isolated from an anemone for a few days is then placed in an aquarium with a *Stoichactis*, it soon sees the anemone and swims close to the tentacles making brief contact. This causes discharge of nematocysts and clinging of the tentacles to the fish, but with a quick jerk the fish escapes. This is repeated many times and within a few minutes, or occasionally as much as three hours, the anemone and the fish become acclimated to one another so that the tentacles no longer cling due to nematocyst discharge when contact is made. The acclimated fish now lives close to its anemone and retreats amongst the tentacles whenever danger threatens. Davenport and Norris (1958) showed that the mucus on the damselfish raises the threshold of discharge of nematocysts whereas the mucus on other fish does not. Schlichter (1968, 1972) has shown that the protective mucus of *Amphiprion bicinctus* is actually derived from the anemone when the fish brushes over the tentacles, and this probably applies to other species of damselfish as well. There is no evidence that the behaviour of the anemone is changed by acclimation of the fish (Mariscal, 1970b). Figure 11.10 shows that fish acclimated to one

Fig. 11.9   The damsel fish *Amphiprion percula* resting amongst the tentacles of the sea anemone *Stoichactis*. (*Based on various sources.*)

anemone are never stung when presented with another anemone of the same species, irrespective of whether or not the new anemone already had a fish living with it. Unacclimated fish were always stung by the anemone, again irrespective of whether or not the anemone already had a fish living with it. Finally, if acclimated fish are wiped before being presented with an anemone, they are always stung. Mariscal further showed that fish acclimated to one species of anemone are not necessarily protected against other species of anemones. Thus fish–anemone associations involve a change in the behaviour of the fish but no change in the behaviour of the anemone. Immunity to nematocyst discharge is brought about by the fish becoming coated with anemone slime, a substance which is adapted to preventing discharge of nematocysts when neighbouring tentacles touch one another or brush against some inanimate object such as a nearby rock. The fish certainly derives protection from this association. The anemone also benefits from the association since *Amphiprion percula* attacks fish which come near including species which normally eat sea anemones, but it may nibble its host's tentacles occasionally (Mariscal, 1970a). Other details of the relationship vary with different species of damselfish and different species of anemone: for example *A. bicinctus* and *A. xanthurus* sometimes feed their anemones but *A. percula* does not; only one individual of *A. xanthurus* normally lives with an anemone, whereas several *A.*

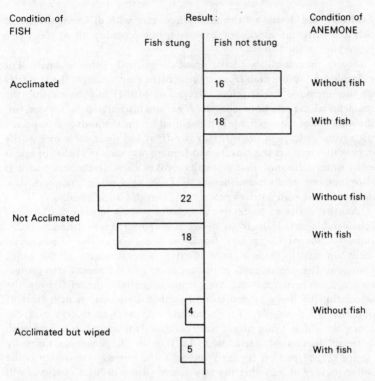

Fig. 11.10 Results of presenting acclimated and unacclimated damsel fish (*Amphiprion xanthurus*) to anemones either with or without associated fish. (*Data from* Mariscal, 1970*b*.)

percula live in a single anemone; and whereas *A. percula* has only one species of host anemone in any one population, *A. xanthurus* may have several species of anemone hosts (Mariscal, 1970*a*, *b*; Gohar, 1948).

The Mediterranean actinian *Anemonia sulcata* also has an anemone fish, *Gobius bucchichii*, belonging to the family Gobiidae. Here too the mucus on the fish inhibits discharge of nematocysts, but there is apparently no period of acclimation. Unlike with *Stoichactis* or *Radianthus* anemones, a new *Anemonia* will accept a fish immediately without any conditioning process (Abel, 1960; Davenport, 1962, 1966*b*), but it is possible that this may be due to the anemone mucus on the fish persisting for a very long time.

Another anemone fish is the labrid *Thalassoma amblycephalus* which retreats from attack under the column or beneath the disc of the anemone, but not between the tentacles (Schlichter, 1970). Nevertheless, it is at least partially immune to its host's nematocysts, since it visits anemones, inspects tentacles, and picks up food particles from them. It is thus a food parasite that also gains protection from the association.

Evidently the details of the association vary with different species of fish and different species of anemone, but probably all of them give protection to the fish.

Some anemones also have associations with other animals. For example, the spider crab *Hyas araneus* often rests amongst the tentacles of the anemone *Tealia felina*. Davenport (1962) has shown that the tentacles of the anemone cling to *Hyas* and discharge nematocysts, but the crab is not damaged nor is it engulfed by the stomodaeum as occurs with other species of crabs. Evidently *Hyas* has some substance in the cuticle that inhibits the closure and feeding response of *Tealia* (or else it lacks some substance that normally evokes these responses) and it is also immune to the nematocysts of *Tealia* and other anemones. Undoubtedly the crab receives protection from this association.

Another series of protective associations which have been studied in considerable detail are those between sea anemones and hermit crabs. Hermit crabs of the genera *Dardanus* and *Pagurus* live in gastropod shells on which there is very often a large anemone of the genus *Calliactis*. The nematocysts of the anemone give the hermit crab protection against predators and the anemone probably benefits from the association by being moved around so that it encounters new areas of potential food supply. The behaviour of the hermit crab and the anemone which bring about and maintain the association vary with different species of participant. In Hawaii, *D. gemmatus* normally carries *C. polypus* on its shell and it is the hermit crab which is the active partner in initiating the association (Ross, 1970). *C. polypus* will attach to almost any hard surface and shows no particular preference for settling on a gastropod shell, either with or without a hermit crab. But if *D. gemmatus*, with no anemone, finds a *Calliactis*, it taps the column and induces the anemone to detach. It then places the anemone on its shell and induces it to reattach firmly. The anemone *Sagartiomorphe guttata* also shows no response to gastropod shells, yet 10 per cent of all *Dardanus venosus* in the Caribbean have this small anemone on the columella of their shell. In this association it is also the hermit crab that is active in transferring the anemone (Cutress and Ross, 1969).

The association between *Pagurus bernhardus* and *Calliactis parasitica* in Europe is superficially very similar, but in this case it is the anemone that is the active partner in initiating transfer (Ross, 1967; Ross and Sutton, 1961*a*). When anemones were placed on glass and then presented with *Buccinum* shells approximately equal numbers transferred to shells with and to shells without hermit crabs in twenty-four hours. It appears to be the periostracum of *Buccinum* rather than the hermit crab that attracts the anemone and induces it to transfer onto the shell. If the hermit crab moves into a larger shell, or if it dies, the anemone remains on the now empty gastropod shell. Thus in this association the crab does not assist in the process of transferring the

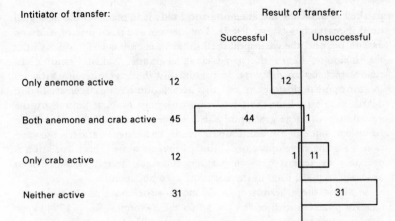

Intitiator of transfer: Result of transfer:

Successful   Unsuccessful

Only anemone active   12   12

Both anemone and crab active   45   44   1

Only crab active   12   1   11

Neither active   31   31

Fig. 11.11   Results of 100 trials of the hermit crab *Dardanus venosus* placed in a tank with the anemone *Calliactis tricolor*. (*Data from* Cutress and Ross, 1969.)

anemone onto the shell, and in fact it is the shell, not the hermit crab, which is attractive to the anemone.

The Caribbean association between *D. venosus* and *Calliactis tricolor* is much more complex. Out of 100 trials in which *D. venosus* was placed in a tank with *C. tricolor*, there were fifty-seven successful transfers of the anemone onto the shell (Fig. 11.11). Sometimes the anemone, sometimes the hermit crab, and sometimes both together were active in initiating transfer, but for transfer to be successful it was normally necessary for the anemone to be active (Cutress and Ross, 1969).

A rather similar situation occurs in *Dardanus arrosor* and *Calliactis parasitica* from the Mediterranean (Ross and Sutton, 1961*b*; Ross, 1967): the hermit crab can merely facilitate transfer, but it cannot detach and reattach the anemone unless the anemone also participates. It was also found that whilst most of the female *D. arrosor* studied actively participated in transfer, most of the males were inactive like *Pagurus bernhardus*. Furthermore, in intraspecific encounters it has been shown that some individuals of *Dardanus arrosor* are dominant over others, and that high ranking *Dardanus* normally possess *Calliactis* whereas low ranking *Dardanus* have no anemone and are unable to induce an anemone to transfer in the presence of a dominant crab (Mainardi and Rossi, 1969). Since *C. parasitica* shows no preference for gastropod shells either with or without hermit crabs, a satisfactory association of hermit crab and anemone is more likely to result if both partners are active in the process of transfer than if only the anemone participates.

A more intimate association is that between *Adamsia palliata* and the hermit crab *Pagurus prideauxi*. This is an obligatory association, neither partner occurring normally alone. When the hermit crab changes its shell it detaches the anemone and holds it in place on the new shell until it reattaches (Ross, 1967). The lobes of the basal disc of *Adamsia* extend beyond the gastropod shell so that the crab is well hidden under the anemone, hence the name cloak anemone. Naked hermit crabs cannot pick up an *Adamsia*, so presumably the anemone only attaches to gastropod periostracum, but this association has not been studied in detail. In areas where the population of hermit crabs is limited by the availability of large gastropod shells *Pagurus prideauxi* may be at an advantage since it can continue to live in small shells yet derive protection from the enveloping pedal disc of the anemone. This association is probably of benefit to both partners since the hermit crab has been observed to place food in the anemone's stomodaeum.

In all of these hermit crab—anemone associations the hermit crab receives protection since few predators eat anemones. Ross (1971) kept *Dardanus arrosor* in a tank with an octopus. He found that all hermit crabs without anemones were eaten in from one to three days. The octopus takes a very long time to extract the hermit crab since this requires a long sustained pull, rather like a starfish opening a bivalve. But whenever the hermit crab had a *Calliactis* on the shell, it was only held by the octopus for fifteen seconds or less, and after several attacks the octopus left it alone. *Pagurus prideauxi*, however, is not fully protected by *Adamsia*. The octopus, although stung by nematocysts, could still get hold of the *Pagurus* and pull it out since it was not possible for the hermit crab to withdraw fully beneath the anemone. Perhaps this association is protective against fish or other predators.

Some species of hermit crabs regularly associate with other species of sessile invertebrates. *Paguristes* sp. from Ghana is often found in gastropod shells with a millepore coral growing on it, whilst other species of hermit crab often have the hydrozoan *Hydractinia* growing over the shell. Millepores have powerful stings, and *Hydractinia* also has nematocysts, so presumably these coelenterates protect the hermit crab against predators. *Hydractinia* can also give protection against shell stealing by other species of hermit crabs. In Texas *Clibanarius vittatus* regularly steals shells from *Pagurus longicarpus* and *P. pollicarpus*, and a hermit crab without a shell is likely to be quickly killed by a predator. However, *Clibanarius* cannot steal a shell that is covered with *Hydractinia* or *Podocoryne* because it gets stung whereas *Pagurus* spp. appear to be unaffected by the nematocysts. Since 20—30 per cent of the *Pagurus* spp. usually have a hydroid on the shell, they are therefore protected against shell stealing by *Clibanarius* (Wright and Matthews, 1973). Other hermit crabs may have their shells overgrown with sponges (*Suberites*) or bryozoans. Sponges are spicular and avoided as food by fish, so this association may also be protective, but the polyzoan associations may perhaps be of benefit to the polyzoan by increasing the area available for food collection but be of no protective value to the crab.

Coelenterates may also be associated with other crustacea. The spider crab *Stenocionops furcata* may carry sponges, algae or anemones on its carapace. One individual was found with twenty-five *Calliactis tricolor* on its carapace and legs (Cutress, Ross and Sutton, 1970). Transfer involves active movements by both crab and anemone. The anemone *Anthothoë paguri* is often found on the carapace or held in the claws of the hermit crab *Diogenes edwardsia*; and the anemone *Bunodeopsis* is often held in the chelae of the crab *Lybia*. *Lybia* simply detaches the anemone and then holds it as a weapon for its own defence (Ross, 1967). All of these associations are probably of protective value to the crustacean. Sponges may similarly be protective though they do not of course have any nematocysts. The crab *Dromia* cuts a piece of the sponge *Ficulina* or *Suberites* and holds it over its carapace with the fourth and fifth legs. After a few weeks the sponge grows into the shape of the carapace. The sponge may camouflage the crab, and in the absence of sponges *Dromia* may hold ascidians, alcyonarians or even algae instead. Polimanti has shown that the sponge is of protective value against octopus since in aquaria *Dromia* covered with a sponge survive whilst naked *Dromia* are attacked and eaten (Dembowska, 1926; Carlisle, 1953). However, it is not clear if the *Dromia* with sponges survive because the octopus cannot find them or because it finds the sponge distasteful.

Hence there are a variety of associations between coelenterates (and sponges) on the one hand and fish, molluscs or crustacea on the other in which the nematocysts of the coelenterate protect the associated commensal animal. The problems of initiating the association and of the commensal avoiding being stung by the host's nematocysts have been solved in different ways in the different associations.

Bird nesting associations

Various species of colonial-nesting birds utilize the defences of some other organism for the defence of the nest against snakes and other predators. As a general rule, small birds which nest solitarily protect the brood against nest predators by concealing the nest so that it is difficult to find. Colonial nesting birds may have some form of group defence which is much more effective a deterrent to predators than is the attack of a single pair of birds (see p. 248) or they may nest in some place which cannot be reached by the predators, for example in trees over water. Some birds normally nest close to the nest of a large bird of prey. Since the raptor normally hunts away from the nest the birds are not attacked by it, nor are they molested by other animals which tend to keep away from the raptor's nest. In one area of northern Nigeria J. F. Walsh (personal communication) found eight out of eleven nests of the red-tailed buzzard *Buteo auguralis* had the red-winged anaplectes (*Anaplectes melanotis*) nesting close by. Elsewhere in West

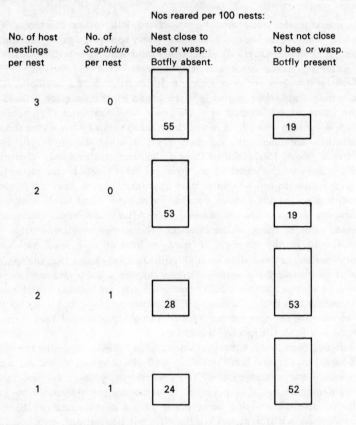

Fig. 11.12 Numbers of young successfully reared in 100 nests of *Cacicus cela* and *Zarhynchus wagleri* in Panama. (*Data from* Smith, 1968.)

Africa the red-winged anaplectes associates with other raptors or nests on its own.

Heuglin's weaver (*Ploceus heuglini*) in southern Ghana commonly nests in trees colonized by the red weaver ant (*Oecophylla longinoda*). The ants run freely over the nests of the bird without attacking the young, yet if a nest falls to the ground the ants immediately attack the nestlings (Grimes, 1973). In northern Ghana Heuglin's weaver often nests in trees with polistid wasps, and other African weavers and neotropical caciques and oropendulas regularly build their nests in the same trees as colonial bees or wasps. The hymenopterans apparently become habituated to the birds and ignore them, but it is likely that other birds, mammals or snakes are attacked. Such an association is of no obvious benefit to the wasp but is of protective value to the birds. An additional benefit to the birds has been demonstrated by Smith (1968). He has shown that cacique and oropendula nests built away from wasp colonies suffer a higher incidence of parasitization by botflies (*Philornis* sp.) than do colonies built close to wasp nests (Fig. 11.12). This is because wasps attack botflies and other insects near their colony. He has also shown that if the nest harbours a brood parasite, the cowbird *Scaphidura oryzivora*, this also reduces the incidence of death from the botfly since the intruding bird pecks and kills botfly larvae on the host nestlings.

Summary

Some animals live in large single species groups. A group of animals in an area is less likely to be encountered by a predator than are the same number of animals dispersed widely over the same area. If a group is attacked, only a few individuals are likely to be killed before the predator is satiated, so the rest will escape. The hunting success of a predator is reduced by the confusion of seeing many animals moving and by individuals scattering so that only one or two can be followed and caught. Groups of some animals can also retaliate against predators more effectively than can single individuals, and the sense organs of the entire group are more likely to detect a predator than are the sense organs of a single animal. Mixed species groups of animals have similar advantages to single species groups but with the added advantage that the sense organs of the different species may be more efficient at detecting predators than are those of a single species.

Some animals live in close association with and derive protection from other well-protected host animals such as colonial Hymenoptera or coelenterates. The processes by which these associations are initiated and maintained are often very complex, and whilst some are harmful to the host (parasitism) others are beneficial to both partners (mutualism).

The evolution of predator—prey systems

Predators superior to the best defence

We have seen in the earlier chapters of this book that there is tremendous selection pressure on prey animals to improve and perfect their anti-predator defences. There will also be selection pressure favouring those predators that are best able to overcome the defences of the prey, so there is a perpetual arms race between prey and predator with the prey constantly improving their defences, the predators constantly devising new or improved techniques for overcoming these defences. No defence gives 100 per cent protection against all predators since selection can never push the improved defences that far. It may even be advantageous to a prey species if defence is not 100 per cent perfect: so long as the healthy individuals are protected the species will surely benefit if those individuals that are deformed, diseased, parasitized or senile are inadequately protected and succumb to predation.

Predators overcoming secondary defences

Molluscs are mostly slow-moving animals that are protected by a thick calcareous shell with an operculum (gastropods) or by two closely fitting shells (bivalves). Predators have overcome this defence in a variety of different ways: the opisthobranch *Aglaja inermis* swallows other opisthobranch gastropods whole and dissolves the soft parts in its stomach (Paine, 1963); plaice and rays crush the shell with batteries of flat teeth; the opisthobranchs *Philine* and *Scaphander* crush the shell with strong gizzard plates; the crab *Calappa* holds shells in its notched chelae and cracks them open like a nutcracker (Shoup, 1968); song thrushes and sea otters smash shells on stones; starfish can give a sustained pull on bivalve shells or gastropod opercula so that the mollusc is forced open sufficiently for the starfish to insert its stomach and digestive juices into the mollusc; and some carnivorous gastropods drill holes through the shell, partly by chemical, partly by mechanical means, and then insert the proboscis to eat out the contents. Hence a large number of predators have overcome the principal defence of typical molluscs.

Like molluscs, glomerid millipedes are also protected by a hard exoskeleton. When disturbed they roll into an impregnable ball (Fig.

10.8) which ants, birds and small rodents are unable either to open or to crush. The banded mongoose, *Mungos mungo*, however, has overcome this defence in the African glomerid *Sphaerotherium*. When it finds a *Sphaerotherium* it holds it in its forepaws and repeatedly throws it between the hindlegs onto a stone until it smashes (Eisner and Davis, 1967; Eisner, 1968).

Predators can sometimes defeat chemical defences too. If they are very hungry they may take prey that is moderately unpalatable but which they would not eat if they were well fed (see p. 79). Some birds such as cuckoos (e.g. *Cuculus clamosus*) and cuckoo-shrikes (e.g. *Campephaga phoenicea*) feed largely on caterpillars and are apparently immune to the urticating properties of hairy caterpillars (Chapin, 1939, 1953). So for these birds the aposematic colours of these caterpillars may actually attract rather than repel, and hence if these birds are common selection may favour hairy caterpillars that are not so conspicuously coloured. Presumably these birds have particularly tough linings to the gut so that the hairs from the caterpillars do no damage.

Other birds specialize in preying on social Hymenoptera. The honey buzzard (*Pernis apivora*) feeds throughout the summer months almost entirely on the larvae of wasps and bees, together with a few adults as well. When raiding a wasps' nest the bird is protected from being stung by the very small and close-fitting feathers on the head, and this armour is supplemented by the judicious use of the small beak in decapitating insects that are too persistent in trying to sting it (Willis, 1972). Thus the honey buzzard has overcome the defences of Hymenoptera by evolving sting-proof feathers and probably also by a partial immunity to the stings.

Bee-eaters (Meropidae) also specialize in preying on Hymenoptera, in this case only on the adults. They do not raid the colony, so, unlike the honey buzzard, they are not normally attacked by a swarm of bees. Instead they pick up bees in flight and carry them to a convenient perch. Then they bang the head of the insect against the perch, and bang and rub the abdomen several times to squeeze the venom out of the sting gland. They give a final one or two head blows, and then swallow the insect head first. Bee-eaters do occasionally get stung, but apparently not seriously, and normally this procedure squeezes out most of the poison. Other non-venomous insects are simply beaten on the head before swallowing — the abdomen is not rubbed. Fry (1969) found that in eleven species of bee-eaters Hymenoptera comprised between 62 and 94 per cent of the food, and honey bees were often the commonest single species taken. The bee-eater's technique of overcoming the prey's defences is very specific: *Melittophagus bullocki* will not eat ponerine ants, which also sting, nor the aposematic butterfly *Danaus chrysippus* which has a different chemical defence.

There are also several species of mammals that feed on social Hymenoptera. Bears, badgers, honey-badgers and wolverines all open up bees nests to feed on honey, and they are apparently not harmed by being stung. But honey merely forms a supplement to their diet: there are other mammals that feed almost entirely on Hymenoptera. In Africa the arboreal pangolins (e.g. *Manis longicaudata*) feed entirely on ants and are apparently not affected by the ants' bites and stings. The larger giant pangolins and the aardvark break open termite mounds and feed almost entirely on *Macrotermes* spp. They are apparently not harmed by the powerful jaws of the termite soldiers which attack them in defence of the colony (Pages, 1970).

Poisonous snakes should in theory by avoided by all except the inexperienced predators yet civets and mongooses often kill snakes, without themselves being bitten, by speed of action and skill (Ewer, 1973). Neither is wholly dependent on snakes for food, but some birds specialize in catching and eating snakes and they may feed almost entirely on them. The secretary bird has long scaly legs which the snake's fangs cannot penetrate, and the short toed eagle has scales extending further up its legs than do other eagles so that it too is protected from snake bites.

Thus for any defence, however perfect it appears to be, one can usually find at least one species of predator that has evolved a means of overcoming it. The disadvantage for the predator is that this can lead to specialization and increased dependence on a narrow range of prey species. If the Hymenoptera or snakes on which the predator feeds become scarce, it may then be unable to capture enough alternative prey to survive in competition with other predators that are specialized in capturing these alternative species of prey.

Predators overcoming primary defences

Predators can overcome primary defences by improved hunting techniques, for example by developing a searching image or searching behaviour pattern for each species of prey. The prey species responds by selection favouring variability in appearance (polymorphism), since this will make it less easy for a predator to develop a searching image or searching behaviour pattern for that particular species of prey, or selection may lead to variability in secondary defensive response (protean behaviour).

In West Africa one of the predators of salticid spiders is the wasp *Pison xanthopus* which sometimes stocks its cells almost entirely with the salticid ant-mimic *Myrmarachne* (see p. 117), so this predator has apparently overcome the defence (mimicry) of its prey. Orb-web spiders are also preyed on by wasps and so selection will favour any defences that reduce the chances of a spider being captured. One of the most elaborate defences must surely be that of *Araneus cereolus* which builds its web at dusk when wasps are no longer hunting and then destroys it completely except for a single thread before dawn. Hence there is no web to act as a visual cue for the hunting wasp, but only a single thread and a couple of leaves bound together with silk in which the spider rests during the day. Nevertheless *Sceliphron* and another species of wasp do prey on *A. cereolus* in Ghana, and they have evidently developed a technique for hunting and finding the spiders despite the absence of a web. *Sceliphron* often hunts near orb-webs searching for the spider's retreat, but it also hunts over foliage searching for spiders without webs.

Moths normally fly at night when visually hunting predators (such as most birds) are unable to catch them. Bats, however, have evolved a special hunting technique which operates efficiently in the dark. They emit short pulses of high frequency sound and if an object such as a tree or an insect is in the vicinity the sound will be reflected back to the bat which is able to pick up and analyse the echo. Thus they are able to avoid objects and hunt and capture prey by means of echo location (sonar) (Griffin, 1958).

A rather different prey-hunting strategy has evolved in some deep-sea fish such as *Pachystomias*. Most deep-sea fish are not sensitive to red light since the only light present in the environment, if any, is blue or violet. The retinal pigment of *Pachystomias*, however, is sensitive to red light, and this fish has light emitting organs close beside the eyes which transmit only red or orange light (Fig. 12.1). These photophores emit red flashes from time to time, like an intermittent searchlight, and so enable the fish to see nearby prey without itself being seen by them (Denton, 1971).

In marine invertebrates the sense of smell (chemoreception) is often important both in hunting for prey and in detection of predators. The starfish *Asterias rubens* searches for its bivalve prey *Mytilus edulis* by orienting towards certain chemicals passed into the water by *Mytilus*. Strangely enough it is only attracted chemically to *Mytilus* between November and May; during the rest of the year it shows no orientation towards *Mytilus*, but whether this is because of a seasonal change in

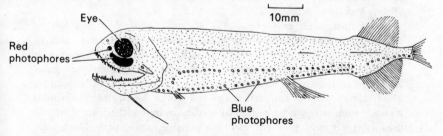

Fig. 12.1 The deep-sea fish *Pachystomias* showing its photophores (heavy circles). The photophores on the belly emit bluish light downwards, similar to those of *Argyropelecus* (Fig. 2.6). The large photophore below the eye emits red light forwards like a searchlight. (*Based on various sources; details of eye and red photophores added from notes provided by* Professor E. J. Denton.)

release of the attractant by the bivalve or because the starfish has a lowered sensitivity at this period is not known (Castilla, 1972). The gastropod *Cassis tuberosa* also detects its prey (sea urchins) by chemoreception, and it shows a distinct preference for some species (e.g. *Echinometra lucunter*) and an avoidance of others (e.g. *Diadema antillarum*) (Hughes and Hughes, 1971). The burrowing starfish *Pisaster brevispinus* also appears to detect prey by chemoreception: from the surface of the sand it is able to detect buried clams and sand-dollars and it then digs down after them (Smith, 1961).

Many elasmobranchs also hunt using the sense of smell. The hammerhead shark (*Sphyrna*) has the nostrils far apart on the flattened T-shaped head. Presumably this enables it to locate food by comparing the intensity of the smell on the two sides, adjusting its position till they are equal, and then moving towards it. Certainly hammerheads are often said to be the first sharks to arrive at bait. Dogfish (*Scyliorhinus canicula*) and rays (*Raia clavata*) can also find food using their olfactory sense organs. When dead fish meat was placed in a buried chamber through which a current of water was passed, dogfish attacked at the outlet of the water current (Fig. 12.2*c*) (Kalmijn, 1971). However, if the buried chamber contained a live plaice (*Pleuronectes platessa*) both dogfish and rays attacked directly above the chamber, not at the water outlet, indicating that they can also hunt using some sense other than chemoreception to detect prey (Fig. 12.2*b*). If the plaice in its chamber was covered with an insulating polyethylene film, no attacks were made (Fig. 12.2*d*). This suggested that the fish might have an electric sense organ that detects muscle action potentials in the prey fish. To prove that this is so, Kalmijn passed a 1 Hz sine wave current at 4 μA amplitude into the aquarium under the sand. The fish then attacked this electrode as if it were a live plaice, even ignoring bait of whiting meat nearby (Fig. 12.2*f*). Hence sharks and rays can detect a prey fish, even when it is resting passively on the sea-bed or hidden under the sand, by means of the electric potentials it produces. The catfish *Ictalurus*, and presumably some other teleosts, also detect hidden prey in this way (Peters and Bretschneider, 1972).

Other fish, such as *Gymnarchus niloticus*, emit electric pulses to detect food and other objects around them. *Gymnarchus* lives in muddy water where visibility is poor and it has very poor eyesight. It produces a series of 3 V electric pulses at a constant frequency of about 280 Hz, and it also has receptors sensitive to electric pulses (Lissmann, 1958, 1963). Thus it can detect any changes in the electric field round its body such as might be caused by a stone or by another fish. *Torpedo* uses its much more powerful electric discharge to stun prey. During the day it rests on the sea-bed, partly buried in sand, and it stuns fish that swim too close above it. At night it cruises over the sea-bed and stuns fish below it on the sand (Pfeiffer, 1961; Belbenoit and Bauer, 1972). The catfish *Malapterurus* and the gymnotid *Electrophorus* also stun

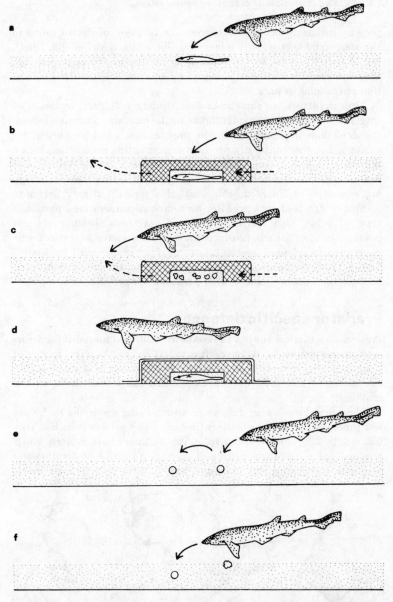

Fig. 12.2   Detection of prey by the dogfish *Scyliorhinus canicula*: (a) the dogfish quickly finds live plaice buried under sand; (b) with the plaice in an agar chamber the fish still attacks just above the plaice, even though the water current (dotted line) comes up elsewhere. Hence the dogfish cannot be hunting by smell; (c) whiting meat is placed in the agar chamber and the dogfish attacks at the water outlet, so here it is hunting by smell; (d) with the agar chamber sealed with a polyethylene film the dogfish cannot detect a buried plaice; (e) the dogfish attacks electrodes which are producing a 1 Hz sine wave current of 4 μA amplitude; (f) although attracted by the smell of buried whiting meat, the dogfish actually attacks the electrodes. (*Redrawn from* Kalmijn, 1971.)

prey by means of electric discharges. The function of electric pulses in the stargazer (*Astroscopus*) is less clear (Pickens and McFarland, 1964), but in *Gnathonemus*, and in some other mormyrid and gymnotid fish, they are used for intraspecific signalling and not, apparently, for location or stunning of prey.

Thus predators can sometimes defeat primary defences by means of specialized hunting techniques. Some of the hunting techniques involve the detection of stimuli which the prey cannot avoid producing: for example live fish constantly produce muscle action potentials which a dogfish can detect, and a flying moth above a certain critical size cannot avoid being detected by a hunting bat. As with predators defeating secondary defences development of a specialized prey detection technique can lead to a predator becoming dependent on a particular type of prey. Such predators are food specialists which could not possibly survive if their favoured prey became scarce or extinct (the consequences of food specialization are further discussed by Emlen, 1968*b*). Hence a balance is maintained such that the predator does not become too common, and neither predator nor prey becomes too rare.

Predator specific defences

Prey species that are subject to serious losses due to specialist predators may themselves evolve defences directed at just one or a few species of predators. Highly specific anti-predator defences occur in many different animals, but they have been studied most intensively in the molluscs.

The normal molluscan defence of withdrawing inside the shell does not protect gastropods or bivalves from specialist predators such as starfish (see p. 226), and so other secondary defences have evolved. When touched by a starfish, a gastropod such as *Nassarius* or *Struthiolaria* responds by flexing and extending the foot so that it leaps away,

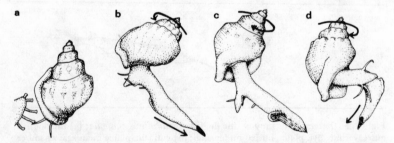

Fig. 12.3   Escape behaviour of the gastropod *Struthiolaria papulosa* on contact with the starfish *Astrostole scabra*: (*a*) the gastropod is touched by the starfish; (*b*) and (*c*) the gastropod elongates the foot and twists the shell violently; (*d*) the foot is flexed to throw the shell in a different direction and the shell is now twisted in the opposite direction. (*Redrawn from* Feder, 1972. Copyright ©1972*b* Scientific American Inc. All rights reserved.)

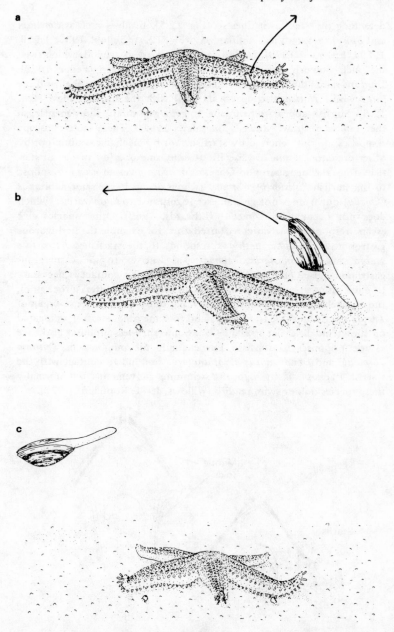

Fig. 12.4   Response of the bivalve *Spisula* to a starfish (*Asterias rubens*): (*a*) the siphon of the buried bivalve is contacted by the starfish; (*b*) the bivalve leaps clear of the sand; (*c*) the bivalve makes several further leaps away from the starfish. (*Redrawn from photographs in* Feder, 1972.)

detaching tube feet as it does so (Fig. 12.3). Bivalves such as *Cardium* and *Spisula* respond in a similar way (Fig. 12.4) (Bullock 1953; Ansell, 1969*a, b*; Feder, 1972). *Haliotis* and *Tegula* are gastropods which twist the shell violently and then crawl away at full speed (Feder, 1972). *Diodora* raises the body and extends the mantle over the shell and the foot, whilst *Natica* also covers its shell with the mantle (Figs. 12.5, 12.6). Starfish appear to be unable to get a grip on the slimy mantle, so the molluscs crawl away (Margolin, 1964). These active escape responses are only elicited by starfish, not by mechanical stimuli or by other predators. Some are specific to only one or a few species of star-fish: thus the neogastropod *Olivella biplicata* gives an escape response to the starfish *Pisaster brevispinus* (which occurs in the same habitat as *Olivella*) but it does not give an escape response to *P. ochraceus* (which does not occur with *Olivella*) (Edwards, 1969). Other species give escape responses to a variety of predators, for example the herbivorous gastropod *Melagraphia aethiops* responds to the starfishes *Astrostole scabra* and *Coscinasterias calamaria* as well as to the carnivorous gastropod *Lepsia haustrum* (Clark, 1958). Usually contact is necessary to elicit an escape response, but as soon as a *Pisaster* enters a rock pool, the topshell *Tegula* responds by crawling out of it. The sense organ of *Tegula* that is sensitive to *Pisaster* (and to some other starfish) is located on the ctenidial leaflet (Szal, 1971). Some shell-less molluscs also have escape responses to starfish. For example the nudibranchs *Tritonia diomedia* and *Dendronotus albopunctatus* respond to contact with the starfish *Pycnopodia* by vigorous swimming movements, but normally these species do not swim readily (Willows, 1971; Robilliard, 1972).

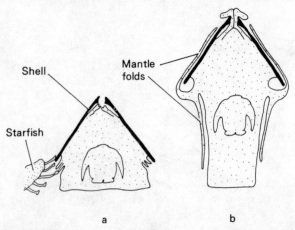

Fig. 12.5 Response of the limpet *Diodora aspersa* to the starfish *Pisaster ochraceus*: (*a*) starfish contacting the limpet shell with its tube feet; (*b*) the limpet has elongated the foot and extended the mantle folds to cover most of the shell so that the tube feet cannot grip it. (*Redrawn with modifications from* Margolin, 1964, Fig. 2.)

a

b

c
Fold of tissue   Exposed shell
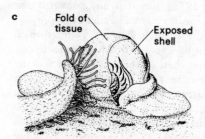

d

Fig. 12.6   Escape behaviour of the gastropod *Natica catena* following contact by the starfish *Asterias rubens*: (*a*) the mollusc detects the starfish; (*b*) the mollusc responds by expanding the foot and erecting a fold of tissue behind the shell; (*c*) this fold of tissue spreads over the shell; (*d*) the shell is completely covered so that the starfish cannot grip the mollusc because of the slimy nature of its surface, and the mollusc is crawling rapidly away. (*Redrawn from* Purchon, 1968, Fig. 146; *after the original by* Dr G. Thorson *drawn by* Poul Winther.)

Thus there are a great variety of molluscs that exhibit predator specific escape responses, but very few predator–prey interactions have been thoroughly examined. Gonor (1965) studied the gastropod *Natica chemnitzii* and its prey *Nassarius luteostoma* in Costa Rica. If a hunting *Natica* crosses the slime trail of a *Nassarius*, it follows it, but often in the wrong direction since it appears to be unable to determine the direction in which the prey was travelling. If it catches up, it may make several attempts to grasp its prey before it is successful. *Nassarius* can detect *Natica* from a distance of 1½–2 cm and responds by accelerated crawling away. When actually touched by *Natica*, *Nassarius* responds by leaping movements. In this case the choice of prey is highly specific, and so is the response of the prey to *Natica*.

The opisthobranch *Aglaja* (= *Navanax*) *inermis* also hunts prey by following slime trails. It ignores trails of prosobranchs but follows those of a wide variety of species of opisthobranch (Paine, 1963, 1965). In nature it preys largely on shelled opisthobranchs (*Bulla gouldiana*, *Haminoea virescens* and *Aglaja* sp.), but it will also take eolids without apparently being affected by the nematocysts which they eject in defence. However, dorids which are protected by spicules (e.g.

Rostanga pulchra) and opisthobranchs with acidic defensive secretions (e.g. *Pleurobranchaea* sp. and *Acteon punctocaelatus*) are normally rejected by *Aglaja*. Occasional individuals of *Aglaja inermis* occur which will eat these normally avoided species of opisthobranch. Presumably these animals will be at a selective advantage to others under conditions of scarcity of the normal species of prey.

Both *Natica* and *Aglaja* follow slime trails, but since they cannot distinguish the direction in which the prey was moving, they must fail to capture 50 per cent of all prey detected. Some gastropods (e.g. *Littorina, Physa* and *Nassarius*) can distinguish the direction of a trail of a conspecific (Wells and Buckley, 1972), but it is not known if any predaceous mollusc can distinguish the direction of a trail of its prey.

Another predaceous gastropod is the naticid *Polinices conicus* which ploughs through sand searching for bivalves on which it feeds. In one area of South Australia it was found that, although *Donacilla* was the commonest bivalve present, very few *Donacilla* shells had actually been drilled by *Polinices* and a much higher percentage of shells of *Mactra* and *Katelysia* had been drilled (Fig. 12.7). Laws and Laws (1972) attribute this to the fact that of these three species of bivalve only *Donacilla* has an active escape response to *Polinices*. As the burrowing *Polinices* approaches, *Donacilla* pops up onto the surface of the sand, and when *Polinices* has passed beneath, it burrows down again.

The defence responses of the bivalve *Pecten maximus* to various species of starfish have been studied by Thomas and Gruffydd (1971). *Pecten* responds to the steroid glycosides on the tube feet of starfish either by swimming, jumping, or by closing the valves. It is clear from Fig. 12.8 that an active locomotory escape response is given more frequently to *Asterias*, which preys largely on molluscs, than to *Porania*, which is not a predatory species. Similar experiments showed that active locomotory responses are frequently given to several species of starfish which prey on molluscs, but they are not often given to starfish which feed on echinoderms or on other foods. One might wonder why, as a precaution, *Pecten* does not respond actively to all starfish, but it is probably a disadvantage to swim or jump unnecessarily since *Pecten* normally lives in a scooped depression in the sea-bed. Swimming or jumping means that it must make a new depression, and so it is more efficient to simply close the valves if touched by a harmless starfish.

Predator specific defences also occur in some coelenterates against eolid nudibranchs: for example when the sea anemone *Anthopleura nigrescens* is attacked by the eolid *Herviella baba*, it detaches its pedal disc from the substrate (Rosin, 1969). When *Actinia equina* is attacked by *Aeolidia papillosa* it responds by crawling away, and if the attack continues it also detaches from the substrate (Edmunds, Potts, Swinfen and Waters, unpublished observations). In the aquarium this is no protection, but in the sea the anemone probably gets carried by water movements some distance from the eolid. The eolid will then continue

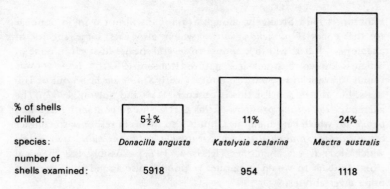

% of shells drilled:	$5\frac{1}{2}$%	11%	24%
species:	*Donacilla angusta*	*Katelysia scalarina*	*Mactra australis*
number of shells examined:	5918	954	1118

Fig. 12.7 Percentage of shells of three species of bivalve which were drilled by the naticid *Polinices conicus* on a South Australian beach. (*Data from* Laws and Laws, 1972.)

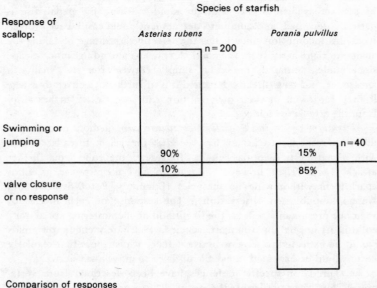

Comparison of responses to the two starfish: $X^2_{(1)} = 107.5$ $P < 0.001$

Fig. 12.8 Responses of the scallop *Pecten maximus* to contact with two species of starfish. (*Data from* Thomas and Gruffydd, 1971.)

to search for anemones, but it is likely that the next one it finds will be a different individual to the one originally attacked. Hence although several different anemones may be attacked and damaged by the eolid, very few are likely to be killed.

The anemones *Stomphia coccinea* from Denmark and *Actinostola* sp. from the Pacific both swim in response to contact by *Aeolidia papillosa*

(Robson, 1966). Strangely enough *Stomphia* will not swim in response to the eolid *Aeolidiella glauca* (which also eats some species of anemones), but it will in response to several species of starfish, none of which is known to attack it in nature (Robson, 1961). *Actinostola* was found to swim in response to contact with *Stomphia* in 53 out of 100 trials, but the reason for this is obscure (Ross and Sutton, 1967). The anemone *Gonactinia prolifera* swims in response to contact by *Aeolidia papillosa*, which eats anemones, but it also swims in response to contact by *Favorinus branchialis* and *Coryphella rufibranchialis*, two eolids which do not eat anemones (Robson, 1971). Possibly the chemical from *Aeolidia* to which anemones respond is also found in a variety of other invertebrates.

By comparing the responses of different species of anemones to attacks by eolids it is possible to see how swimming and detachment may have evolved. Anemones such as *Anemonia sulcata* respond to contact with, or a bite from, *Aeolidia* by crawling away from the point of contact as fast as they can. In rapid crawling the pedal disc is partially detached, and complete detachment could easily have evolved from this locomotion since it has the selective advantage of taking the anemone right away from the eolid. A crawling anemone cannot escape since, unless satiated, the eolid simply crawls after it. Finally in anemones that have already detached it is of further selective advantage if they can swim of their own volition and so get even further away from the attacking eolid.

Metridium senile and *Sagartia elegans* are two anemones that have a different response to attack by *Aeolidia*: they eject threads covered with nematocysts (acontia) which often deter the eolid from further attack. *Tealia felina*, however, appears to have no response to eolids apart from withdrawing its tentacles (Edmunds, Potts, Swinfen and Waters, unpublished observations). The response of each species of anemone presumably reflects the likelihood of encountering a predatory eolid in its normal environment: species which rarely encounter eolids would be expected to have no active defence whilst those that regularly do so would be expected to eject acontia or to move away actively.

Few predator specific defences have been described from vertebrates, but they undoubtedly occur. The scorpion fish (*Scorpaena guttata*) flees from a hunting octopus (*Octopus bimaculatus*), but it stays put and raises its spines if approached by predatory fish, by man, or if it sees a resting octopus (Taylor and Chen, 1969). The octopus is one of the few predators that can successfully tackle a scorpion fish.

The African ground squirrel jumps vertically into the air in response to rustling noises in the grass. This is the only defence that is likely to succeed against a snake that is about to strike (Ewer, 1966), but unfortunately it has not actually been observed as a response to a live snake. Larger mammals also respond differently to different species of predators, but they have not often been studied with this in mind.

$$\frac{a \mid b}{c \mid d}$$

Plate 7

(a) Deimatic behaviour of young white-faced owl (*Otus leucotis*). Note the feathers are fluffed out to give apparent increase in size and to accentuate black facial marks, and the bill is snapped repeatedly. Legon, Ghana.

(b) Deimatic display of the neotropical hawkmoth *Leucorampha*. Note the false eyes. N. Smythe, Canal Zone.

(c) Peacock butterfly (*Nymphalis io*) on *Buddleja davidii*. Harlow, England.

(d) Hawkmoth *Platysphinx constrigilis* exposing eyespots after being poked. Legon, Ghana.

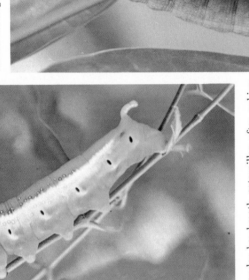

Plate 8
(a) Oleander hawkmoth (*Deilephila nerii*) on *Salacia baumannii* in normal posture. Akosombo, Ghana.

(b) Oleander hawkmoth caterpillar after poking. Note the swollen thorax with false eye now visible. Akosombo, Ghana.

(c) Hawkmoth caterpillar (*Hippotion eson*): brown morph on *Morinda lucida*, with head withdrawn after being disturbed. Legon, Ghana.

(d) Green morph of *Hippotion eson* on *Cissus quadrangularis*. Legon, Ghana.

Thomson's gazelle flee or move away from wild dogs but not from other predators at 500–1 000 m distance. The 'flight distance' is 100–300 m in response to cheetah and lion, 50–100 m in response to hyaena, and 5–50 m in response to jackals (Walther, 1969). There is much variation in the flight distance according to age, sex and social status of the gazelle, and according to whether the predator is actively hunting or not, and if singly or in a pack. Nevertheless, the series does reflect the prey-capture capabilities of the predators and the escape potential of the prey. Jackals probably cannot run down a healthy adult gazelle and rarely prey on them, hence for a gazelle to flee from a jackal is a waste of energy unless it is very close – jackals do prey on adult gazelles occasionally (H. and J. van Lawick-Goodall, 1970) but it is not known if the prey individuals were healthy or diseased. At the other extreme, wild dogs are noted for their tremendous stamina and they can continue a chase for several miles, much further than can a lion or cheetah. Hence the gazelle responds to wild dogs whilst they are still a long way off.

The meercat (*Suricata suricatta*), when above ground, regularly scans the sky for predatory hawks or eagles. If it sees one, it gives an alarm call and all meercats in the area freeze or run to their burrows. The response to ground predators such as dogs is quite different and involves an intimidatory display (Ewer, 1963), so here too the response of the prey depends on what sort of predator is present.

Birds too may show highly specific responses to certain predators. Black-headed gulls actively attack carrion crows (*Corvus corone*) which prey on their eggs and chicks (Kruuk, 1964), and they also attack kestrels if these are hovering near the nesting colony. But they very rarely attack a peregrine falcon: instead they fly up from the colony in tight formation and then fly fast with many zig-zags and irregular movements. Since the kestrel is a predator of chicks whilst the peregrine attacks adult birds, the two responses are obviously adaptive, but it is not easy for an inexperienced human observer to distinguish a peregrine from a kestrel. The birds also occasionally make an error and fly up in formation if a kestrel appears suddenly, but they usually recognize the kestrel as soon as it hovers.

Mallard ducklings exhibit fear responses (running, crying and wing flapping) when presented with a hawk model but not when presented with a duck model (Melzack, 1961). This response is innate, but it quickly habituates so that after many repeated presentations a hawk model no longer elicits the complete fear response. However, the ducklings still orient towards the hawk, and probably if it changed in size (by approaching nearer) instead of placidly moving across the sky, they would respond fully.

Although many of these predator specific defences are innate they can probably be modified by experience. Paulson (1973) has suggested that small birds and mammals may, through experience, learn the colour pattern and outline of commonly seen predatory birds so that they respond more quickly and hence reduce the hunting success of the predator. Selection will then favour those predators which have a different colour pattern since the prey will not immediately recognize them as predators, and hence the hunting success of these predators will be improved. Evidence to support this theory is that most open country predators of mammals (buzzards of the genus *Buteo*) and predators of small birds (hawks of the genus *Accipiter*) are polymorphic in colour, whereas forest buzzards (where prey will have little chance of seeing a predator before an attack) and hawks which prey on lizards and insects are usually monomorphic. Hence the prey may be selective agents favouring the establishment and maintenance of polymorphism in a predator, and we can regard colour polymorphism in these birds as being a predator response to efficient predator detection systems in the prey.

A more elaborate response to a particular predator is shown by the rotifer *Brachionus* which develops long spines on the carapace in response to secretions in the water from the predatory rotifer *Asplanchna* (Gilbert, 1967). These spines reduce the likelihood of *Brachionus* being swallowed by *Asplanchna* (see p. 185).

Predator-specific defences also occur in insects. The young of digger wasps are often heavily parasitized by bee-flies, miltogrammine flies, mutillids (velvet ants) and cuckoo wasps, and some wasps have defences directed at these parasites (Evans, 1966). Some digger wasps temporarily close their burrows after each visit, and this probably reduces the incidence of parasitization, but other species build false accessory burrows near to the entrance of the true nesting burrow. Evans thinks that originally these false burrows may simply have been quarries for material to stop up the true burrow, but that selection then favoured insects that built false burrows since some parasites laid eggs in these instead of in the true nesting burrows. Thus, *Bembix amoena* quarries material in order to close the true burrow temporarily, and in so doing she digs one or more false burrows. Parasites have been observed to enter and oviposit in both the false and the true burrows. Obviously the incidence of parasitization is likely to be reduced if some of the parasites' eggs are wasted in empty burrows. Some digger wasps such as *Philanthus coronatus* and *Sphex argentatus* not only dig false burrows but also rebuild them if they are destroyed. There is considerable variation in behaviour in different species. Thus *Sphex argentatus* builds two or three false burrows as soon as the true burrow has reached the stage of being temporarily closed, but *Bembix sayi* only builds its single deep false burrow after the true burrow is completed and blocked up for the last time (Fig. 12.9). Proof that false burrows reduce the incidence of parasitization is not easy to obtain, but Tsuneki found that colonies of *Sphex argentatus fumosus*, which build false burrows, had a much lower incidence of parasitization than colonies of *S. flammitrichus*,

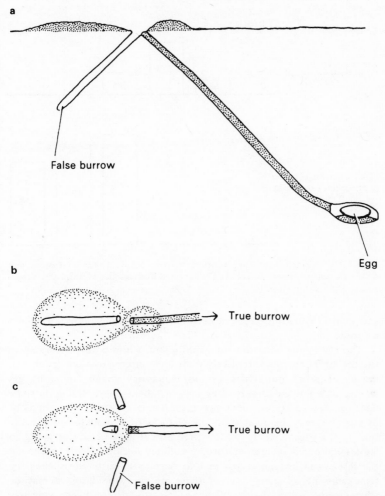

Fig. 12.9 Diagram to show true nesting burrows and false burrows of *Bembix* wasps: (*a*) side view of burrow of *Bembix sayi* with false burrow shown in white, true burrow filled with sand (stippled), and excavated mounds of sand (stippled); (*b*) top view of *Bembix sayi* burrows; (*c*) top view of *Bembix amoena* burrows. (*Redrawn from* Evans, 1966. Copyright 1966 by the American Association for the Advancement of Science.)

which do not build false burrows (Fig. 12.10). One of the colonies examined contained both species and the incidence of parasitization was very different, but since the total number of cells examined in all colonies was only 110 for *S. argentatus* and 45 for *S. flammitrichus*, this is not conclusive. Presumably there will be conflicting selection pressure for wasps building false burrows and hence reducing the incidence of parasitization, and for wasps building more nesting burrows

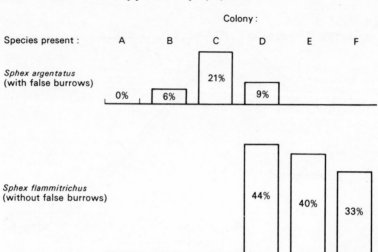

Fig. 12.10   Incidence of parasitization in colonies of *Sphex argentatus* and *Sphex flammitrichus*. Only colony D contained both species of wasp. (*Data from* Evans, 1966, *after* Tsuneki.)

and hence laying more eggs rather than wasting time on building false burrows.

The egg laying behaviour of Lepidoptera is also perhaps adapted to the predators and parasites likely to be encountered by the eggs of each species of insect. Butterflies and moths may lay a large number of eggs in one or a few batches, or they may lay them singly, flying on some distance before laying successive eggs. These contrasting methods of oviposition may be related to the chances of a predator or parasite finding the eggs. If there are few predators it may be advantageous to lay a large batch since predators are unlikely to find a single batch, and the insect can lay more eggs in one place than scattered utilizing the same amount of energy. But if there are many predators it may be advantageous to lay the eggs singly but scattered over a very wide area so that there will be increased chances of a few escaping detection. *Danaus chrysippus* eggs suffer very heavy predation from ants, but as they are laid singly a few usually hatch successfully. Some nymphalid butterflies from the neotropical region protect their eggs from chalcids and other egg parasites and predators even more effectively. *Euptychia hermes* and *E. renata* search for a larval food plant, then fly a short distance away and glue the eggs to dead leaves, twigs, or to some other plant, whilst *E. penelope* drops its eggs on the ground beside the larval food plant (Singer *et al.*, 1971). It is likely that predators and parasites, which normally search on the larval food plant, may be unable to find these eggs. Presumably the loss of eggs and larvae due to other predators finding them, or to the newly hatched larvae being unable to find their food plant, is less than the number normally lost due to predation on the food plant, otherwise the habit would not have evolved.

Insects may also have chemical defences directed against specific predators. Many of the chemical defences described on pp. 191–5 are directed against ants since in many parts of the world, especially in the tropics, ants are the most serious predator likely to be encountered by an insect.

Predator specific chemical defences also occur in vertebrates. For example the skin secretion of the red-spotted newt (*Notophthalmus viridescens*) appears to be directed against blood sucking leeches. When exposed to attack by the leech *Batracobdella phalera* they received significantly fewer attacks than did other species of newts (*Ambystoma* and *Triturus* spp.) (Pough, 1971). In addition 7.0 per cent of field caught *Ambystoma maculata* ($n = 1\ 580$) had leeches attached to them, but only 0.1 per cent of *Notophthalmus* from the same place had leeches ($n = 23\ 074$). Since leeches thrash around in water containing skin secretions of *Notophthalmus* but not in water containing skin secretions of *Ambystoma*, it is clear that *Notophthalmus* contains a skin secretion which repels leeches.

Integrated defence systems

We have seen that no one defence can give protection against all predators an animal may encounter, and further that some defences are directed against just one species of predator. The result is that many animals have several different defence mechanisms. If the first line of defence (e.g. crypsis) is breached, then a second defence comes into operation (e.g. flight, a bluff display, or a genuine warning display). If this also fails to deter a predator, the prey animal may attack the predator, or it may be distasteful and so be rejected. Thus many animals have three or more lines of defence which come into operation in sequence (Kettlewell, 1959). Because each species of prey is likely to be preyed on by several different predators with different hunting methods, some of these defences are likely to be specifically directed at particular species of predators.

The caterpillars of the sphingid *Errinyis ello* have two important predators in Jamaica, the wasp *Polistes crinitus* and the anole lizard *Anolis lineatopus* (see pp. 56–7). The primary defence of the caterpillar is crypsis, but there are four colour forms. Polymorphism is probably a prey response to predators that hunt by the searching image method, but since neither the lizards nor the wasp use coloration in their search, there must be a further predator involved, probably a bird. The wasp only searches amongst the leaves whilst the lizards only search on the trunk and large branches; but both of these predators take any caterpillars they find irrespective of colour. Hence predation

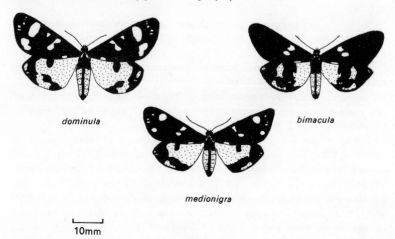

dominula

bimacula

medionigra

```
⌊___⌋
10mm
```

Fig. 12.11   Three morphs of the scarlet tiger moth *Panaxia dominula*. Form *dominula* is the commonest with the heterozygote *medionigra* and the very rare homozygote *bimacula* occurring in some populations. The forewings are black with yellow and cream spots, the hindwings are red with black spots. (*Redrawn from photographs in* Ford, 1955, Plates 11 and 14.)

pressure by the wasp should force caterpillars to rest during the day on the trunk, but predation pressure by the lizard should work in the other direction and drive them back amongst the leaves. Hence the caterpillars are polymorphic in choice of resting place as well as in colour.

Another caterpillar with a complex defence system is *Papilio demodocus*. The young larvae are bird-dropping mimics which rest on the upper surfaces of leaves, but the last instars are black and green and rest on stems or petioles (see Plate 2*d* and p. 54). Both young and old caterpillars hunch up when poked and eject from the back of the head a Y-shaped organ (the osmeterium). This organ is coloured red or orange and emits a nauseous smell. However, it is not clear at what predators this is directed: it does not deter red weaver ants (*Oecophylla*), and some birds certainly eat the caterpillars, though it may possibly make them less acceptable if alternative food is available.

The scarlet tiger moth *Panaxia dominula* is a brightly coloured day-flying moth which lives in scattered colonies (Fig. 12.11). At rest it is often cryptic, resting on flowers or vegetation, with the dark black of the forewings blending with dark shadows on leaves and the yellow spots resembling buds or flecks of sunlight. When it flies the scarlet on the hindwings may be a flash colour since it disappears on landing. Finally, if the resting insect is disturbed it exposes the red and black hindwings in a deimatic display and it may exude a drop of fluid from the cervical glands on the thorax (Ford, 1964). The insects are rejected as food by a variety of species of reptiles, amphibians, birds and mammals. Thus the scarlet tiger has several defences against predation.

It is also polymorphic in colour (Fig. 12.11), but there is no evidence that the rarer morphs are at either a selective advantage or disadvantage to the common morph with respect to visually hunting predators.

Most small species of praying mantis are cryptic, but when disturbed they run or fly away. Larger species are also cryptic, but they have several methods of secondary defence (Crane, 1952; Edmunds, 1972). *Polyspilota aeruginosa* for example may run, fly, or give a deimatic display, or it may slash at the attacker. It can also feign death if persistently roughly handled, but it soon recovers and attempts to escape actively. The bright colours on the abdomen may also act as flash colours since they are hidden when the animal comes to rest after flying. *Tarachodes afzellii* has a similar defensive repertoire: it is normally a bark mimic (Fig. 12.12), but when disturbed it runs (both sexes) or flies (male only), or gives a deimatic display. However, if it is approached by an ant it behaves differently. If hungry, it strikes and captures it, but if well fed it strikes with the forelegs but keeps the tibial claws flexed against the femur. The result is that the ant is knocked to the ground, away from the mantid, but is otherwise unhurt. This is obviously a defence specially adapted to the requirements of a bark-living species which must encounter many more ants than it can eat and which could easily be killed by them.

Even small apparently defenceless insects have anti-predator defensive systems. Aphids are often protected by the presence of ants which keep away many predatory insects, but they also have several escape responses when they are approached by a predator. An important predator of the sycamore aphid (*Drepanosiphum platanoides*) is the bug *Anthocoris nemorum*. *Anthocoris* detects aphids by contact with its antennae or rostrum, but the aphids see the *Anthocoris* approaching and take evasive action. If the predator is small the most frequent response of a large aphid is to kick it and so to delay an attack whilst the aphid removes its stylets from the host plant and walks away. The most usual response to large *Anthocoris* is to walk away (Russel, 1972). Aphids also occasionally respond by swivelling on their stylets so that they can continue feeding but with their body no longer in the path of the predator, or they can drop vertically from the plant. Similar defensive responses are shown by aphids to ladybirds. Although very simple, these escape responses are often successful so that 90—100 per cent of adult aphids escape predation in this way. Young aphids, however, are very often unsuccessful in their attempts to escape.

It is interesting to note that very few aphids respond to a predator by dropping yet this is always a successful method of escape. One disadvantage of dropping is that the animal may have great difficulty in again finding a suitable plant on which to feed, particularly if it is immature and has no wings. Nevertheless, for some small Hemiptera dropping is the only defence likely to be successful. Leston (1973) describes how the cocoa-capsid *Distantiella theobroma* drops when

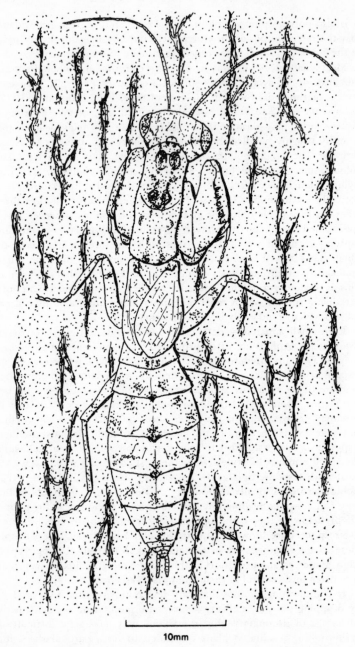

Fig. 12.12   Female mantid *Tarachodes afzellii* on the bark of a tree. Notice the forelegs held slightly open to reduce lateral shadow and the head tucked under so that the mouthparts are hidden. (*Reproduced from* Edmunds, 1972, Fig. 3.)

disturbed by red weaver ants (*Oecophylla longinoda*). As they fall, adult capsids disengage the forewings from the margin of the scutellum and couple them with the hindwings, and they then fly. This preparation for flight can be done before take-off (for example when they fly spontaneously), but it takes time, and when attacked by an ant it is more important to get out of the way quickly than to couple the wings and fly. Leston showed that out of ninety-three capsids attacked by ants thirty-nine (42 per cent) were killed and fifty-four escaped successfully by dropping. He further showed that capsids can only survive where weaver ants are rare or absent, since where the ants are common, the capsids are sure to be encountered and either killed or driven elsewhere.

Eolid nudibranchs are protected by nematocysts which they derive from their food (coelenterates) and these can be ejected when the animal is itself attacked. But eolids are occasionally eaten by fish, so evidently this protection is not perfect, and it may be that it is in response to these predators that many species of eolid have developed glandular defences as well (Edmunds, 1966a). Just as nematocysts are ejected from the tips of the papillae, which are the first parts of the animal to be touched by a predator, so the defensive glands are also found close to the tips of the papillae in species of *Eubranchus*, *Trinchesia* and *Catriona* (Fig. 9.5). It is not known, however, if the glands are defensive against fish or against invertebrate predators.

The complexity of the defence system is well illustrated in the orb-web spiders of the genera *Argiope* and *Araneus*, but as is often the case, we know very little about the predators to which the various defences are directed. Many species of *Araneus* spend most of the day in a retreat to one side of the web and they only come out when a signal on their guide line informs them that a prey is trapped. Nevertheless, most species are cryptically coloured green or brown so that when they are in the centre of the web they are not conspicuous. Spider-hunting wasps, however, may use the web as a visual cue in their search for a spider, and with persistent search they often find the retreat (which typically consists of two or three leaves spun together with silk). The spider may then drop to the ground and rest motionless, hidden under a leaf or a blade of grass, but wasps are very adept at capturing the spider as she falls to the ground. A different strategy used by species such as *Araneus cereolus*, is to build the web at dusk when wasps are going to roost and to destroy it again before dawn so that all that remains is one or two threads. This means that the spider must obtain all her food at night-time, but she is well hidden during the day. Nevertheless, some wasps do succeed in finding her and some individuals even specialize in capturing such species of spider (see p. 228). The large tropical orb-web spider *Caerostris albescens* only sits in the centre of her web at night-time and during the day she rests motionless on the branch of a tree (Plate 6d). Since her shape and coloration blend perfectly with the bark she presumably derives protection from visually hunting predators in this way.

Argiope rests by day in the centre of her web where she can quickly run and capture any insect that blunders into the web. But although this gives her an advantage over *Araneus* in the speedy capture of temporarily entangled prey, she is at a disadvantage with respect to conspicuousness to predators. Her legs are disruptively banded dark and pale, so their outline is not very conspicuous. If disturbed, the creamy white *Argiope flavipalpis* may drop vertically to the ground, changing colour to a dull brown as she falls, and then rest motionless and cryptic on the soil (J. Edmunds, in preparation). Alternatively she may make vigorous pumping movements at the hub of the web so that it vibrates rapidly backwards and forwards. Her outline is then obscured so that a predator may be baffled as to her exact position, or it may perhaps be intimidated by the movements. In addition young spiders build circular platforms or devices of thick white webbing at the hubs, and they rest behind this (Robinson and Robinson, 1970; Ewer, 1972). Here they are concealed from one side by the white device, and they are cryptic from the other side since the body of the spider is silvery white as well. When poked (as by a bird) the young spider, with incredible speed, darts through the web beside the device and takes up position on the opposite side from the attacking stimulus. Finally, if all other defences fail, the spider may attack the aggressor with her chelicerae (jaws). Some of these defences may perhaps be directed at wasps, some at birds, and some perhaps at other predators such as mantids or other spiders. In addition most species of *Argiope* build a zig-zag white webbing device in the form of a diagonal line or cross: / or X. The function of these devices is probably to reduce the probability of predators finding the spider, but the evidence for this is discussed more fully on pp. 37—40.

A very complex integrated defence system is found in the sandwich tern. This bird nests in colonies which are usually sited in the midst of a much larger colony of black-headed gulls or of arctic terns. The host species (e.g. the black-headed gull) relies for protection of its eggs and young on the principle of safety in numbers. There are so many eggs in a small area that a predator is unlikely to take more than a small proportion of those present. In addition the eggs and chicks are camouflaged, the parent birds remove broken egg shells, and they defaecate away from the colony so that there are no obvious visual clues to the presence of chicks in the area. If a predator such as a crow does enter the colony it is attacked by large numbers of gulls, not just one or two, and consequently it tends to make only brief visits to the colony to search for food and then to get away quickly. The sandwich tern relies for protection of its eggs and chicks on the defences of its host. Since the terns nest in the midst of the host colony the chances are that a marauding crow or fox will find food soon after it enters the host colony, and the host gulls will intimidate and perhaps drive off the predator before it gets near the terns' nests. The eggs of the terns are also liable to be attacked by the host gulls, but the parent terns sit tight on their nests when gulls come near and fend them off with jabs of the bill.

Rand (1967) has pointed out that selection will favour 'aspect diversity', or differentiation of appearance, in two species occupying the same habitat. If the sandwich tern's eggs closely resemble those of the host, then a predatory crow which has a searching image for the eggs of the host may include those of the tern in this image, and hence prey extensively on these as well. However, predators that hunt by searching

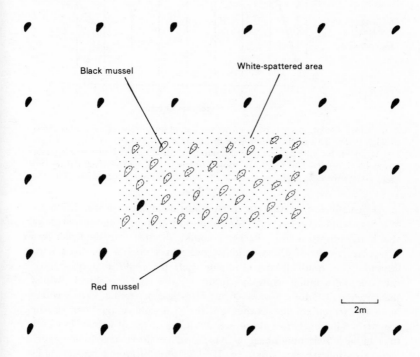

Fig. 12.13 Experiment designed to simulate predation by crows on sandwich terns. Thirty red mussel shells were laid out on shingle to resemble black-headed gull eggs. The central (stippled) area was spattered with white paint to resemble the area occupied by sandwich terns. Two red shells were placed in this spattered area to encourage crows to visit it to some extent. After two days during which only red shells were exposed to predation, black spattered shells were presented in the central area, initially at low frequency, but increasing in number each day. This resembles the egg laying behaviour of sandwich terns. The results of the numbers of shells eaten each day are presented in Fig. 12.14. (*Redrawn with modifications from* Croze, 1970, Fig. 45.)

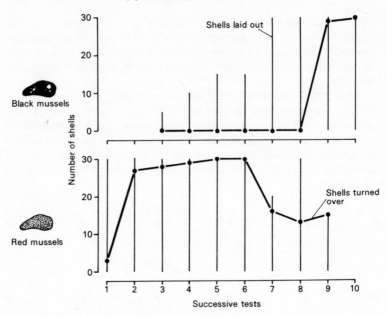

Fig. 12.14   Predation by carrion crows on baited mussel shells laid out as shown in Fig. 12.13. The vertical lines give the numbers of mussels laid out, and the points joined by a heavy line give the numbers turned over. Red mussels (simulating black-headed gull eggs) are shown below, black mussels (simulating tern eggs) are shown above. (*Redrawn with modifications from* Croze, 1970, Fig. 46.)

image usually restrict the search to a particular area (Croze, 1970; see also Smith and Dawkins, 1972). The sandwich tern takes advantage of this behaviour to protect its eggs. Unlike its host, the sandwich tern defaecates freely within its colony and leaves broken egg-shells lying around. The result is that the sandwich tern colony is quite conspicuous, with much white spattering on the sand and shingle, whilst the larger gull colony around it has no white spattering. Crows which hunt by searching image amongst the gull colony rarely cross the boundary into the tern colony because it is a different habitat in which their gull-egg searching image is inappropriate. Croze devised an experiment to simulate this situation. He laid out thirty red mussel shells baited with meat, 4 m apart, on an area of shingle, but with a shell-free area in the centre (Fig. 12.13). This shell-free area was spattered with white paint to resemble the faeces spattered shingle of the sandwich tern colony. The crows quickly learned to find the red mussels and on most days found almost all of them (Fig. 12.14). When the crows had discovered the red mussels, black mussels spattered with white paint and baited with meat were introduced in the central white spattered area. They were introduced at low frequency, but their numbers were increased each day to simulate the egg laying behaviour of the terns.

Previous experience had shown that crows quickly find and build up a searching image for any new food, but in this case it took seven trials before the crows found the black mussels. Next day only black mussels were laid out and all thirty were found by the crows. Hence it appears that the black mussels were protected in the midst of the red ones because the searching image of the predators was not just for 'red', but for 'red on a particular area of clean shingle'. Marking the shingle with white paint prevented the crows from searching there (at least to some extent), and so prevented them from finding the black mussels.

This experiment helps to explain a number of otherwise puzzling features about the sandwich tern. It demonstrates the advantage of modifying the habitat by defaecation and failure to remove egg shells. It further explains why the terns pair up away from the colony and start to lay eggs well after egg-laying in the host species has started — predators are then already conditioned to the appearance of 'eggs on clean shingle' in the area, and they may then overlook the second type of 'eggs on spattered shingle'. Further, both eggs and chicks of the tern resemble guano-spattered stones so are slightly different in appearance to the eggs and chicks of the host. Finally the tern families leave the colony only a week after the eggs hatch thereby allowing a minimum of time for predators to find them and hence to form a searching image for 'tern eggs on spattered shingle' as well as for 'gull eggs on clean shingle'.

Early warning as a defence

For species which are palatable it will be of advantage if they can detect their predators before the predators detect them, and if they can initiate their active defence (flight) before, or as soon as possible after, the predator has noticed them. Here a conflict is apparent: it will be of advantage to the prey to flee before the predator starts to give chase, but only if the predator was about to give chase anyway. If the predator had not actually detected the prey, a precipitate flight might draw attention to the prey rather than facilitate escape. Nevertheless, on balance it is probably an advantage to the prey to be aware of a predator before the predator is aware of the prey: it enables the prey to adjust its behaviour according to that of the predator, rather than to be forced into panic flight which may be the only means of escape if surprised at close quarters.

Insectivorous bats detect and locate prey insects by means of sonar (Griffin, 1958). For an insect that relies for predator detection on sight and vibration receptors, by the time it has detected an approaching bat there is nothing it can do to escape except to make erratic flight movements so that the bat may miss it at the first attempt at interception. But some insects have receptor organs that are sensitive to ultrasound. Many arctiid, noctuid and geometrid moths respond to ultrasound by diving to the ground either passively or by powered flight, whilst others make erratic flight movements. These all make sense in terms of escape behaviour from a bat which is already homing in on the insect (see p. 145 and Fig. 6.2). Some species can apparently detect ultrasound at very low intensity (i.e. when the bat is still very far away), and these respond by turning away and flying fast out of the range of the bat's sonar. Obviously if they can detect the bat before it detects them, the best escape is to move out of the line of flight of the bat (Roeder, 1962, 1965). In addition some arctiids can produce ultrasonic clicks which may possibly interfere with the bat's sonar. If a bat about to capture a projected mealworm is exposed to these clicks, it swerves away. Possibly the clicks are a warning that the arctiid is unpalatable, or perhaps they are a deimatic reaction. Hawkmoths (Sphingidae) lack tympanal organs, but species of *Celerio* (subfamily Choerocampinae) detect ultrasonic sounds by means of a special organ in the pilifer (part of the labrum), and these hawkmoths also show vigorous escape responses when subjected to ultrasonic sounds (Roeder, Treat and Vandeberg, 1968, 1970). The lacewing, *Chrysopa carnea*, also has a tympanum sensitive to ultrasound to which it responds by ceasing flight and so diving to the ground (Miller, 1971).

Dunning (1968) investigated the responses of bats to arctiid moths which produced sounds. In one series of experiments he found that the bat *Myotis lucifugus* withdrew significantly more often to clicking than to silent moths of the genera *Halysidota* and *Haploa*, indicating that the clicks are deimatic warning sounds (Fig. 12.15). Presumably clicks from different species of distasteful arctiids are examples of müllerian mimicry. However, only one out of five moths responded to the presence of a bat by clicking. This low proportion may perhaps be an artefact of the experimental situation in which moths were presented on the floor of a cage rather than flying. Dunning also showed that many species of arctiid and ctenuchid moths produce clicks in response to the sonar pulses of a bat. They can respond when the bat is 2 m distant, far enough for an attacking bat to hear the click and to evade the moth.

The moth *Pyrrharctia isabella* appears to be less distasteful than are species of *Halysidota* or *Haploa*. Dunning found that clicking *Pyrrharctia* caused bats to withdraw but silent ones were more likely to be ignored or even eaten (Fig. 12.15). He suggested that clicks in this species may be mimicking the clicks of the much more distasteful moths such as *Haploa* (batesian mimicry). One of the bats in these experiments learned to distinguish the clicks of *Pyrrharctia* from those of other arctiids and so ate them regularly.

Social mammals also have an early warning defence, although in this case their sense receptors may be no better than those of the predator. With many animals in the group it is not necessary for all to be equally alert to danger all of the time for the group to be able to respond the

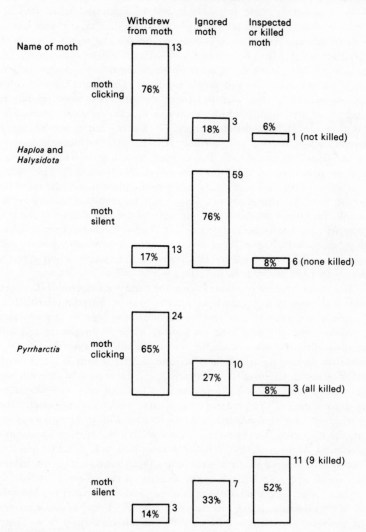

Fig. 12.15 Responses of bats (*Myotis lucifugus*) to silent and to clicking arctiid moths. (*Data from* Dunning, 1968.)

moment one of their number detects a predator. Group vigilance is common amongst small rodents and carnivores and also among larger ungulates and even ostriches. One member of the group may be a sentinel on the alert for danger (e.g. baboons), or each individual may raise its head, alert, from time to time whilst feeding. The moment one animal detects a predator it gives the alarm, either as a call or a movement, and all the group are immediately alerted. Thomson's gazelle only occasionally raise their heads and look around when they are feeding, but the herds are normally so large that at any one instant several animals are almost certain to be on the watch for danger (Walther, 1969). Gazelle and topi rely on sight to detect predators, but waterbuck and elephant rely more on smell. Birds rely on sight and give an alarm call when they detect a predator. This alerts birds of other species as well as their own to danger. Alarm calls in many species of bird are similar: high pitched whistles or squeaks on a single note with no phase or intensity differences and no clear beginning or end, characteristics which make it very difficult for a predator to locate the source of the sound. Mobbing calls, by contrast, have a wide frequency range and are thus easy for other birds to locate (Marlar, 1957). This means that other small birds can easily come to the place to join the mobbing group, but a hunting predator cannot easily locate a bird which gives an alarm call. Ostriches often associate with wildebeest and zebra and it is thought that the combined senses, particularly sight and smell, of all three animals may give them early warning of predators. Giraffe and zebra can also be alerted by tick birds on their backs which perceive danger by sight before their host does (see p. 207).

Having detected a predator before the predator is aware of it, a prey animal is then in a dilemma, as pointed out by Smythe (1970). If it ignores the predator, this will be successful so long as the predator continues past the prey without sensing it, or if the predator is not hunting. But the prey animal must watch all the time in case the predator decides to attack, and this means that it cannot get on with other essential activities. Alternatively it can move quietly away till out of sensory contact with the predator – but this may be even more dangerous since if the predator changes its direction of movement, the prey animal may actually blunder into contact with it. The prey animal can simply watch the predator, or even walk towards it, as Thomson's gazelle do (Walther, 1969), but this wastes time that could be spent in feeding. Mobbing similarly wastes time although it does draw the attention of all the prey animals in the vicinity to the whereabouts of the predator. The prey animal can also freeze and remain cryptic, but this too wastes time; or, it can run away quickly, which may actually attract the predator's attention to a prey it had not previously seen. Many animals in fact appear to do what at first sight seems the most dangerous thing: they approach the predator and then turn to flee, but they flee rather slowly with much apparently unnecessary movement, and exposing conspicuous colour patches. Gazelle 'stot' or run in a jerky, jumping way, exposing the white rump patch (Fig. 12.16). Rabbits run exposing the white tail. It seems likely that it will benefit the prey animal to test the intentions of the predator as quickly as possible. In this way the prey is either chased at once, or it is able to return to feeding, instead of wasting time in watching the predator.

Fig. 12.16   Stotting gait of a Thomson's gazelle, adopted when approached by a predator. (*Redrawn from* Walther, 1969.)

Hence Smythe thinks that fast running prey may goad the predator into attacking them under conditions most favourable to their own escape. As soon as the predator gives up the chase, the prey animal can continue to feed and the predator will be likely to turn to chase some other, less alert, individual. Gazelle stot just before and directly after fleeing at full speed, at just the times when one would predict stotting should occur if it is designed to attract a predator to give chase. If the predator can be induced to give chase when so far from the prey that there is little chance of it catching up, the defence will have worked. The final stot at the end of the flight is to test if the predator is really giving up or is about to give chase again — rather like the grayling butterfly exposing its eyespots, in case a predator is around, to direct any attacks where little harm can be done. Stotting is also contagious, so it may also distract the predator from the animal it was initially chasing and so confuse it (Walther, 1969).

Observations on a South American rodent, the mara (*Dolichotis patagonicum*), support Smythe's theory. When approached directly by a predator (man), the mara gallops directly away. When approached at an angle, it freezes — the man may then pass by without noticing it. But when approached at an angle so that the predator may pass within about 20 m, the mara stots away, stops, watches, and then stots again, apparently inducing the predator to give chase.

Some other animals may also goad predators into chasing them, and Young (1971) has suggested that the brilliant blue colours and bobbing flight of *Morpho* butterflies may induce pursuit in this way. *Morpho amathonte* is a very fast flier and it can probably evade most birds that give chase. It is possible that birds that have chased several butterflies unsuccessfully may learn not to pursue insects of that particular colour pattern and behaviour. This hypothesis is not easy to prove, but in one area Young found that 80 per cent of less brilliant species of *Morpho* had beak marks on the wings (fifty-two out of sixty-three *M. granadensis* and fifteen out of nineteen *M. peleides*) indicating that they had been captured by a bird but had escaped, whilst nought out of thirty-one of the brilliant *M. amathonte* had beak marks. In addition, fragments of the wings of the first two species were found under places where birds roosted, but no traces were found of *M. amathonte*. Probably such butterflies might be better protected if they were cryptic, but bright colours are important in courtship, and the conflicting selection pressures of sexual selection and predator selection may lead to different results in quite closely related species. Lindroth (1972) has drawn attention to a similar phenomenon in flea beetles (Chrysomelidae). Some of these are brightly coloured but are not aposematic nor apparently mimetic. However, when they are approached, they jump so quickly that a bird is likely to be baffled and unable to follow them. It is possible that birds learn, after a few unsuccessful attacks, that they cannot associate insects of this colour with food, and so they avoid them. There are even some carabid beetles (*Lebia* spp.) which mimic the bright colours of flea beetles (*Disonycha* spp.) and live closely associated with them (Fig. 12.17). Presumably these mimics derive protection because birds which have learned that it is futile to attack *Disonycha* will also refrain from attacking them.

Another type of early warning defence occurs in many species of shoaling fish. If a predator snaps up one fish, the doomed fish releases a warning chemical ('alarm substance' or 'alarm pheromone') into the water which induces a fright response in the remaining fish. Although one fish is killed, the response is of obvious survival value to the remaining fish in the shoal so that the predator is likely to capture only this one individual. Most fish which show a fright reaction of this type have special cells in the epidermis which are responsible for the secretion of the alarm substance (Pfeiffer, 1962; Pfeiffer and Lemke, 1973; Reutter and Pfeiffer, 1973). They occur in several families of ostariophysid fish especially shoaling cyprinids, but also in Characidae and Siluroidea. The fright reaction of minnows (*Phoxinus laevis*) involves rapid swimming in all directions away from the stimulus source, but the fright reaction of *Gobio fluviatilis* is to rest motionless and cryptic on the bottom of the aquarium. Verheijen and Reuter (1969) point out that alarm substance could alarm conspecific members of a shoal against predation by a different species of fish, or it could alarm them against cannibalism by conspecifics. They found that when large roach (*Leuciscus rutilus*) or minnows preyed on small roach there was no alarm reaction, but when pike preyed on small roach there was an alarm reaction. They further

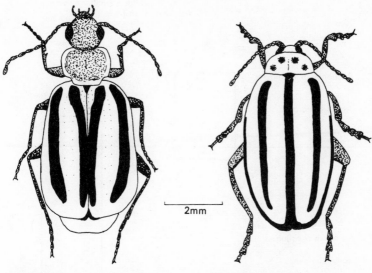

Lebia vittata *Disonycha alternata*

Fig. 12.17 The carabid beetle *Lebia vittata* which mimics the chrysomelid flea beetle *Disonycha alternata*. Both are mostly black and yellow with red-brown head and thorax in *Lebia*, and red-brown legs in *Disonycha*. (Lebia *redrawn from* Lindroth, 1971; Disonycha *original*.)

placed blinded minnows in a tank and exposed them to water from a tank in which a pike had eaten a small roach, or to water from a tank in which a minnow had eaten a small roach. They found that alarm reactions were given significantly more frequently to the pike water than to the minnow water (Fig. 12.18). Since inexperienced minnows show no alarm reaction to pike odour, this result is, at first sight, difficult to explain. However, it was found that when a pike snaps a small roach it breaks the skin with its sharp teeth, and this probably releases alarm substance. But when a minnow catches a small roach it swallows it intact with little or no damage to the skin, and so no alarm substance is released. Therefore it is clear that alarm substance is a defence against predation by other species of fish.

An alarm reaction similar to that of minnows occurs in tadpoles of the toad *Bufo bufo*, whilst the snail *Helisoma nigricans* buries into the mud when exposed to crushed tissue of a conspecific. Juice from an injured sea urchin (*Diadema antillarum*) also causes an alarm reaction in other urchins — they rise on their ventral spines and crawl rapidly downstream. They show no reaction to crushed tissues of some other sea urchins although there is a strong response to *Lytechinus variegatus*. This is probably a defence against the predaceous gastropod *Cassis* which feeds on sea urchins; *Diadema* flees on contact with *Cassis*, but

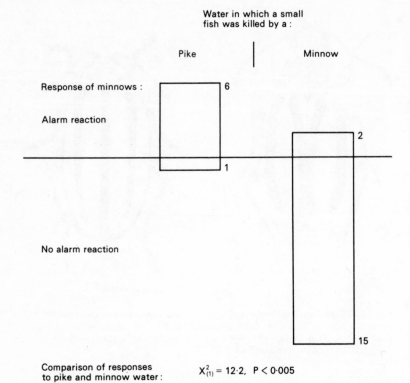

Fig. 12.18 Alarm reaction of blinded minnows to water in which a pike has killed a small fish and to water in which a minnow has killed a small fish. (*Data from* Verheijen and Reuter, 1969.)

shows no response to the harmless gastropod *Fasciolaria* (Snyder and Snyder, 1970).

Alarm pheromones occur widely in ants and they may be released either when an ant is injured or killed, or by voluntary action at certain times. The alarm pheromones are secreted by the mandibular gland, by Dufour's gland at the tip of the abdomen, and also by the poison gland at the tip of the abdomen. A variety of chemicals are produced by any one species of ant and their functions are very complex so that Wilson and Regnier (1971) propose that the entire complex of alarm pheromones be called the alarm-defence system. Alarm pheromones in ants are highly volatile terpenes, alkanes and ketones. In the Formicinae, both Dufour's gland and the mandibular gland may be very large. The response to alarm pheromones varies in different species and depends on the size of the colony. Species which have very small or diffuse colonies respond by 'panic alarm' behaviour which involves scattering and removing some of the brood to safety; whilst species with larger, more compact colonies respond by 'aggressive alarm' behaviour in which they actively attack the aggressor. Primitive formicine ants have small mandibular and Dufour's glands, and most of them respond to alarm pheromones by panic alarm. From these primitive formicines have evolved ants such as *Formica sanguinea* and *F. subintegra* with small mandibular glands and large Dufour's glands, and with large colonies and aggressive alarm behaviour. *F. subintegra* is one of the slave-raider ants which sprays the colony it is attacking with alarm pheromones. This causes fellow raiders to attack, but the defending colony (*Formica subsericea*) responds by panic alarm. The result is that there is much less mortality than would occur if the defending ants fought back. The raiders succeed in removing many of the defender's young, but at the same time the defending colony is not entirely wiped out and may save some of its brood (Regnier and Wilson, 1971).

Another group of formicine ants have retained small Dufour's glands but most of them have well-developed mandibular glands, and these also form large colonies with aggressive alarm behaviour. *Acanthomyops* is one such genus which, because it tends root-living Homoptera, cannot abandon its colony when attacked without losing its main source of food. Hence it forms large colonies which stand firm and fight off aggressors.

Some stingless bees also have alarm pheromones of defensive value, and these also appear to cause panic alarm in some species and aggressive alarm in others. *Lestrimelitta limao* is a meliponine bee which robs food from the nests of stingless bees (*Trigona testaceicornis, Melipona quadrifasciata*, and other species). When a scout *Lestrimelitta* finds a suitable colony for raiding it may be killed by the guard bees and this injury causes release of an alarm pheromone called citral from its mandibular gland. Citral causes the defending bees to panic and attempt to escape whilst it attracts further *Lestrimelitta* bees to the attack. There are, however, some species of *Trigona* in which citral releases aggressive alarm behaviour, but these are not normally attacked by *Lestrimelitta* (Blum, 1970; Blum *et al.*, 1970). Thus, as with the slave-raiding ants, the alarm pheromone in stingless bees has a variety of functions related both to defence and attack.

Some non-social insects also produce alarm pheromones. For example, when aphids are attacked they release droplets of fluid from abdominal processes called cornicles. This fluid contains a chemical whose odour induces neighbouring aphids to move away or even to drop off the substrate (Bowers *et al.*, 1972; Kislow and Edwards, 1972). The pheromone thus gives warning of a predator to neighbouring insects even though the aphid attacked is likely to be killed.

Conflict of defensive systems with other essential systems

As predators become more adept at capturing prey, selection will favour any prey animals that are better protected than their fellows, and so will lead to highly efficient protective mechanisms. But an animal may not be able to evolve as perfect a defensive system as is theoretically possible because this may conflict with other essential activities. In Section 1 I pointed out that anachoretic and cryptic animals cannot remain in the best protected place and posture all the time since they have to move around in order to feed and reproduce. Many such animals are only active at night when visually hunting predators are less numerous and less able to locate them than during the day. Animals that are active during the day face the problem of communicating with conspecifics without making themselves conspicuous to predators. Crickets, grasshoppers and cicadas are normally cryptic to visually hunting predators, but advertise their presence to conspecifics by song. Many passerine birds are also cryptic and they communicate with conspecifics by song. The reed warbler (*Acrocephalus scirpaceus*) lives in dense masses of reeds whilst the sedge warbler (*A. schoeno-baenus*) lives in more open habitats where it is more vulnerable to visually hunting predators. In the reed warbler song is used to demarcate territory, attract females, and to perform other social functions, but the impenetrable habitat prevents potential predators which hear the song from finding the bird. In the sedge warbler, however, song is only used to attract females. In the more open habitat of this species song probably results in increased predation, so males sing much less often, and the song performs only one function (Catchpole, 1973). Other birds (such as the chaffinch) and some reef fish communicate with conspecifics by the judicious display of normally concealed colours. The reef fish *Chaetodon lunula* is normally counter shaded and cryptic, but when two individuals are cleaning each other or being aggressive, they change colour so as to enhance colour contrast. At the end of the encounter the contrasting colours fade and the fish become cryptic again (Hamilton and Peterman, 1971). Thus animals can communicate with conspecifics but avoid being detected by predators either by restricting communication to brief but well-defined times, or by communicating in a different sensory modality to that used by the hunting predator.

Robinson (1973) has evidence suggesting that marmosets (*Saguinus geoffroyi*) use the legs of insects as a visual clue in the detection of prey. He found that the marmosets ignore plain sticks but attack sticks with insect legs attached to them. Furthermore, they could find dead stick insects with the legs projecting but they could not find dead stick insects with the legs fixed in a cryptic, life-like position. This suggests that selection should favour grasshoppers that have small legs. But this conflicts with the secondary defences of grasshoppers which require large and powerful hindlegs for kicking with the tibial spines, jumping, and flying (which starts as a jump to give rapid initial acceleration away from the predator). The hindlegs can be folded to some extent, but they are still rather large and obvious. Hence they are often coloured with disruptive patterns (Fig. 6.3). In the most perfectly concealed grass mimics the hindlegs are reduced in size and although they can still function for jumping, the jump is much less powerful and less effective than in other grasshoppers (Fig. 4.22).

Stick-mimicking insects face similar conflicting selection pressures (Robinson, 1969a; Edmunds, 1972). If they have long legs it enables them to rock gently from side to side or to 'teeter' backwards and forwards. In this way they can move slowly but mask the movement since the rocking or teetering resembles a twig swaying in the wind. This is of advantage in defence (in phasmids, long-legged bugs, and spiders such as *Pholcus*) and also in attack when stalking prey (in mantids such as *Danuria*). Rocking may also enable a mantid with a narrow head and eyes close together to get two bearings on a prey insect and hence to increase the probability of a strike being successful. But long legs are conspicuous, and so the conflicting selection pressures produce a compromise solution depending on the relative intensities of these pressures in each species. The mantids *Pyrgomantis* and *Catasigerpes* have short legs and do not teeter or rock, whilst *Danuria* has long legs and does. Long legs also give the mantid a greater strike distance, and although conspicuous, legs can be concealed to some extent by frills or disruptive colour marks as in *Phyllocrania* and *Pseudocreobotra*.

Mantids may also be recognized by their predators by the conspicuous head and foreleg profile in typical praying posture. This is concealed in stick-mimics by protraction of the forelegs (*Danuria, Hoplocorypha* and *Angela*) (Fig. S1.1), or by extending them laterally like two side branches of a twig (*Stenovates*) (Fig. 4.21). But in neither of these positions is the mantid in a suitable position for prey-capture, so normally it rests with the forelegs partially flexed, and only adopts the fully cryptic posture when it is disturbed. This requirement to feed places restrictions on the perfection of the stick-mimicry in mantids, but it does not apply to the same extent to phasmids since these insects do not use the forelegs in obtaining food, and many of them feed mainly by night. The phasmid *Pterinoxylus spinulosus* has a very bizarre resting posture with the body attenuated like a stick, the forelegs continuing the body profile and touching a branch, and the second and third legs flexed to resemble broken twigs (Fig. 7.4). Leaf mimics have equally bizarre resting postures in which they are totally unable to perform any other activity at all, for example the tettigonid *Mimetica mortuifolia* (Fig. 12.19), but leaf-mimicking mantids often rest in a posture which is suitable for prey-capture. Flattened bark mimics are also maladapted to feeding when in the best concealed position since the head and mouthparts must be either angled forwards (as in *Theopompella*) or reflexed backwards (as in *Tarachodes*) to give efficient contour concealment. When such mantids feed, the head is adjusted so

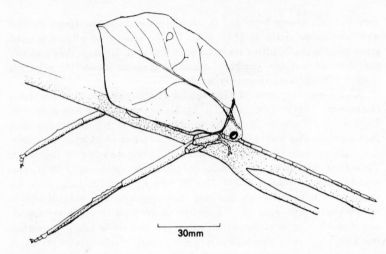

Fig. 12.19   Resting position of the leaf-mimicking grasshopper *Mimetica mortui-folia* with twig-like hindlegs projecting below it. (*Redrawn from* Robinson, 1969*b*, Fig. 10.)

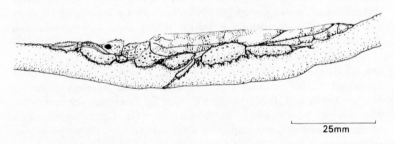

Fig. 12.20   Lichen-mimicking phasmid *Prisopus berosus* resting on a branch. (*Redrawn from* Robinson, 1969*b*, Fig. 16.)

that the mouthparts are at right angles to the body and temporarily the insect is not in the best concealed position. Lichen-mimicking phasmids and tettigonids remain in the most cryptic position all day when not feeding (Fig. 12.20).

There may also be a conflict between predation pressure for one defensive mechanism and predation pressure for another defensive mechanism. *Cleora repandata* is a moth that normally rests during the day on tree trunks in the British Isles. In industrial areas there is a melanic form *nigra* which has become very abundant, presumably because it is the better camouflaged morph in such areas. But in northern Scotland there is another melanic morph controlled by another gene, the *negrina* form, and this may comprise 10 per cent of the population in areas remote from industry and pollution. Kettlewell has evidence to suggest that the normal form is the better protected on pine tree trunks because it is cryptic, but the *negrina* form is better protected when the insects fly during the day because it is very difficult to follow in flight in the dark gloom of the pine forest. Since ants regularly disturb resting moths there is a high probability that a moth will have to fly at some time during the day and therefore be exposed to predation by birds which hawk after flying insects. The frequency of the *negrina* form is thus the result of selection pressures related to degree of disturbance of resting moths and frequency of hawking birds in the area (Ford, 1964).

The situation appears to be even more complex in the moths *Amathes glareosa* and *Lasiocampa quercus* in which some populations in northern Scotland also have high frequencies of melanics (Kettlewell *et al.*, 1969, 1971, and earlier). The melanic form of *A. glareosa* is better protected on peaty soil than is the typical form, and the melanic oak eggar (*L. quercus*) is less heavily preyed on by gulls than is the typical form when resting on heather. However, since the typicals persist in the populations they must have some as yet unknown advantage to compensate for this selective predation which results in a balanced polymorphism.

The colour polymorphism of spittlebugs (*Philaenus spumarius*) can also perhaps be explained in terms of conflicting selection pressures. Most colour forms are cryptic and their polymorphism may simply be due to predators learning the colour patterns of the commoner morphs so that rarer ones come to have a selective advantage relative to them (apostatic selection). But the form *marginella* is black and cream and very conspicuous. Thompson (1973) suggests that, as with Lindroth's flea beetles, some predators learn to avoid the spittlebug because its agility at jumping means that they come to associate the colour pattern with lack of reward. Furthermore, *marginella* is restricted to the long-lived females whilst the cryptic males live only a few days, and it is possible that the dark colour enables *marginella* to warm up quicker than do other forms so that it has a much more rapid escape movement.

A similar conflict between the requirements of two defensive mechanisms is probably quite common since perfection of some form of crypsis is likely to reduce the efficiency of active escape movements and to make deimatic displays less conspicuous and intimidating than they might otherwise be. In addition it may be of value to be cryptic with respect to one predator but conspicuous with respect to another. *Danaus chrysippus* caterpillars have narrow black, white and yellow transverse bands on the body. There are also three pairs of horns which are black with red bases. The last instar caterpillar is very conspicuous to the human eye and one could argue that its pattern is aposematic. But sometimes it feeds on non-poisonous food plants and so is actually palatable, and during the first instars it is unlikely to have accumulated sufficient cardenolides to be emetic when eaten by a bird. Because the black, white and yellow bands are very narrow, particularly in the earlier instars, the small caterpillars are actually not easy to see, and the bands give it cryptic colouring. The horns too are rudimentary and not red at this stage. Even the later, larger caterpillars are not in fact conspicuous from a distance; they often rest underneath leaves and from several metres distance the black and white bands cannot be resolved by a vertebrate eye. In other words they are probably cryptic from a distance, but the later instars are aposematic when seen close to. The black and white stripes of zebras may similarly be cryptic from a distance, particularly in hazy conditions or at night-time. When a predator is closer, the stripes are conspicuous (though not aposematic), and they may induce it to give chase too soon to ensure capture — though this is speculation.

Predator selection for crypsis may also conflict with reproductive selection as in the three-spined stickleback. At Lake Wapato in North-West America the male stickleback is dimorphic in coloration, its throat being either red or black (Semler, 1971). There are more red-throated males in deep water than in shallow water (Fig. 12.21), and in addition the proportion of reds declines from 12.8 per cent in mid May to 5.3 per cent by the end of June. When trout are given pairs of stickleback prey, one with a painted red throat, the other with a painted dull grey throat, Moodie (1972) found that they take significantly more of the red-throated fish (forty-two red against eighteen dull; $\chi^2_{(1)} = 15.29$, $p < 0.005$ for one series of experiments). Presumably predatory fish take more of the red than of the dark-throated sticklebacks in Lake Wapato too. This would account for the seasonal decline in the proportion of red males captured and also for the higher frequency of reds in deeper water than in shallow water since they are much less conspicuous in the darker deeper places. The factor which enables red males to survive at all is that given a choice of mates, a female prefers a male with a red throat to one with a black throat (Fig. 12.22). Hence sexual selection favours the red males whilst predator selection favours the black males. The result depends on the intensities of the two selection pressures, and this varies in different lakes and at different places in the same lake. At Mayer Lake the result of predation pressure is that only 14 per cent of the fish have red throats, and all have larger spines and larger body size than do sticklebacks from nearby streams where predation is less intense (Moodie, 1972).

In the Conner Creek river system (North-West America), black-throated males comprise 90—100 per cent of the population except at the mouth of the river where 20 per cent are red-throated and where there are also many hybrids. This distribution correlates with that of the predatory mudminnow, *Novumbra hubbsi*, which occurs throughout the river system except at the mouth. Mudminnows, however, prey only on young sticklebacks, not on adults, and the young of both red and black-throated males are identical in coloration. McPhail (1969)

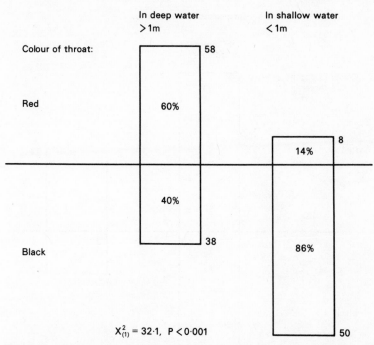

Fig. 12.21 Frequencies of red- and black-throated male sticklebacks (*Gasterosteus aculeatus*) in deep and shallow water in Lake Wapato, Washington State, U.S.A. (*Data from* Semler, 1971.)

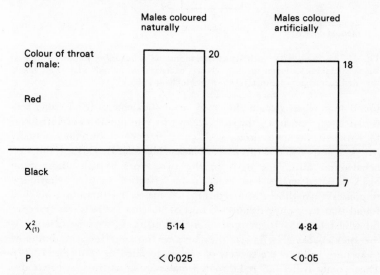

Fig. 12.22 Choice of red- or black-throated males by female sticklebacks. (*Data from* Semler, 1971.)

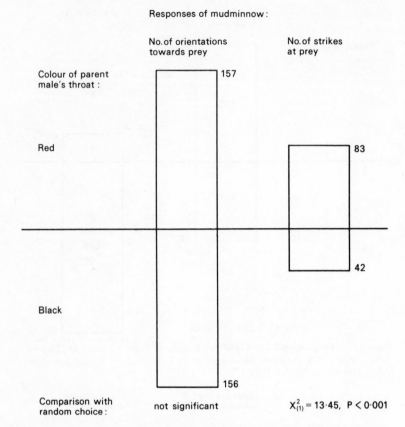

Fig. 12.23   Frequency with which mudminnows (*Novumbra hubbsi*) struck at larval sticklebacks when given a choice of larvae from red- and from black-throated parent populations. (*Data from* McPhail, 1969.)

found that when given a choice of larval sticklebacks from either red-throated or from black-throated males, the mudminnows oriented as if to strike at the two forms with equal frequency, but they actually struck at reds twice as often as at blacks (Fig. 12.23). This was due to a behavioural difference between the two types of fish: those from black-throated males had much more effective evasive behavioural responses than those from red-throated males. Furthermore, McPhail found that for young fish 50 per cent of all strikes at reds were success-ful whilst only 5–10 per cent of strikes at blacks were successful. As the sticklebacks grew in size they became too large for a mudminnow to swallow, and so the success of strikes declined to zero (Fig. 12.24). Thus young fish from red-throated males are at a strong selective dis-advantage compared with young from black-throated males. There is also evidence that mudminnows are attracted to red adult males rather

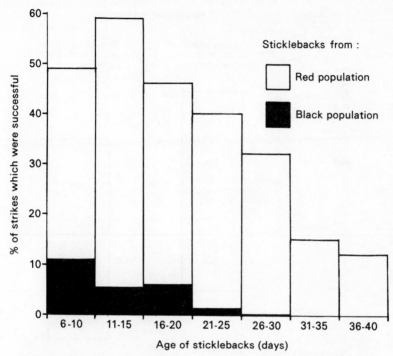

Fig. 12.24 Percentage of strikes by mudminnows (*Novumbra hubbsi*) on larval sticklebacks of different ages from red- and from black-throated parent populations. (*Data from* McPhail, 1969, Fig. 9.)

than to black males (Fig. 12.25), so they are likely to be well placed to prey on young fish as soon as these leave the nest. The result of this powerful predator selection pressure is that male sticklebacks are almost always black wherever mudminnows occur, but since female sticklebacks prefer to mate with red rather than with black-throated males, the reds persist elsewhere.

The general conclusion from these and other studies on sticklebacks (see also Hagen and Gilbertson, 1972) is that the colour of the males and the morphology of both sexes (e.g. size, spine length, number of lateral plates) vary in different populations according to the different selection pressures acting in each area, but in most populations one of the most important selection pressures is that exerted by predators.

Conflict between predator selection for crypsis and reproductive selection for brightly coloured males is probably widespread in vertebrates. In the West African lizard *Agama agama*, the male is grey with an orange head and orange and black tail, whilst the female is cryptically brown with orange on the flanks. A common predator is the shikra hawk, *Accipiter badius*, and one might guess that male agamids are much more easily seen and caught than females. The males are very

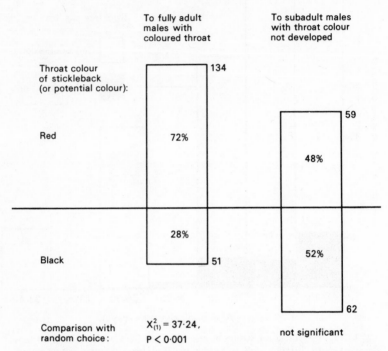

Fig. 12.25 Frequency with which mudminnows (*Novumbra hubbsi*) associate with male sticklebacks when given a choice between red- and black-throated fish. (*Data from* McPhail, 1969.)

alert to danger except when engaging in territorial combats with rivals. Such combats, with each lizard manoeuvring so as to slash its opponent with its tail, may last for several hours, and the combatants are thus very vulnerable to predation. However, during combats, but not at other times, the orange colour on the head of the male is replaced by a mottled brown appearance so that the animal becomes cryptic instead of conspicuous.

Predator–prey systems and ecology

The reciprocal effects of selection of prey for improved anti-predator defences and of predators for improved strategies for capturing prey should not be thought to imply that predators inevitably control the numbers of prey animals. In the butterfly *Danaus chrysippus* (and probably in many other species too) the population is almost certainly regulated by parasites — in Sierra Leone by tachinids (Owen, 1971), but in southern Ghana rather more by the hymenopteran *Apanteles*.

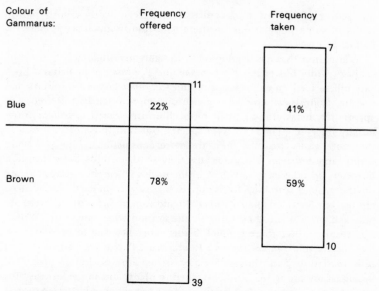

Colour of Gammarus:

Frequency offered

Frequency taken

Blue 22% 41%

Brown 78% 59%

Comparison of numbers taken with numbers surviving: $X^2_{(1)} = 3.96$, $P < 0.05$

Fig. 12.26 Frequencies of blue and brown *Gammarus lacustris* taken by a duckling. (*Data from* Hindsbro, 1972.)

Danaus larvae that are parasitized by *Apanteles* develop a paler and less conspicuous coloration than do normal caterpillars. One would expect it to be of advantage to the species if caterpillars were rendered more conspicuous as a result of being parasitized, since then they would be more likely to be found by a predator and hence fewer parasites would survive. On the other hand, selection would also favour those parasites which alter the host's colour the least since then more parasites would escape predation by birds. It is known that parasites can alter the colour of their host to their advantage: for example, the amphipod *Gammarus lacustris* is often parasitized by cystacanths of the acantho-cephalan *Polymorphus*, and about half of all parasitized *Gammarus* become blue instead of brown (Hindsbro, 1972). The blue *Gammarus* are more positively phototropic than the brown, and since they are also more conspicuous they are more likely to be seen and caught by surface feeding ducks (Fig. 12.26). In this case the parasite benefits from the heavier predation on blue *Gammarus* since ducks are the final host for the *Polymorphus* cystacanth. Since *Polymorphus* can alter the colour of its host to its advantage, it is reasonable to suppose that *Apanteles* may be able to cause a similar change in the coloration of its host *Danaus chrysippus*. However, in *Danaus* the situation is further complicated by the fact that the caterpillars may sometimes be emetic to birds, depend-ing on their food plant and on the bird's tolerance of cardenolides (see pp. 104–11), and by the fact that a rare, green highly cryptic morph exists in some populations, so there is obviously much scope here for further research.

A situation that has been more thoroughly investigated is that of the cabbage white butterflies *Pieris rapae* and *P. brassicae* in Britain. Both caterpillars feed on cabbages, but whilst *P. rapae* larvae are solitary and cryptic, *P. brassicae* larvae are gregarious, more brightly coloured, and apparently distasteful to small birds. Blue tits (*Parus coeruleus*) prey heavily on young larvae of both species, but find those of *P. brassicae* the more easily because they are more conspicuous. Blue tits avoid eating large *brassicae* larvae because they are distasteful. Song thrushes, however, do take large larvae of both species when the caterpillars are on the ground searching for a pupation site, and on earth *P. brassicae* are better concealed than *P. rapae*. In addition, if *P. brassicae* larvae are attacked on a plant they drop to the ground where they are cryptic. The result is that there is much higher mortality due to predation on late instars of *P. rapae* than of *P. brassicae*. Pupae are also subject to bird predation and those of *P. brassicae* are the better protected because they are more firmly attached to plants and so are less easy for small birds to remove. But it is not at all certain that bird predation is important in controlling the numbers of either species of butterfly. About 12 per cent of *P. rapae* and 80 per cent of *P. brassicae* larvae are parasitized by *Apanteles* spp. This percentage difference could be due to the gregariousness of *P. brassicae* attracting a higher incidence of parasitization than the solitary, scattered individuals of *P. rapae* which are less easy for a parasite to find and are therefore better protected, but Baker (1970) has proposed an alternative explanation. He suggests that *Apanteles* has a definite preference for *brassicae*: because of the higher survival of *brassicae* larvae with respect to bird predation, selection favours those parasites that choose to lay their eggs on this species. A complication is that there are two species of *Apanteles*, *A. rubecula* which leaves the caterpillar at the fourth instar, and *A. glomeratus* which leaves just before pupation. Since *P. rapae* suffers less predation from birds during the early instars, *A. rubecula* is probably adapted to parasitizing this species. *A. glomeratus* is probably adapted to parasitizing *brassicae* which survives better during the last instars. To attempt to sort out the factors which actually control the population of either insect in this complex situation is very difficult. However, it is clear that there is a series of anti-predator adaptations involving coloration (cryptic or aposematic, or even both at the same time to different predators), and behaviour (gregariousness), and there are also strategies on the part of the parasite for coping with the situation (different emergence times related to bird predation and perhaps selection of *brassicae* as a host because it survives better than *rapae*).

Equally complex anti-predator adaptations have no doubt evolved in other species too, but where these are behavioural or ecological they are not so easy to discover. Morton (1971) points out that the clay-coloured robin (*Turdus grayi*) breeds towards the end of the dry season in Central America when there is much less food available for nestlings and fledglings than there is later on in the rains. Weights of nestlings and fledglings were lower at the peak breeding season than they were later on. But he found that nest predation was much heavier in the rainy season than earlier, so that early breeders have a 42 per cent chance of fledging young whilst later breeders have only a 15 per cent chance of fledging their young. Hence predation pressure is more important than food supply in determining when the birds breed, and early breeding in this species can be regarded as an anti-predator adaptation.

A predator–prey interaction in a freshwater ecosystem has been described by Zaret (1972). Cladocera often have projections on the carapace which may be of protective value against predation. In Gatun Lake, Panama, *Ceriodaphnia cornuta* may be horned or hornless. Each form is parthenogenetic and normally breeds true. It was found that in open water 89–98 per cent of the population is hornless, but close inshore horned morphs are more numerous so that 48–86 per cent of the population is hornless. Zaret found that the atherinid fish *Melaniris chagresi* prey on *Ceriodaphnia*, and in laboratory trials it was found to take more hornless than horned animals. He further compared the frequency of horned *Ceriodaphnia* in the stomachs of fish caught in the

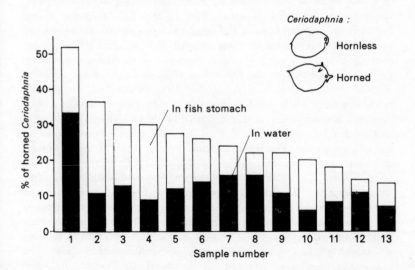

Fig. 12.27 Frequencies of horned *Ceriodaphnia cornuta* in thirteen paired samples from Gatun Lake, Panama. Samples from stomachs of the fish *Melaniris chagresi* are shown in white, samples from the lake near the fish are in black. (*Data from* Zaret, 1972.)

lake with that of cladocerans in the surrounding water: in every case there was a higher frequency of horned in the water sample than in the fish stomach, indicating that the fish were preying selectively on horn-less *Ceriodaphnia* (Fig. 12.27). The reason why hornless individuals persist in the population is that they have a higher fecundity producing on average 9.1 eggs in their lifetime compared with only 4.3 eggs for horned ones. In other species of Cladocera temperature and viscosity of the water are known to affect the frequency of horned and hornless morphs, but there are a few other species known in which the horns or spines are probably anti-predatory in function. Rather similar predator—prey interactions occur in sticklebacks and their predators (see pp. 264—7) and in the rotifers *Brachionus* and *Asplanchna* (see pp. 185—6), but the details of the interaction and of the inheritance of the prey species' spines or horns differ.

The two most comprehensive studies on predation in vertebrates are by Mech and others on the wolf (summarized in Mech, 1970) and by Kruuk, Schaller and their colleagues on the Serengeti predators (summarized by Kruuk, 1972, and Schaller, 1972). The situation is simplest in the case of the wolf since it is the only important predator in the ecosystem, and in many areas there is often only one or two species of principal prey. On Isle Royale, moose are the principal prey of the wolf, and there are numerical data on predation of adult moose which give information on the effectiveness of anti-predator defences. In a long series of observations by Mech, 131 moose were detected by wolves, most of them by scent (Fig. 12.28). Eleven of the moose detected the wolves at about the same time and moved out of sensory contact with the wolves, so these escaped because of an efficient early warning defence. A further twenty-four moose stood their ground as the wolves approached, and the wolves soon gave up and moved away without actually attacking them. So these twenty-four successfully intimidated the predators. Whether this is a genuine warning or a bluff display is a moot point, a moose can certainly kick viciously at wolves with both fore and hindfeet, but on the other hand wolves can and do kill and eat moose. The remaining ninety-six of the 131 moose turned and fled: forty-three of these escaped because of superior speed and stamina since the wolves gave up the chase, but the other fifty-three were caught up. Twelve of these turned at bay and threatened the wolves who soon gave up and wandered away, whilst the remaining forty-one moose continued running with the wolves running alongside them. Even then, thirty-four moose outran the wolves — a running moose can still kick viciously and so the wolves hesitated to attack. The remaining seven were attacked on the rump, flanks and nose, and though they retaliated by kicking and trampling, six were killed whilst the seventh escaped seriously wounded (probably it died or was killed by the wolves a few days later). Thus the success rate of hunts in which wolves actually confronted or gave chase to moose is only six or seven

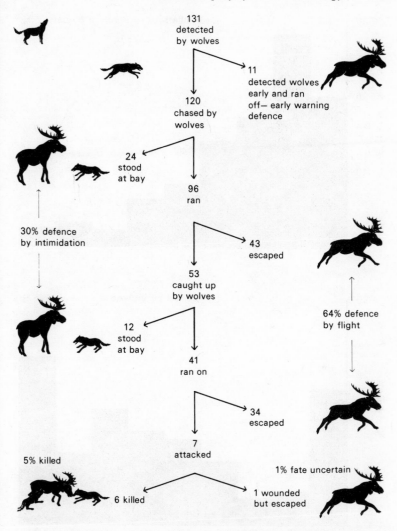

131
detected
by wolves

11
detected wolves
early and ran
off— early warning
defence

120
chased by
wolves

24
stood
at bay

96
ran

30% defence
by intimidation

43
escaped

53
caught up
by wolves

12
stood
at bay

41
ran on

64% defence
by flight

34
escaped

7
attacked

5% killed

1% fate uncertain

6 killed

1 wounded
but escaped

Fig. 12.28   Fate of 131 moose detected by wolves on Isle Royale, U.S.A. For further explanation see text. (*Data from* Mech, 1970.)

out of 120 (5—6 per cent). Sixty-four per cent of the moose escaped by flight, and 30 per cent successfully defended themselves. Wolves also attack moose calves, but unless there are several wolves attacking, the mother often fends off the attack successfully.

Wapiti, bison and musk ox can also retaliate successfully against wolves using their hooves or horns. The defensive ring of musk-ox is probably specially adapted to warding off wolves: such a herd flees from man, but not from dogs, which are domesticated wolves. Other

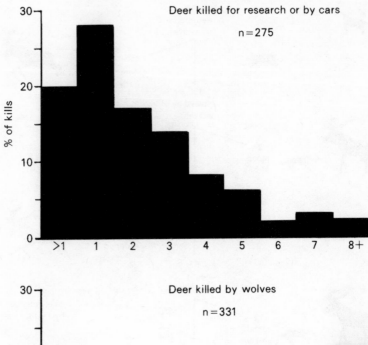

Fig. 12.29 Frequencies of different age classes amongst white-tailed deer killed in Ontario, Canada. (*Data from* Mech, 1970.)

prey of wolves have no retaliatory defence and must rely on speed and stamina to escape once the pursuit has started. Of nine chases by wolves on white-tailed deer followed by Mech, all of the prey escaped, but sixteen of thirty-five chases tracked in the snow by Kolenosky (1972) were successful (46 per cent). The difference in success may be related partly to different tracking methods: by following snow tracks some chases which were short and in which the prey escaped could easily have been overlooked. This criticism also applies to other hunting success rates of North American carnivores deduced from snow tracks and summarized by Schaller (1972). White-tailed and mule deer both escape by speed, but dall sheep escape by climbing up rocky cliffs. Caribou (or reindeer) herds are often chased by wolves as if 'tested'. Wolves apparently set a herd running in order to pick out any individual which lags behind the rest, and then they concentrate on this animal. This testing of prey implies that wolves prey more on weaker, diseased animals than on healthy ones, and evidence from wolf kills supports this view. On Isle Royale, only a quarter of the adult moose population (20—35 per cent) are more than eight years old whereas nearly all adult moose killed by wolves (91 per cent) were found to be more than eight years old. Wolves also take many calves — in a sample of eighty moose killed by wolves, 94 per cent were either calves or adults over eight years old. In addition many of the moose killed by wolves had diseased jaws, hydatid cysts or other evidence of being unhealthy such as low bone-marrow fat content. In the case of deer, many more of those killed by wolves were old (four to eight years) than occurred in the population as judged by a sample shot by man or killed by motor cars (Fig. 12.29). The conclusion from these and other studies is that wolves take more older and diseased animals than would be expected on the hypothesis that they kill a random sample of the population, and hence that healthy prey stand a good chance of escaping predation.

In the Serengeti ecosystem of East Africa the situation is far more complex since there are six important predators, hyaena, lion, wild dog, leopard, cheetah and jackal. There are three very abundant species of prey, wildebeest, zebra and Thomson's gazelle, together with several other species which are also numerous including Grant's gazelle, giraffe, eland, topi, buffalo and warthog. There is variation in the hunting methods of the different predators, and the species of prey have each evolved different defences directed at one or more of the predators. Lion and leopard capture most of their prey by stalking until near enough to charge or spring, whereas the other predators are coursers, capturing most prey only after a fast and often long chase.

The most abundant small herbivore in the Serengeti is Thomson's gazelle whose reactions to predators have already been described on pp. 239 and 254. Once a predator has started to chase a Thomson's gazelle, its only defence, except possibly to the jackal, is flight, initially in a straight line or a large circle, but with erratic zig-zags when the predator is very close. In addition a female whose fawn is being chased may attack jackals and attempt to distract a hyaena by running close to it. It does not attempt to attack a predator larger than a jackal, and no doubt such an attack would very likely end in the death of the mother as well as the fawn. Kruuk followed forty-three hunts of Thomson's gazelle by spotted hyaenas of which fourteen (33 per cent) were successful. There

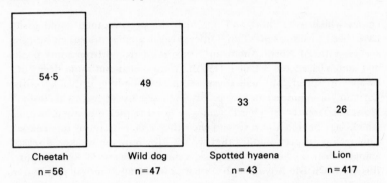

Fig. 12.30   Percentage of hunts which were successful (hunting success rates) on Thomson's gazelle (excluding very young fawns), by predators in the Serengeti, East Africa. The data for lions may include some attacks on very young fawns, but if these were excluded the success rate would probably be even lower. (*Data from* Kruuk, 1972 and Schaller, 1972.)

was a slightly lower success rate for adult than for young gazelle, but the difference is not significant. By contrast the two other coursing predators, cheetah and wild dog, have a higher success rate of about 50 per cent, whilst lions have a low success rate (26 per cent) (Fig. 12.30). No figures are available for the other two predators of Thomson's gazelle, the leopard and the jackal. These figures must be treated with caution since the final success rate calculated will depend on the age of the prey and on the criteria used to judge what constitutes a genuine hunt. The hunting success rates of wild dog and cheetah on very young gazelle fawns are 95 and 100 per cent ($n = 22$ and 31 respectively). If these are added to the figures given in Fig. 12.30 the overall success rate for wild dogs is 64 per cent ($n = 69$) and for cheetah it is 70 per cent ($n = 87$). Wild dogs are also very successful at hunting other prey: Schaller found that their overall success rate for all hunts in the Serengeti was 89 per cent ($n = 112$) whilst in the Ngorongoro Crater it was 86 per cent ($n = 29$) (Estes and Goddard, 1967). H. and J. van Lawick-Goodall (1970) observed a success rate of only 43 per cent (thirty-nine out of ninety-one hunts) in the Serengeti, so evidently they used different criteria to judge what constitutes a genuine hunt, or possibly they omitted from their figures captures of very young fawns. Nevertheless the high success rate of capturing gazelle by coursing predators contrasts with the low success by the stalking lion. Schaller found that lion were very efficient at stalking, and in 63 per cent of the hunts they were able to get sufficiently close to the quarry to be in a position to charge or spring. However, only 26 per cent of hunts resulted in capture and this was due to the fact that the gazelle are very agile at evading the final rush, sometimes even jumping away over the approaching lion. One would like to know if the age distribution of prey captured by each predator corresponds with that in the gazelle population as a whole, or if some of the predators are selectively taking more of the older and diseased animals. Unfortunately the evidence is not conclusive, but the following tentative conclusions can be suggested. Hyaenas, which are generalized coursing predators and also highly efficient scavengers, take mainly older prey not in prime condition, and the 33 per cent success rate represents the ability of the hyaenas to judge the condition of the prey gazelle and hence their chances of capturing it. Wild dog and cheetah are highly specialized coursing predators, both particularly adapted to preying on gazelle rather than on larger herbivores. They both take gazelle of different age classes roughly in proportion to their frequency in the population, although with a slight bias towards very old or very young ones. The 50 per cent success rates for cheetah and wild dog demonstrate how efficient these predators are at capturing gazelle. Thomson's gazelle have a much greater flight distance to wild dog than they do to cheetah, hence the wild dog are forced to chase gazelle under conditions more favourable to the prey's escape than is the case for cheetah. The lion is a stalking predator and has a high success rate at approaching gazelle, but the low capture rate (26 per cent) is entirely due to the efficiency of the gazelle's defence by erratic jumping. Even old gazelle appear to be equally adept at evading lion so that the segment of the population taken by lion is more or less random with respect to age. Similar conclusions probably apply to the leopard but there is little data available for this stalking predator, except that it differs from the lion in taking many more fawns. Finally the jackal (*Canis aureus* and *C. mesomelas*) is really a predator on small animals including rodents and gazelle fawns, but it is also a very fast runner and can occasionally chase and capture a fully adult gazelle (H. and J. van Lawick-Goodall, 1970).

Wildebeest are much larger animals than Thomson's gazelle but they too lack any retaliatory defence towards their predators, principally lion, hyaena and wild dog. A mare with foal will attack wild dogs and hyaenas, but she does not actively defend herself when she is on her own. Like Thomson's gazelle, adult wildebeest rely for defence on speed and stamina. Out of forty-nine chases observed by Kruuk (1972), eighteen (37 per cent) were successful. Of the thirty-one unsuccessful chases, twenty-four animals escaped by flight, either outrunning the predators or entering the clan territory of a different pack of hyaenas so that its pursuers gave up the chase rather then face attack from the neighbouring clan (Fig. 12.31). The remaining seven ran into a herd of animals and the pursuing hyaenas either lost their quarry amongst the herd or switched to chasing a different individual. This is defence due to living in a large group; even though there is no retaliation against hyaenas, there is advantage to the wildebeest simply due to the mass of animals confusing predators. These figures are not directly comparable with Mech's data for wolves hunting moose: wolves usually detect moose by scent and so the 5—6 per cent killed represents the kills per

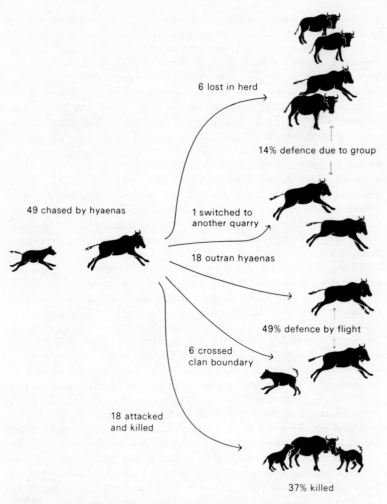

Fig. 12.31   Fate of forty-nine adult wildebeest hunted by spotted hyaenas in East Africa. For further explanation see text. (*Data from* Kruuk, 1972.)

100 encounters between wolf and moose. With hyaenas and wildebeest, as with wolves and caribou, the predators often charge a herd and then watch the running animals for some time before selecting a particular individual to chase. Hence before the chase starts the predator 'tests' the prey and singles out an individual which perhaps shows signs of being slower or in some other way easier to capture than the rest. Sometimes dozens of wildebeest may be tested in this way before a serious chase is begun. With moose the comparable figures for those tested may be the fifty-three which were actually caught up by wolves. Six or seven of these fifty-three were killed, giving a success rate of

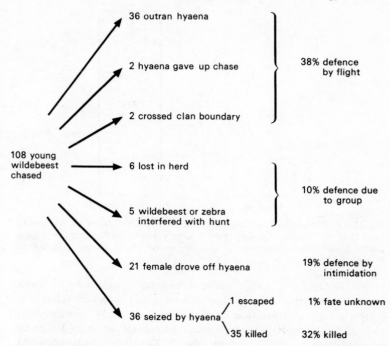

Fig. 12.32   Fate of 108 young wildebeest chased by one or more spotted hyaenas in East Africa. (*Data from* Kruuk, 1972.)

11–13 per cent – still appreciably less than that for hyaenas and wildebeest, but then the latter have no defence once caught up by their pursuers.

For young wildebeest hyaenas usually hunt alone or in pairs, not in packs. Thirty-two per cent of 108 such hunts were successful, and in 38 per cent of the hunts the prey escaped by flight (Fig. 12.32). In 19 per cent of the cases the mother successfully defended her calf; defence by the mother was always successful against only one hyaena ($n = 20$), but always failed to protect the calf when two or more hyaenas were attacking ($n = 19$), hence the hunting success on wildebeest calves depends very much on how many hyaenas are hunting (Fig. 12.33). The effect of hyaena predation on the wildebeest population is very serious; Kruuk estimated that three-quarters of all calves are killed within a few months of birth in Ngorongoro, mostly by hyaenas. Consequently one of the best defences of wildebeest is synchronized breeding so that for a month or two the market is flooded. Predators then eat their fill on the vulnerable calves, but these are too numerous for all to be killed at this time. Synchronized breeding also occurs in Thomson's gazelle although a few fawns may be born at other seasons. These few are cryptic when resting so many escape detection by predators. Wildebeest calves are never cryptic and are always obvious to a hunting predator, so there is

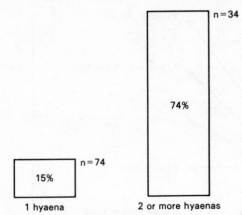

Fig. 12.33   Percentage of successful hunts of wildebeest calves by one and by more than one hyaena. The difference is largely due to the fact that the cow can fend off one hyaena successfully, but she cannot drive off two. (*Data from* Kruuk, 1972.)

very strong selection against wildebeest calving outside the peak breeding season. Another defence of wildebeest is that cows with calves have a greater flight distance than have cows without calves, and they may also dodge through a herd of adult wildebeest so as to be partly screened from a hunting predator (Fig. 12.34). Hence, although wildebeest appear to be defenceless when attacked by hyaenas, wild dogs or lions, they actually possess several indirect defences related to their herding behaviour. A vulnerable prey animal can often escape predation by associating with healthy (non-vulnerable) prey animals of its kind.

Hyaenas probably take more diseased wildebeest than healthy ones (Kruuk, 1972). Evidence for this hypothesis is the method of 'testing' prey, and the observation that the age distribution of hyaena—killed wildebeest is similar to that of wildebeest which died of natural causes (i.e. old age or disease). Lions usually hunt by stalking their prey and they kill a quite different cross-section of the wildebeest population from that killed by hyaenas. Nevertheless their hunting success rate (32 per cent out of fifty-nine hunts) is very similar to that of hyaenas (Schaller, 1972), but since it is difficult to judge what constitutes a genuine hunt this similarity may have no great significance.

The other large Serengeti herbivores can all defend themselves actively. Buffalo and eland are rarely killed by hunting dogs or hyaenas. Kruuk recounts how a buffalo warded off a pack of hyaenas with its horns, and of how eland cows ran to the defence of an isolated calf and drove off a pack of hyaenas. Eland use both hooves and horns to fend off hyaenas (Fig. 12.35), and this group defence is similar to the defensive ring of musk oxen described earlier. Lions do stalk and kill buffalo, but usually several lions are involved in the hunt, and buffalo have been known to attack and kill a lion.

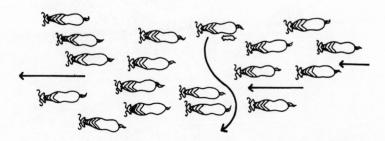

Fig. 12.34 Defensive behaviour of cow wildebeest in a herd when passing close to a resting hyaena. As the herd moves on she dodges through the animals so that she and her calf are on the side away from the hyaena. (*Redrawn from* Kruuk, 1972, Fig. 40. © 1972 by the University of Chicago.)

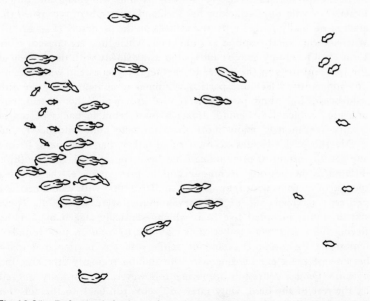

Fig. 12.35 Defensive behaviour of eland when threatened by a pack of spotted hyaenas. The cows with calves keep well back whilst the cows without calves stand forwards facing the hyaenas. The hyaenas (on the right) are about 100 m from the group of cows with calves. (*Redrawn from* Kruuk, 1972, Fig. 51. © 1972 by the University of Chicago.)

Fig. 12.36   Defensive behaviour of zebra when hunted by a pack of hyaenas. Notice that the foals are in the centre of a closely bunched herd of mares and yearlings. The stallion lags behind and makes occasional attacks on the pursuing hyaenas. (*Redrawn from* Kruuk, 1972, Fig. 46. © 1972 by the University of Chicago.)

Zebra also defend themselves actively. A mare will charge and bite at hyaenas or wild dogs attacking her foal, and when a herd is pursued the stallion normally makes repeated attacks on the predators (Fig. 12.36). Whereas wildebeest respond as individuals when they are chased, zebra form up into family groups and gallop away slowly with the stallion at the rear and foals in the centre of the herd. Hyaenas and wild dogs do not apparently select one particular animal to chase, as they do with wildebeest, but they pursue the entire group and then attack any individual which lags behind. Hyaenas hunt zebra in packs of ten to twenty-five animals, many more than normally hunt wildebeest, and clearly the pack 'decides' to hunt zebra rather than wildebeest before any specific group of prey animals has been approached. Despite these differences in hunting technique and in prey defence, the hunting success of hyaenas with zebra is very similar to that with wildebeest: 34 per cent (sixteen out of forty-seven hunts) were successful. These sixteen hunts included five foals which eventually lagged behind the fleeing herd and were pulled down by the hyaenas. Kruuk found it impossible to correlate escape of zebra with any particular defence because of the close herding with the stallion regularly attacking the hyaenas. Once a zebra has been caught, however, it is normally ignored by the rest of the herd. Only rarely do zebra run back to the aid of a fellow individual that has been caught and attacked: H. and J. van Lawick-Goodall (1970) describe one dramatic example in which ten zebra returned to the aid of a mare, a yearling and a foal who had been surrounded by wild dogs, and then led them all to safety.

Lion and wild dog also have a low success rate in hunting zebra, 27 per cent out of fifty-six hunts for lion, and 17 per cent out of thirty hunts for wild dog, but again it is not possible to correlate this with any particular defence of the prey. There is no convincing evidence that hyaena take more adult zebra that are old or diseased but lions definitely appear to take rather more very old ones than one would expect if they took them at the same frequency with which they occur in the population.

In a complex ecosystem like the Serengeti where there are six or more important predatory mammals and a similar number of common prey species, the predator—prey interactions are very complex. Some features of the prey species defences can be seen to be directed specifically at one particular species of predator: for example the female gazelle with fawn who attacks a jackal, distracts a hyaena, but retreats from a larger carnivore; or the evasive jumps of gazelle when pursued by predators which are clearly adapted towards avoiding a lion or leopard, but are useless against packs of hyaenas or wild dogs. On the other hand, the predators' hunting strategies may be directed at one particular species of prey, for example the hyaenas hunting method for zebra. Apparently maladaptive behaviour, such as failure to defend young against some predators, may be related to the complexity of the situation: it may not be possible for a prey population to evolve the best predator—specific defences against all the different predators in the area, particularly if the predators hunting techniques vary with different individuals (lions) or different packs (hyaenas). For example, Kruuk found that one clan of hyaenas in Ngorongoro preyed largely on zebra whilst a neighbouring clan preyed largely on wildebeest. Hence different clans can become specialists on different species of prey, but the prey animals moving throughout the area must be adapted to the entire predator spectrum, not to just a part of it. Nothing is known as to what parts of a herbivore mammals defensive responses are innate and directed at general conditions in the area, and what parts are learned and can thus be modified by local conditions, such as by specialist clans of predators in a particular area.

Information on hunting methods and hunting success rates are not available for many other animals, but the lynx has hunting success rates on showshoe hares varying from 42 per cent in Newfoundland to 24 and 9 per cent in Alberta. The variation may be related to different ecological conditions such as snow cover and abundance of prey (Nellis and Keith, 1968). The sparrow falcon (*Falco sparverius*) has a success rate of 33 per cent (*n* = 54) (Sparrowe, 1972), but the gyr falcon (*Falco rusticolus*) has a success rate of 77 per cent (seven out of nine attacks) (Bengtson, 1971). The gyr falcon hunts by attempting to surprise its prey, and many potential prey are never actually attacked because it is evident to the falcon that they are aware of its presence and so will escape by diving (if on water) or by evasive flight tactics. Thus the high success rate in this case reflects a specialized hunting strategy that detects and ignores alert prey and only attacks those with inefficient early warning defence systems. An experienced tiger is also reported to ignore prey that has already detected it and to hunt only those prey which are not aware of its presence (Schaller, 1967).

It is clear from an examination of these predator–prey systems that a balance must be achieved. It is true that selection will favour those predators with the most efficient hunting strategy and those prey with the best defences, but both predator and prey have survived millions of years till now, so they must have achieved a balance. If the anti-predator defences become too efficient then the predator will either become extinct or turn to an alternative source of food. If the hunting strategy becomes too efficient then the prey either becomes extinct or so rare that the predator has to find an alternative source of food or itself go extinct. The situation is delicately balanced in some of the simpler arctic systems where one predator feeds principally on one species of prey, for example the wolf on the vast caribou herds in Alaska and northern Canada. Here the wolf population is probably adjusted to the level that can exist on the surplus old and diseased caribou, supplemented by a glut of young at the appropriate season, whilst most of the healthy caribou are able to escape. Such simple systems can oscillate, and these can be simulated in the laboratory. For example, a culture of the bacterium *Klebsiella aerogenes* with the ciliate *Tetrahymena pyriformis* as its predator oscillates violently for several generations with first the prey and then the predator approaching extinction. However, this situation eventually stabilized with both species coexisting. The reason for the change was that the prey bacteria had evolved a new habit, not present in the original population, of resting on the walls of the culture vessel where they could not be removed by the ciliate (Ende, 1973). Thus violently oscillating predator–prey systems can evolve into more stable systems. In the case of voles and lemmings, whose populations also oscillate, the breeding success of the predators is determined by the numerical abundance of the prey population. The predators never become sufficiently numerous to be able to control the numbers of the much more rapidly breeding prey because of the violent population crashes of the prey species. The causes of these population fluctuations are still not clear: probably the ultimate factor involved is food shortage or deterioration of habitat, but the population actually falls before this limit is reached as a result of genetically controlled changes in the birth rate, death rate, and behaviour of the voles (Krebs *et al.*, 1973). Predators may prey very heavily on voles in certain years but they never actually control the population. Similarly, there is no reliable evidence to suggest that predators control the numbers of mule deer on the Kaibab Plateau (North America) (Caughley, 1970). It is only in rather special circum-stances that predators can temporarily control prey numbers. For example, at Tilden Park, California, during a plague year of meadow voles (*Microtus californicus*), Pearson (1964) calculated that predators killed 88 per cent of the population between June 1961 and spring 1962, eventually switching to other species as the vole population declined. However, reproductive activities in the vole also ceased at this time so that predation, however slight, was bound to lower the population. A second example of predators controlling a prey population is of lions reducing a herd of wildebeest in Nairobi National Park from 1 780 in 1961 to 253 in 1966 (summarized in Schaller, 1972). However, by 1967 the wildebeest population was so low that the predators had turned to alternative prey, and there were signs that the wildebeest population was on the increase.

In the tropics there are usually more species of predator and more alternative prey, so the situation is much more complex. In the past many prey species may well have become extinct because they could not defend themselves adequately from some of these predators, and conversely some of the predators that did not become specialized on particular prey may have been unable to capture sufficient to survive. In the Serengeti the greatest competition between predators will be for the smaller species of prey, the gazelles (Schaller, 1972). A Thomson's gazelle provides small reward in terms of food per kill, and hence if it is to survive on gazelles alone a predator must have a high hunting success rate, or else use a hunting method which uses up little time and energy. For lion and hyaena, if gazelle are scarce they can always turn to alternative, larger, prey. But for cheetah or jackal this is not possible since neither can kill an adult wildebeest or buffalo. Because of this competition between predators for the available prey, each species has evolved specializations. Two of the predators, the lion and the hyaena, are specialists at tackling large prey, but whilst the lion is a generalized stalking predator and can kill buffalo as well as wildebeest, zebra and gazelle, the hyaena is a coursing predator which is also a highly efficient bone-crushing scavenger. The remaining four predators are small prey specialists. The leopard is a stalker which can take a variety of species whilst the other three are all coursers. Both black-backed and golden jackals are really scavengers and generalized predators of smaller game, although they can take adult gazelle or the newborn young of larger herbivores. Wild dog are highly efficient coursing predators which can also, because of their social habits, tackle wildebeest and zebra as well as gazelle although the hunting success rate for zebra is very low. Cheetah are the coursing specialists par excellence and rely for over 90 per cent of their kills on gazelle. One of the implications of this speciali-zation is that if for any reason gazelles were to become scarce the cheetah might well become extinct since it is unlikely to be able to adapt to capturing larger or better defended species such as zebra. On the other hand, assuming that hyaenas can only capture the 10 per cent of the gazelles that are the oldest and most diseased, then cheetah could probably still catch a few more gazelle which could outrun a hyaena, because cheetahs are more highly specialized at capturing gazelles — hence the hyaenas would have to find alternative prey.

One would like to know if any or all of these predators exert any limiting effect on the populations of herbivores, but the evidence is not conclusive. In the case of wildebeest, it appears that in the Serengeti the population is limited more by food shortage and disease than by predation, and here the population has responded by reducing its fecundity so that 96 per cent of cows do not calve until their third year (Schaller, 1972). On the other hand, in Ngorongoro there is much heavier predation and no evidence for large-scale mortality due to disease or starvation. Here the population has responded by increasing its fecundity so that 75 per cent of cows calve in their second year. Information on the other prey species is very incomplete, but it appears that a complex of factors all play a part in regulating the numbers of the herbivore populations. Probably predation, disease and intrinsic factors (such as reduced fertility under certain conditions) normally keep the population below the level at which deterioration of habitat, and hence starvation, occur. This is highly speculative, but it should be clear that in any environment the predator–prey system must achieve a balance such that both predators and prey coexist. On occasions when the environment changes or is disturbed, a new predator–prey balance will evolve, either with the same or with different species in the system.

Finally it can be argued that predator selection benefits the prey species since it weeds out those that are vulnerable to illness or disease or which are in any way deformed. In the absence of predation a species population might expand and there would be no mechanism for eliminating sublethally infected individuals which could then transmit disease to the rest of the population. In fact one of the defences of animals against parasites may well be that parasitized animals suffer a higher mortality due to predation than do healthy animals, and if this kills the parasite it will benefit the population.

Summary

A study of predator–prey interactions suggests that there is an arms race between predator and prey. The predators are constantly evolving more efficient methods of detection and capture of prey, whilst the prey are evolving more efficient methods of detecting predators at a distance and more effective primary and secondary defences. Some defences are specific to one or a few species of predators, but selection must act on the entire defensive system of an animal, not just on one or two defences. Improvement of some defences may conflict with the efficiency of other defences or of other essential activities of the animal so that a compromise is achieved. Selection cannot perfect a species defensive system such that all individuals escape predation, and it is probably of advantage to a species if diseased, senile or otherwise sub-optimal animals succumb to predation so long as the healthy ones survive. Predator—prey systems have evolved a balance such that both predator and prey populations survive in equilibrium, albeit an equilibrium that may involve considerable fluctuations in numbers. Such predator—prey systems are comparatively simple in temperate, arctic or island ecosystems where there are few species of predator and prey, but they can be very complex in the tropics where species diversity is much greater.

Section 5
Conclusions

In this book I have attempted to look at the anti-predator defences of animals with particular emphasis on the effect of each defence on the predators. Some of the more important conclusions have already been given in the chapter summaries and in the summaries at the end of Sections 3 and 4. It is also possible to take an historical view and consider the evolution of defensive systems within each class or phylum of animals, for example in the Mollusca.

Molluscs have evolved from slow-moving soft bodied animals of similar bodily organization to that found today in the Turbellaria. They probably lived on the surface of the sea-bed and were protected from predators by glandular secretions, but they must have experienced very heavy predation from the benthic carnivores of precambrian seas. Selection would have favoured any individuals which had a thicker cuticle, and it is likely that the molluscan shell evolved in this way as an anti-predator defence. The defensive advantage of the shell must have been very considerable since it brought with it numerous disadvantages such as the difficulty of concealing an animal with a shell, restriction of the surfaces available for gaseous exchange and for discharge of renal and reproductive products, and the adverse effect on speed of crawling which a large shell must have entailed. This last disadvantage was to some extent overcome by coiling of the shell so that it became less cumbersome. Predators could still prey on molluscs, however, if they could attack the soft body underneath the shell. One prey response to this attack-strategy was to develop a powerful suction-pad foot so that the mollusc could pull the shell down around it and be very difficult to dislodge from a rock. This can be observed in chitons, *Neopilina*, and in various prosobranch, opisthobranch, and pulmonate gastropod limpets today. Primitive gastropods underwent torsion and evolved an operculum; they could then withdraw fully into the shell and at the same time plug the opening with the calcareous operculum. This may not have been the original selective advantage of torsion, as was suggested by Garstang (see the discussions of torsion by Ghiselin, 1966, and Thompson, 1967), but undoubtedly the ability to retract fully in this way was of very considerable protective value against predation and probably enabled gastropods to spread into many predator-rich habitats from which they had previously been excluded.

The success of gastropods resulted in their forming an abundant potential food supply for any predator that could overcome the defences of the shell and the operculum. As pointed out on p. 226, numerous predators have evolved means of preying on gastropods, and this in turn imposed tremendous selective advantage on any individuals that could effectively counter these new predator-strategies. Cowries and cone shells have evolved very narrow apertures so that it is difficult for a predator to insert its mouthparts, whilst muricids have evolved very thick shells with long and thick spines which break up the outline of the shell, make it difficult to swallow, and also make it more difficult for drilling gastropod predators to bore through it. Cowries camouflage the shell with the expanded, papillate mantle which also secretes sulphuric acid, and many other gastropods, such as the pulmonate limpet *Siphonaria*, have defensive glandular secretions on the mantle or the sides of the foot. In addition some gastropods have evolved highly specific anti-predator escape responses towards starfish and predatory gastropods in which they may leap or swim away. Some species even detect such a predator whilst it is still some distance away so that they can take evasive action before an attack has actually been launched (see pp. 232–7).

The shell has several disadvantages, as already mentioned, and it is also a firm surface which starfish can grip when they are attacking a gastropod. Hence provided that the mollusc is well protected in other ways, it has sometimes been of advantage to reduce or completely lose the shell. *Natica* extends the mantle over the shell when approached by a starfish, but the limpet *Scutus* has the mantle permanently covering the shell so that it is impossible for a starfish to grip it and pull it off a rock. Many opisthobranch molluscs have reduced the shell and have elaborated other defences: *Aplysia* is cryptic with glandular defensive secretions; many dorids are cryptic and spicular, and some have acidic or other defensive secretions; some sacoglossans and eolids have autotomizable papillae which contain defensive glands and (in eolids) nematocysts so that their defences are concentrated in the least important part of the body which is also the first part likely to be contacted by a predator (Edmunds, 1966a, b, 1968a).

Thus whilst the shell is probably the most important single character responsible for the success of the gastropods, there are many predators which have overcome this defence, and hence living gastropods have evolved a variety of other defences towards these predators.

This brief outline of the evolution of the defence systems of gastropods could be repeated for other groups of animals. In the case of annelids the compartmentalized coelom probably evolved because it enabled primitive annelids to burrow and so to escape from crawling predators on the sea-bed (Clark, 1964), and from this initial stock the present wide diversity of different crawling, burrowing, swimming and tubicolous annelids have evolved. Amongst the insects, wings have provided an efficient escape mechanism, but large insects have conspicuous wings and cannot fly quickly the moment they are attacked. Hence they have evolved other defences; for example in large mantids there may be flash colours, deimatic responses and thanatosis, and very often the wings have been reduced or completely lost in the female since they are a conspicuous recognition mark for predators (Edmunds, 1972).

It is obvious that one could elaborate in this way on the evolution of defensive systems for many groups of animals. But such an analysis only considers a part of the evolutionary history — it ignores the problems faced by an animal in obtaining its own food and in reproducing successfully. Selection pressures for efficient defence and for efficient food collection or reproduction may conflict, and all three need to be considered in any account of the evolution of a major group of animals.

For a population or a species to survive it must satisfy three requirements:

(a) It must reproduce successfully, hence selection will favour high fecundity and/or methods of caring for eggs and young.

(b) It must obtain sufficient food both to provide energy for its own survival and also to enable it to reproduce, hence selection will favour adaptations which help an animal to obtain food more efficiently either by becoming more specialized on a particular food resource, or more generalized on the type of food it can take.

(c) It must avoid being killed either by predators, parasites or disease on the one hand, or by harmful environmental conditions, such as cold, on the other.

These three requirements are interrelated since the defensive system of an animal may give protection both against predators of the adult and predators of the eggs or young; hence it increases the chances of individual survival (c) and also of reproducing successfully (a).

The defensive system of most animals has two components, primary defences which operate regardless of whether there is a predator in the vicinity, and secondary defences which operate when a predator encounters a prey animal. The requirements of any one successful defensive mechanism may conflict with those of another, or with the animal's method of obtaining food or of reproducing, and each species has evolved a unique solution to these conflicting selection pressures. This makes it very difficult to make any generalizations on anti-predator defences, but one can ask why it is that some species are highly specialized in their defences whilst others are not. For example the mantid *Sphodromantis* is simply camouflaged whilst *Phyllocrania* is a specialized leaf mimic; house flies have no obvious primary defence whilst syrphids are often wasp mimics; and minnows are unspecialized counter shaded fish whilst *Lobotes* and *Argyropelecus* have highly elaborate camouflages. Presumably part of the answer to the question is that each species has been acted on by slightly different selection pressures and so has responded in different ways. The most highly specialized species are probably on the road to extinction since they can only evolve better defences by still further specialization, and they cannot readapt to new environments. By contrast the species with less specialized defences have the possibility of evolving more efficient defences in any one of several different directions, so that they have the potentiality for surviving under different ecological environments.

It is of interest that most of the highly specialized defensive adaptations are found in areas of environmental stability such as a tropical forest, a coral reef, the pelagic region of the oceans or the abyss. By contrast there are few specialized defences in animals of the arctic, in deserts, in estuaries or in littoral habitats, all areas subject to extremes of variation of some physical parameter of the environment such as temperature, or salinity. One reason for this is simply that there are more species of animal in the stable environments compared with the unstable ones, hence, numerically, there will be more of the more specialized animals present. But part of the reason must also be that in unstable environments an animal must strive against the environment: it may well be killed by the harsh environmental conditions, and so must become physiologically specialized if it is to survive. Such physiological specialization may conflict with anatomical anti-predator specializations. In a stable environment there is much weaker selection pressure towards physiological specialization due to the environment, but an animal faces much stronger competition from all the other animals which can also survive there. Hence there will be much greater selection pressure to evade predation and to evolve highly elaborate anti-predator defences in a stable environment than in an unstable environment.

Bibliography

ABEL, E. F. (1960) 'Liaison facultative d'un poisson (*Gobius bucchichii* Steindachner) et d'une anémone (*Anemonia sulcata* Penn.) en Méditerranée', *Vie Milieu*, 11, 517–31.

ALCOCK, J. (1969) 'Observational learning by fork-tailed flycatchers (*Muscivora tyrannus*)', *Anim. Behav.*, 17, 652–7.

ALCOCK, J. (1970a) 'Punishment levels and the response of black-capped chickadees (*Parus atricapillus*) to three kinds of artificial seeds', *Anim. Behav.*, 18, 592–9.

ALCOCK, J. (1970b) 'Punishment levels and the response of white-throated sparrows (*Zonotrichia albicollis*) to three kinds of artificial models and mimics', *Anim. Behav.*, 18, 733–9.

ALCOCK, J. (1971) 'Interspecific differences in avian feeding behaviour and the evolution of Batesian mimicry', *Behaviour*, 40, 1–9.

ALDER, J. and HANCOCK, A. (1864) 'Notice of a collection of nudibranchiate Mollusca made in India by Walter Elliot, Esq., with descriptions of several new genera and species', *Trans. zool. Soc. Lond.*, 5, 113–47.

ALEXANDER, A. J. (1958a) 'On the stridulation of scorpions', *Behaviour*, 12, 339–52.

ALEXANDER, A. J. (1958b) '*Peripatus*: fierce little giant', *Anim. Kingd.*, 61, 122–5.

ALEXANDER, A. J. (1959) 'A survey of the biology of scorpions of South Africa'. *Afr. wild Life*, 13, 99–106.

ALEXANDER, A. J. (1960) 'A note on the evolution of stridulation within the family Scorpionidae', *Proc. zool. Soc. Lond.*, 133, 391–9.

ALEXANDER, A. J. and EWER, R. F. (1959) 'Observations on the biology and behaviour of the smaller African polecat', *Afr. wild Life*, 13, 313–20.

ALLEN, J. A. (1972) 'Evidence for stabilizing and apostatic selection by wild blackbirds', *Nature, Lond.*, 237, 348–9.

ALLEN, J. A. and CLARKE, B. (1968) 'Evidence for apostatic selection by wild passerines', *Nature, Lond.*, 220, 501–2.

AMEYAW-AKUMFI, C. E. (1971) *The Biology of Two Littoral Species of Hermit Crab* Clibanarius chapani *Schmitt and* Clibanarius senegalensis *Chevreux and Bouvier*. MSc. thesis, University of Ghana.

ANNANDALE, N. (1905) 'Notes on some oriental geckos in the Indian Museum, Calcutta, with descriptions of new forms', *Ann. Mag. nat. Hist. (Ser. 7)*, **15**, 26–32.

ANSELL, A. D. (1969*a*) 'Leaping movements in the Bivalvia', *Proc. malac. Soc. Lond.*, **38**, 387–99.

ANSELL, A. D., (1969*b*) 'Defensive adaptations to predation in the Mollusca', *Proceedings of Symposium on Mollusca*, Part 2, 487–512; Marine Biological Association of India.

APLIN, R. T., BENN, M. H. and ROTHSCHILD, M. (1968) 'Poisonous alkaloids in the body tissues of the cinnabar moth (*Callimorpha jacobaeae* L.)', *Nature, Lond.*, **219**, 747–8.

ARNOLD, D. C. (1953) 'Observations on *Carapus acus* (Brünnich), (Jugulares Carapidae)', *Pubbl. Staz. zool. Napoli*, **24**, 152–66.

ARNOLD, R. (1969) 'The effects of selection by climate on the landsnail *Cepaea nemoralis* (L.)', *Evolution, Lancaster, Pa.*, **23**, 370–8.

BAERENDS, G. P. (1941) 'Fortpflanzungsverhalten und Orientierung der Grabwespe *Ammophila campestris* Jur', *Tijdschr. Ent.*, **84**, 68–275.

BAKER, R. R. (1970) 'Bird predation as a selective pressure on the immature stages of the cabbage butterflies, *Pieris rapae* and *P. brassicae*', *J. Zool., Lond.*, **162**, 43–59.

BARNOR, J. L. (1972) *Studies on Colour Dimorphism in Praying Mantids*. MSc. thesis, University of Ghana.

BATE, C. M. (1973) 'The mechanism of the pupal gin trap. I. Segmental gradients and the connections of the triggering sensilla', *J. exp. Biol.*, **59**, 95–108.

BATESON, W. (1890) 'The sense-organs and perceptions of fishes; with remarks on the supply of bait', *J. mar. biol. Ass. U.K.*, **1**, 225–56.

BAYER, F. M. (1963) 'Observations on pelagic mollusks associated with the siphonophores *Velella* and *Physalia*', *Bull. mar. Sci. Gulf Caribb.*, **13**, 454–66.

BEDFORD, G. O. and CHINNICK, L. J. (1966) 'Conspicuous displays in two species of Australian stick insects', *Anim. Behav.*, **14**, 518–21.

BEEBE, W. (1953) 'A contribution to the life history of the euchromid moth, *Aethria carnicauda* Butler', *Zoologica, N.Y.*, **38**, 155–60.

BELBENOIT, P. and BAUER, R. (1972) 'Video recordings of prey capture behaviour and associated electric organ discharge of *Torpedo marmoratus* (Chondrichthyes)', *Mar. Biol.*, **17**, 93–9.

BELJAJEFF, M. M. (1927) 'Ein Experiment über die Bedeutung der Schutzfärbung', *Biol. Zbl.*, **47**, 107–13.

BELL, T. R. D. and SCOTT, F. B. (1937) *The Fauna of British India Including Ceylon and Burma. Moths Vol. 5. Sphingidae*. London, Taylor and Francis.

BENFIELD, E. F. (1972) 'A defensive secretion of *Dineutes discolor* (Coleoptera: Gyrinidae)', *Ann. ent. Soc. Am.*, **65**, 1324—7.

BENGTSON, S-A. (1971) 'Hunting methods and choice of prey of gyrfalcons *Falco rusticolus* at Myvatn in northeast Iceland', *Ibis*, **113**, 468—76.

BENSON, W. W. (1971) 'Evidence for the evolution of unpalatability through kin selection in the heliconiinae (Lepidoptera)', *Am. Nat.*, **105**, 213—26.

BENSON, W. W. (1972) 'Natural selection for Müllerian mimicry in *Heliconius erato* in Costa Rica', *Science, N.Y.*, **176**, 936—9.

BISHOP, J. A. (1972) 'An experimental study of the cline of industrial melanism in *Biston betularia* (L.) (Lepidoptera) between urban Liverpool and rural North Wales', *J. Anim. Ecol.*, **41**, 209—43.

BISSET, G. W., FRAZER, J. F. D., ROTHSCHILD, M. and SCHACHTER, M. (1960) 'A pharmacologically active choline ester and other substances in the garden tiger moth, *Arctia caja* (L.)', *Proc. R. Soc., B*, **152**, 255—62.

BLEST, A. D. (1957*a*) 'The function of eyespot patterns in the Lepidoptera', *Behaviour*, **11**, 209—56.

BLEST, A. D. (1957*b*) 'The evolution of protective displays in the Saturnioidea and Sphingidae (Lepidoptera)', *Behaviour*, **11**, 257—309.

BLEST, A. D. (1963*a*) 'Relation between moths and predators', *Nature, Lond.*, **197**, 1046—7.

BLEST, A. D. (1963*b*) 'Longevity, palatability and natural selection in five species of New World saturniid moth', *Nature, Lond.*, **197**, 1183—6.

BLEST, A. D. (1964) 'Protective display and sound production in some New World arctiid and ctenuchid moths', *Zoologica, N.Y.*, **49**, 161—81.

BLEST, A. D., COLLETT, T. S. and PYE, J. D. (1963) 'The generation of ultrasonic signals by a New World arctiid moth', *Proc. R. Soc., B*, **158**, 196—207.

BLUM, M. S. (1970) 'The chemical basis of insect sociality', in: *Chemicals Controlling Insect Behaviour*, ed. M. Beroza, 61—94, New York and London, Academic Press.

BLUM, M. S., CREWE, R. M., KERR, W. E., KEITH, L. H., GARRISON, A. W. and WALKER, M. M. (1970) 'Citral in stingless bees: isolation and functions in trail-laying and robbing', *J. Insect Physiol.*, **16**, 1637—48.

BOER, M. H. DEN (1971) 'A colour polymorphism in caterpillars of *Bupalus piniarius* (L.) (Lepidoptera: Geometridae)', *Neth. J. Zool.*, **21**, 61—116.

BOURLIÈRE, F. (1955) *The Natural History of Mammals*. London, Harrap.

BOWERS, W. S., NAULT, L. R., WEBB, R. E. and DUTKY, S. R.

(1972) 'Aphid alarm pheromone: isolation, identification, synthesis', *Science, N.Y.*, 177, 1121–2.

BOYCOTT, B. B. (1958) 'The cuttlefish — *Sepia*', *New Biology*, 25, 98–118.

BOYCOTT, B. B. and YOUNG, J. Z. (1950) 'The comparative study of learning', *Symp. Soc. exp. Biol.*, 4, 432–53.

BRAAMS, W. G. and GEELEN, H. F. M. (1953) 'The preference of some nudibranchs for certain coelenterates', *Archs néerl. Zool.*, 10, 241–64.

BREDER, C. M. (1949) 'On the behaviour of young *Lobotes surinamensis*', *Copeia* (1949), 237–42.

BREDER, C. M. (1963) 'Defensive behaviour and venom in *Scorpaena* and *Dactylopterus*', *Copeia* (1963), 698–700.

BRISTOWE, W. S. (1958) *The World of Spiders*. London and Glasgow, Collins.

BROCK, V. E. and RIFFENBURGH, R. H. (1960) 'Fish schooling: a possible factor in reducing predation', *J. Cons. perm. int. Explor. Mer*, 25, 307–17.

BROWER, J. V. Z. (1958a) 'Experimental studies of mimicry in some North American butterflies. Part 1. The monarch, *Danaus plexippus*, and viceroy, *Limenitis archippus archippus*', *Evolution, Lancaster, Pa.*, 12, 32–47.

BROWER, J. V. Z. (1958b) 'Experimental studies of mimicry in some North American butterflies. Part 2. *Battus philenor* and *Papilio troilus*, *P. polyxenes* and *P. glaucus*', *Evolution, Lancaster, Pa.*, 12, 123–36.

BROWER, J. V. Z. (1958c) 'Experimental studies of mimicry in some North American butterflies. Part 3. *Danaus gilippus berenice* and *Limenitis archippus floridensis*', *Evolution, Lancaster, Pa.*, 12, 273–85.

BROWER, J. V. Z. (1960) 'Experimental studies of mimicry. 4. The reactions of starlings to different proportions of models and mimics', *Am. Nat.*, 94, 271–82.

BROWER, J. V. Z. (1963) 'Experimental studies and new evidence on the evolution of mimicry in butterflies', *Int. Congr. Zool.*, 16, 4, 156–61.

BROWER, L. P. (1962) 'Evidence for interspecific competition in natural populations of the monarch and queen butterflies, *Danaus plexippus* and *D. gilippus berenice* in south central Florida', *Ecology*, 43, 549–52.

BROWER, L. P. (1963) 'The evolution of sex-limited mimicry in butterflies', *Int. Congr. Zool.*, 16, 4, 173–9.

BROWER, L. P. (1969) 'Ecological chemistry', *Scient. Am.*, 220(2), 22–9.

BROWER, L. P., ALCOCK, J. and BROWER, J. V. Z. (1971) 'Avian feeding behaviour and the selective advantage of incipient mimicry', in: *Ecological Genetics and Evolution*, ed. R. Creed, 261—74, Oxford and Edinburgh, Blackwell.

BROWER, L. P. and BROWER, J. V. Z. (1956) 'Cryptic coloration in the anthophilous moth *Rhododipsa masoni*', *Am. Nat.*, 90, 177—82.

BROWER, L. P. and BROWER, J. V. Z. (1962a) 'Experimental studies of mimicry. 6. The reaction of toads (*Bufo terrestris*) to honeybees (*Apis mellifera*) and their dronefly mimics (*Eristalis vinetorum*)', *Am. Nat.*, 96, 297—307.

BROWER, L. P. and BROWER, J. V. Z. (1962b) 'The relative abundance of model and mimic butterflies in natural populations of the *Battus philenor* mimicry complex', *Ecology*, 43, 154—8.

BROWER, L. P. and BROWER, J. V. Z. (1964) 'Birds, butterflies, and plant poisons: a study in ecological chemistry', *Zoologica, N.Y.*, 49, 137—59.

BROWER, L. P. and BROWER, J. V. Z. (1965) 'Experimental studies of mimicry. 8. Further investigation of honeybees (*Apis mellifera*) and their dronefly mimics (*Eristalis* spp.)', *Am. Nat.*, 99, 173—87.

BROWER, L. P., BROWER, J. V. Z. and COLLINS, C. T. (1963) 'Experimental studies of mimicry. 7. Relative palatability and Müllerian mimicry among neotropical butterflies of the subfamily Heliconiinae', *Zoologica, N.Y.*, 48, 65—83.

BROWER, L. P., BROWER, J. V. Z. and CORVINO, J. M. (1967) 'Plant poisons in a terrestrial food chain', *Proc. natn. Acad. Sci. U.S.A.*, 57, 893—8.

BROWER, L. P., BROWER, J. V. Z. and WESTCOTT, P. W. (1960) 'Experimental studies of mimicry. 5. The reactions of toads (*Bufo terrestris*) to bumblebees (*Bombus americanorum*) and their robberfly mimics (*Mallophora bomboides*), with a discussion of aggressive mimicry', *Am. Nat.*, 94, 343—55.

BROWER, L. P., McEVOY, P. B., WILLIAMSON, K. L. and FLANNERY, M. A. (1972) 'Variation in cardiac glycoside content of monarch butterflies from natural populations in eastern North America', *Science, N.Y.*, 177, 426—9.

BROWER, L. P., POUGH, F. H. and MECK, H. R. (1970) 'Theoretical investigations of automimicry, 1. Single trial learning', *Proc. natn. Acad. Sci. U.S.A.*, 66, 1059—66.

BROWER, L. P., RYERSON, W. N., COPPINGER, L. L. and GLAZIER, S. C. (1968) 'Ecological chemistry and the palatability spectrum', *Science, N.Y.*, 161, 1349—51.

BULLOCK, T. H. (1953) 'Predator recognition and escape responses of some intertidal gastropods in presence of starfish', *Behaviour*, 5, 130—41.

BURNS, J. M. (1966) 'Preferential mating versus mimicry: disruptive selection and sex-limited dimorphism in *Papilio glaucus*', *Science, N.Y.*, 153, 551—3.

BURTON, M. (1969) *Animal Partnerships*. London, Warne.

BURTT, E. (1951) 'The ability of adult grasshoppers to change colour on burnt ground', *Proc. R. ent. Soc. Lond., A*, 26, 45—8.

BUSTARD, H. R. (1968) *'Pygopus nigriceps* (Fischer): a lizard mimicking a venomous snake', *Br. J. Herpet.*, 4, 22—4.

BUSTARD, H. R. (1969) 'Defensive behaviour and locomotion of the Pacific boa, *Candoia aspersa*, with a brief review of head concealment in snakes', *Herpetologia*, 25, 164—70.

CAIN, A. J. (1971) 'Colour and banding morphs in subfossil samples of the snail *Cepaed*', in: *Ecological Genetics and Evolution*, ed. R. Creed, 65—92, Oxford and Edinburgh, Blackwell.

CARLISLE, A. I. (1953) 'Observations on the behaviour of *Dromia vulgaris* Milne Edwards with simple ascidians', *Pubbl. Staz. zool. Napoli*, 24, 142—51.

CARPENTER, G. D. H. (1921) 'Experiments on the relative edibility of insects, with special reference to their coloration', *Trans. R. ent. Soc. Lond.*, 54, 1—105.

CARPENTER, G. D. H. (1938) 'Audible emission of defensive froth by insects with an appendix on the anatomical structures concerned in a moth by H. Eltringham', *Proc. zool. Soc. Lond., A*, 243—52.

CARPENTER, G. D. H. (1941) 'The relative frequency of beak-marks on butterflies of different edibility to birds', *Proc. zool. Soc. Lond.*, 111 (A), 223—31.

CARPENTER, G. D. H. (1949) *'Pseudacraea eurytus* (L.) (Lep. Nymphalidae): a study of a polymorphic mimic in various degrees of speciation', *Trans. R. ent. Soc. Lond.*, 100, 71—133.

CARPENTER, G. D. H. and FORD, E. B. (1933) *Mimicry*. London, Methuen.

CARRINGTON, R. (1963) *The Mammals*. New York, Time Incorporated.

CARTER, M. A. (1967) 'Selection in mixed colonies of *Cepaea nemoralis* and *Cepaea hortensis*', *Heredity*, 22, 117—39.

CASTILLA, J. C. (1972) 'Responses of *Asterias rubens* to bivalve prey in a Y-maze', *Mar. Biol.*, 12, 222—8.

CATCHPOLE, C. K. (1973) 'The functions of advertising song in the sedge warbler (*Acrocephalus schoenobaenus*) and the reed warbler (*A. scirpaceus*)', *Behaviour*, 46, 300—20.

CAUGHLEY, G. (1970) 'Eruption of ungulate populations, with emphasis on himalayan thar in New Zealand', *Ecology*, 51, 53—71.

CAULLERY, M. (1952) *Parasitism and Symbiosis*. London, Sidgwick and Jackson.

CESNOLA, A. P. DI (1904) 'Preliminary note on the protective value of colour in *Mantis religiosa*', *Biometrika*, 3, 58—9.

CHANCE, M. R. A. and RUSSELL, W. M. S. (1959) 'Protean displays: a form of allaesthetic behaviour', *Proc. zool. Soc. Lond.*, 132, 65—70.

CHAPIN, J. P. (1939) 'The birds of the Belgian Congo. Part 2', *Bull. Am. Mus. nat. Hist.*, 75, 1—632.

CHAPIN, J. P. (1953) 'The birds of the Belgian Congo. Part 3', *Bull. Am. Mus. nat. Hist.*, 75A, 1—821.

CHRISTENSEN, A. M. and McDERMOTT, J. J. (1958) 'Life-history and biology of the oyster crab, *Pinnotheres ostreum* Say', *Biol. Bull. mar. biol. Lab.*, Woods Hole, 114, 146—79.

CLARK, R. B. (1964) *Dynamics in Metazoan Evolution.* Oxford, Clarendon Press.

CLARK, W. C. (1958) 'Escape responses of herbivorous gastropods when stimulated by carnivorous gastropods', *Nature, Lond.*, 181, 137—8.

CLARKE, B. (1960). 'Divergent effects of natural selection on two closely-related polymorphic snails', *Heredity*, 14, 423—43.

CLARKE, B. (1962a) 'Balanced polymorphism and the diversity of sympatric species', in: *Taxonomy and Geography*, ed. D. Nichols, 47—70, London, The Systematics Association.

CLARKE, B. (1962b) 'Natural selection in mixed populations of two polymorphic snails', *Heredity*, 17, 319—45.

CLARKE, B. (1969) 'The evidence for apostatic selection', *Heredity*, 24, 347—52.

CLARKE, C. A., DICKSON, C. G. C. and SHEPPARD, P. M. (1963) 'Larval color pattern in *Papilio demodocus*', *Evolution, Lancaster, Pa.*, 17, 130—7.

CLARKE, C. A. and SHEPPARD, P. M. (1966) 'A local survey of the distribution of industrial melanic forms in the moth *Biston betularia*, and estimates of the selective values of these in an industrial environment', *Proc. R. Soc., B*, 165, 424—39.

CLARKE, C. A. and SHEPPARD P. M. (1972a) 'Genetic and environmental factors influencing pupal colour in the swallowtail butterflies *Battus philenor* (L.) and *Papilio polytes* L', *J. Ent., A*, 46, 123—33.

CLARKE, C. A. and SHEPPARD, P. M. (1972b) 'The genetics of the mimetic butterfly *Papilio polytes* L', *Phil. Trans. R. Soc., B*, 263, 431—58.

CLARKE, W. D. (1963) 'Function of bioluminescence in mesopelagic organisms', *Nature, Lond.*, 198, 1244—7.

CLYNE, D. (1969) 'A spider that mimics the green tree ant', *N. Qd Nat.*, 36, 3—4.

COOK, E. F. (1962) 'A study of food choices of two opisthobranchs, *Rostanga pulchra* McFarland and *Archidoris montereyensis* (Cooper)', *Veliger*, 4, 194—6.

COPPINGER, R. P. (1969) 'The effect of experience and novelty on avian feeding behaviour with reference to the evolution of warning coloration in butterflies. Part 1: reactions of wild caught adult Blue Jays to novel insects', *Behaviour*, 35, 45—60.

COPPINGER, R. P. (1970) 'The effect of experience and novelty on avian feeding behaviour with reference to the evolution of warning coloration in butterflies. II. Reactions of naive birds to novel insects', *Am. Nat.*, 104, 323—35.

COTT, H. B. (1940) *Adaptive Coloration in Animals.* London, Methuen.

COTT, H. B. (1947) 'The edibility of birds: illustrated by five years experiments and observations (1941—1946) on the food preferences of the hornet, cat and man; and considered with special reference to the theories of adaptive coloration', *Proc. zool. Soc. Lond.*, 116, 371—524.

CRANE, J. (1952) 'A comparative study of the innate defensive behaviour in Trinidad mantids (Orthoptera, Mantoidea)', *Zoologica, N.Y.*, 37, 259—93.

CRANE, J. M. (1969) 'Mimicry of the gastropod *Mitrella carinata* by the amphipod *Pleustes platypa*', *Veliger*, 12, 200.

CREED, E. R. (1966) 'Geographic variation in the two-spot ladybird in England and Wales', *Heredity*, 21, 57—72.

CREED, E. R. (1971) 'Industrial melanism in the two-spot ladybird and smoke abatement', *Evolution, Lancaster, Pa.*, 25, 290—3.

CROWCROFT, P. (1966) *Mice all Over.* London, Foulis.

CROZE, H. (1970) 'Searching image in carrion crows', *Z. Tierpsychol.*, supplement 5, 1—86, Paul Parey, Berlin and Hamburg.

CURIO, E. (1965) 'Die Schutzanpassungen dreier Raupen eines Schwärmers (Lepidopt., Sphingidae) auf Galapagos', *Zool. Jb. Syst.*, 92, 487—522.

CURIO, E. (1970a) 'Die Messung des Selektionswertes einer Verhaltens-weise', *Verh. dt. zool. Ges.*, 64, 348—52.

CURIO, E. (1970b) 'Die Selektion dreier Raupenformen eines Schwärmers (Lepidopt., Sphingidae) durch einen *Anolis* (Rept., Iguanidae)', *Z. Tierpsychol.*, 27, 899—914.

CURIO, E. (1970c) 'Validity of the selective coefficient of a behaviour trait in hawkmoth larvae', *Nature, Lond.*, 228, 382.

CUTRESS, C. E. and ROSS, D. M. (1969) 'The sea anemone *Calliactis tricolor* and its association with the hermit crab *Dardanus venosus*', *J. Zool., Lond.*, 158, 225—41.

CUTRESS, C. E., ROSS, D. M. and SUTTON, L. (1970) 'The association of *Calliactis tricolor* with its pagurid, calappid, and majid partners in the Caribbean', *Can. J. Zool.*, 48, 371—6.

DALES, R. P. (1966) 'Symbiosis in marine organisms', in: *Symbiosis*, Vol. 1, ed. S. M. Henry, 299—326, New York and London, Academic Press.

DALY, J. W. and MYERS, C. W. (1967) 'Toxicity of Panamanian poison frogs (*Dendrobates*): some biological and chemical aspects', *Science, N.Y.*, 156, 970—3.

DAVENPORT, D. (1962) 'Physiological notes on actinians and their associated commensals', *Bull. Inst. océanogr. Monaco*, No. 1237, 1—15.

DAVENPORT, D. (1966a) 'Echinoderms and the control of behaviour in associations', in: *Physiology of Echinodermata*, ed. R. A. Boolotian, 145—56, New York, Interscience.

DAVENPORT, D. (1966b) 'Cnidarian symbiosis and the experimental analysis of behaviour', *Symp. zool. Soc. Lond.*, 16, 361—72.

DAVENPORT, D. and NORRIS, K. S. (1958) 'Observations on the symbiosis of the sea anemone *Stoichactis* and the pomacentrid fish *Amphiprion percula*', *Biol. Bull. mar. biol. Lab.*, *Woods Hole*, 115, 397—410.

DAY, J. H. (1967) *A Monograph on the Polychaeta of Southern Africa. Part 2. Sedentaria.* London, British Museum (Natural History).

DEMBOWSKA, W. S. (1926) 'Study on the habits of the crab *Dromia vulgaris* M.E.', *Biol. Bull. mar. biol. Lab.*, *Woods Hole*, 50, 163—78.

DEMPSTER, J. P. (1971) 'The population ecology of the cinnabar moth, *Tyria jacobaeae* L. (Lepidoptera, Arctiidae)', *Oecologia*, 7, 26—67.

DENTON, E. J. (1970) 'On the organization of reflecting surfaces in some marine animals', *Phil. Trans. R. Soc.*, B, 258, 285—313.

DENTON, E. J. (1971) 'Reflectors in fishes', *Scient. Am.*, 224(1), 64—72.

DENTON, E. J. and NICOL, J. A. C. (1965) 'Reflexion of light by external surfaces of the herring, *Clupea harengus*', *J. mar. biol. Ass. U.K.*, 45, 711—38.

DENTON, E. J. and NICOL, J. A. C. (1966) 'A survey of reflectivity in silvery teleosts', *J. mar. biol. Ass.*, *U.K.*, 46, 685—722.

DIJKGRAAF, S. (1963) 'The functioning and significance of the lateral-line organs', *Biol Rev.*, 38, 51—105.

DIMOCK, R. V. and DAVENPORT, D. (1971) 'Behavioural specificity and the induction of host recognition in a symbiotic polychaete', *Biol. Bull. mar. biol. Lab.*, *Woods Hole*, 141, 472—84.

DINGLE, H. and CALDWELL, R. L. (1969) 'The aggressive and territorial behaviour of the mantis shrimp *Gonodactylus bredini* Manning (Crustacea: Stomatopoda)', *Behaviour*, 33, 115—36.

DINSMORE, J. J. (1973) 'Foraging success of cattle egrets, *Bubulcus ibis*', *Am. Midl. Nat.*, 89, 242—6.

DITMARS, R. L. (1953) *The Reptiles of North America.* New York, Doubleday.

DIX, T. G. (1969) 'Association between the echinoid *Evechinus chloroticus* (Val.) and the clingfish *Dellichthys morelandi* Briggs', *Pacif. Sci.*, 23, 332—6.

DOESBURG, P. H. V. (1968) 'A revision of the New World species of *Dysdercus* Guerin Meneville (Heteroptera, Pyrrhocoridae)', *Zool. Verh., Leiden*, No. 97, 1—215.

DONNE, J. (1624) *Devotions upon Emergent Occasions, and Severall Steps in my Sickness.* London, Thomas Jones.

DUNCAN, C. J. and SHEPPARD, P. M. (1965) 'Sensory discrimination and its role in the evolution of Batesian mimicry', *Behaviour*, 24, 269–82.

DUNNING, D. C. (1968) 'Warning sounds of moths', *Z. Tierpsychol.*, 25, 129–38.

EDMUNDS, J. and EDMUNDS, M. (1974) 'Polymorphic mimicry and natural selection: a reappraisal', *Evolution, Lancaster, Pa.*, in press.

EDMUNDS, M. (1966a) 'Protective mechanisms in the Eolidacea (Mollusca Nudibranchia)', *J. Linn. Soc. (Zool.)*, 46, 27–71.

EDMUNDS, M. (1966b) 'Defensive adaptations of *Stiliger vanellus* Marcus, with a discussion on the evolution of "nudibranch" molluscs', *Proc. malac. Soc. Lond.*, 37, 73–81.

EDMUNDS, M. (1968a) 'Acid secretion in some species of Doridacea (Mollusca, Nudibranchia)', *Proc. malac. Soc. Lond.*, 38, 121–33.

EDMUNDS, M. (1968b) 'On the swimming and defensive response of *Hexabranchus marginatus* (Mollusca, Nudibranchia)', *J. Linn. Soc. (Zool.)*, 47, 425–9.

EDMUNDS, M. (1969a) 'Polymorphism in the mimetic butterfly *Hypolimnas misippus* L. in Ghana', *Heredity*, 24, 281–302.

EDMUNDS, M. (1969b) 'Evidence for sexual selection in the mimetic butterfly *Hypolimnas misippus* L', *Nature, Lond.*, 221, 488.

EDMUNDS, M. (1971) 'Opisthobranchiate Mollusca from Tanzania (suborder: Doridacea)', *Zool. J. Linn. Soc.*, 50, 339–96.

EDMUNDS, M. (1972) 'Defensive behaviour in Ghanaian praying mantids', *Zool. J. Linn. Soc.*, 51, 1–32.

EDWARDS, D. C. (1969) 'Predators on *Olivella biplicata*, including a species–specific predator avoidance response', *Veliger*, 11, 326–33.

EDWARDS, J. S. (1960) 'Insect assassins', *Scient. Am.*, 202(6), 72–8.

EHRLICH, P. R. and RAVEN, P. H. (1967) 'Butterflies and plants', *Scient. Am.*, 216(6), 105–13.

EIBL-EIBESFELDT, I. (1952) 'Nahrungserwerb und Beuteschema der Erdkröte (*Bufo bufo* L.)', *Behaviour*, 4, 1–36.

EIBL-EIBESFELDT, I. (1961) 'The fighting behaviour of animals', *Scient. Am.*, 205(6), 112–21.

EIBL-EIBESFELDT, I. (1970) *Ethology the Biology of Behaviour.* New York, Holt, Rinehart & Winston.

EIBL-EIBESFELDT, I. and EIBL-EIBESFELDT, E. (1968) 'The workers bodyguard', *Animals*, 11, 16–17.

EISENBERG, J. F. and GOULD, E. (1970) 'The tenrecs: a study in mammalian behaviour and evolution', *Smithson. Contr. Zool.*, 27, 1–137.

EISENTRAUT, M. (1927) 'Beitrag zur Frage der Farbanpassung der Orthopteren an die Färbung der Umgebung', *Z. Morph. Ökol. Tiere*, 7, 609–42.

EISNER, T. (1968) 'Mongoose and millipedes', *Science, N.Y.*, **160**, 1367.

EISNER, T. and DAVIS, J. A. (1967) 'Mongoose throwing and smashing millipedes', *Science, N.Y.*, **155**, 577—9.

EISNER, T., HENDRY, L. B., PEAKALL, D. B. and MEINWALD, J. (1971) '2,5-Dichlorophenol (from ingested herbicide?) in defensive secretion of grasshopper', *Science, N.Y.*, **172**, 277—8.

EISNER, T., KAFATOS, F. C. and LINSLEY, E. G. (1962) 'Lycid predation by mimetic adult cerambycidae (Coleoptera)', *Evolution, Lancaster, Pa.*, **16**, 316—24.

EISNER, T., KLUGE, A. F., CARREL, J. E. and MEINWALD, J. (1971*a*) 'Defence of phalangid: liquid repellent administered by leg dabbing', *Science, N.Y.*, **173**, 650—2.

EISNER, T., KLUGE, A. F., IKEDA, M. I., MEINWALD, Y. C. and MEINWALD, J. (1971*b*) 'Sesquiterpenes in the osmeterial secretion of a papilionid butterfly, *Battus polydamas*', *J. Insect Physiol.*, **17**, 245—50.

EISNER, T. and MEINWALD, J. (1966) 'Defensive secretions of arthropods', *Science, N.Y.*, **153**, 1341—50.

ELTRINGHAM, H. (1913) 'On the urticating properties of *Prothesia similis*', *Trans. ent. Soc. Lond.* (1913), 423—7.

EMLEN, J. M. (1968*a*) 'Batesian mimicry: a preliminary theoretical investigation of quantitative aspects', *Am. Nat.*, **102**, 235—41.

EMLEN, J. M. (1968*b*) 'Optimal choice in animals', *Am. Nat.*, **102**, 385—9.

EMMEL, T. C. (1972) 'Mate selection and balanced polymorphism in the tropical nymphalid butterfly, *Anartia fatima*', *Evolution, Lancaster, Pa.*, **26**, 96—107.

EMSLEY, M. G. (1966) 'The mimetic significance of *Erythrolamprus aesculapii ocellatus* Peters from Tobago', *Evolution, Lancaster, Pa.*, **20**, 663—4.

ENDE, P. van den (1973) 'Predator—prey interactions in continuous culture', *Science, N.Y.*, **181**, 562—4.

ENE, J. C. (1962) 'Parasitisation of mantid oothecae in West Africa', *Int. Congr. Ent.*, **11**, 4, 725—7.

ERGENE, S. (1950*a*) 'Untersuchungen über farbanpassung und farbwechsel bei *Acrida turrita*', *Z. vergl. Physiol.*, **32**, 530—51.

ERGENE, S. (1950*b*) 'Wählen Heuschrecken ein homochromes Milieu?', *Dt. zool. Z.*, **1**(2), 122—32.

ERGENE, S. (1951) 'Hat homochrome Färbung Schutzwert?', *Dt. zool. Z.*, **1**(3), 187—95.

ERGENE, S. (1953) 'Weitere Untersuchungun über die biologische Bedeutungder Schutzfärbung', *Mitt. zool. Mus. Berl.*, **29**, 127—33.

ESTES, R. D. and GODDARD, J. (1967) 'Prey selection and hunting behaviour of the African wild dog', *J. Wildl. Mgmt*, **31**, 52—70.

EUW, J. VON, FISHELSON, L., PARSONS, J. A., REICHSTEIN, T.

and ROTHSCHILD, M. (1967) 'Cardenolides (heart poisons) in a grasshopper feeding on milkweeds', *Nature, Lond.*, **214**, 35—9.

EUW, J. VON, REICHSTEIN, T. and ROTHSCHILD, M. (1968) 'Aristolochic acid-I in the swallowtail butterfly *Pachlioptera aristolochiae* (Fabr.) (Papilionidae)', *Israel J. Chemy*, **6**, 659—70.

EVANS, H. E. (1966) 'The accessory burrows of digger wasps', *Science, N.Y.*, **152**, 465—71.

EWER, D. W. (1957) 'Notes on acridid anatomy. IV. The anterior abdominal musculature of certain acridids', *J. ent. Soc. sth. Afr.*, **20**, 260—79.

EWER, R. F. (1963) 'The behaviour of the meerkat, *Suricata suricatta* (Schreber)', *Z. Tierpsychol.*, **20**, 570—607.

EWER, R. F. (1966) 'Juvenile behaviour in the African ground squirrel, *Xerus erythropus* (E. Geoff.)', *Z. Tierpsychol.*, **23**, 190—216.

EWER, R. F. (1968) *Ethology of Mammals*. London, Logos Press.

EWER, R. F. (1972) 'The devices in the web of the West African spider *Argiope flavipalpis*', *J. nat. Hist.*, **6**, 159—67.

EWER, R. F. (1973) *The Carnivores*. London and New York, Weidenfeld & Nicholson.

FARQUHARSON, C. O. (1921) 'Five years observations (1914—1918) on the bionomics of southern Nigerian insects, chiefly directed to the investigation of lycaenid life-histories and to the relation of Lycaenidae, Diptera, and other insects to ants', *Trans. ent. Soc. Lond.*, (1921), 319—448.

FEDER, H. M. (1972) 'Escape responses in marine invertebrates', *Scient. Am.*, **227**(1), 93—100.

FEIR, D. and SUEN, J. (1971) 'Cardenolides in the milkweed plant and feeding by the milkweed bug', *Ann. ent. Soc. Am.*, **64**, 1173—4.

FICKEN, R. W., MATTHIAE, P. E. and HORWICH, R. (1971) 'Eye marks in vertebrates: aids to vision', *Science, N.Y.*, **173**, 936—9.

FISHELSON, L. (1960) 'The biology and behaviour of *Poekilocerus bufonius* Klug, with special reference to the repellent gland (Orth. Acrididae)', *Eos, Madr.*, **36**, 41—62.

FORD, E. B. (1945) *Butterflies*. London, Collins.

FORD, E. B. (1955) *Moths*. London, Collins.

FORD, E. B. (1964) *Ecological Genetics*. London, Methuen.

FORD, H. A. (1971) 'The degree of mimetic protection gained by new partial mimics', *Heredity*, **27**, 227—36.

FRASER, J. (1962) *Nature Adrift*. London, Foulis.

FRAZER, J. F. D. and ROTHSCHILD, M. (1962) 'Defence mechanisms in warningly-coloured moths and other insects', *Int. Congr. Ent.*, **11**, 3, 249—56.

FRETTER, V. and GRAHAM, A. (1962) *British Prosobranch Molluscs*. London, Ray Society.

FRICKE, H. W. (1970) 'Ein mimetisches Kollektiv — Beobachtungen an Fischschwärmen, die Seeigel nachahmen', *Mar. Biol.*, **5**, 307—14.

FRICKE, H. W. and HENTSCHEL, M. (1971) 'Die Garnelen-Seeigel-Partnerschaft — eine Untersuchung der optischen Orientierung der Garnele', *Z. Tierpsychol.*, **28**, 453—62.

FRY, C. H. (1969) 'The recognition and treatment of venomous and non-venomous insects by small bee-eaters', *Ibis*, **111**, 23—9.

FRYER, J. C. F. (1913) 'Pupal coloration in *Papilio polytes* Linn.', *Trans. ent. Soc. Lond.* (1913), 414—19.

FUSEINI, B. A. (1972) *The Biology of Cotton Stainers* (Dysdercus *spp.* Heteroptera; Pyrrhocoridae) *in southern Ghana.* MSc. thesis, University of Ghana.

GANS, C. (1961) 'Mimicry in procryptically colored snakes of the genus *Dasypeltis*', *Evolution, Lancaster, Pa.*, **15**, 72—91.

GANS, C. (1964) 'Empathic learning and the mimicry of African snakes', *Evolution, Lancaster, Pa.*, **18**, 705.

GANS, C. and RICHMOND, N. D. (1957) 'Warning behaviour in snakes of the genus *Dasypeltis*', *Copeia* (1957), 269—74.

GEIST, V. (1966) 'The evolution of horn-like organs', *Behaviour*, **27**, 175—214.

GELPERIN, A. (1968) 'Feeding behaviour of the praying mantis: a learned modification', *Nature, Lond.*, **219**, 399—400.

GHISELIN, M. (1966) 'The adaptive significance of gastropod torsion', *Evolution, Lancaster, Pa.*, **20**, 337—48.

GIBSON, R. W. (1971) 'Glandular hairs providing resistance to aphids in certain wild potato species', *Ann. appl. Biol.*, **68**, 113—19.

GIESEL, J. T. (1970) 'On the maintenance of a shell pattern and behaviour polymorphism in *Acmaea digitalis*, a limpet', *Evolution, Lancaster, Pa.*, **24**, 98—119.

GILBERT, J. J. (1967) '*Asplanchna* and posterolateral spine production in *Brachionus calyciflorus*', *Arch. Hydrobiol.*, **64**, 1—62.

GILBERT, L. E. (1971) 'Butterfly—plant coevolution: has *Passiflora adenopoda* won the selectional race with heliconiine butterflies?', *Science, N.Y.*, **172**, 585—6.

GILL, J. S. and LEWIS, C. T. (1971) 'Systemic action of an insect feeding deterrent', *Nature, Lond.*, **232**, 402—3.

GOHAR, H. A. F. (1948) 'Commensalism between fish and anemone (with a description of the eggs of *Amphiprion bicinctus* Rüppell)', *Publs mar. biol. Stn Ghardaqa*, **6**, 35—44.

GONOR, J. J. (1965) 'Predator—prey reactions between two marine prosobranch gastropods', *Veliger*, **7**, 228—32.

GOTTO, R. V. (1969) *Marine animals. Partnerships and other associations.* London, English Universities.

GOTWALD, W. H. (1972) 'Analogous prey escape mechanisms in a pulmonate mollusk and lepidopterous larvae', *Jl N.Y. ent. Soc.*, **80**, 111—13.

GRIFFIN, D. R. (1958) *Listening in the Dark.* New Haven, Yale University.

GRIMES, L. G. (1973) 'The breeding of Heuglin's masked weaver and its nesting association with the red weaver ant', *Ostrich*, 44, 170—5.

GUTHRIE, R. D. and PETOCZ, R. G. (1970) 'Weapon automimicry among mammals', *Am. Nat.*, 104, 585—8.

HAEFELFINGER, H. R. (1969) 'Pigment and pattern in marine slugs', *Documenta Geigy, Nautilus*, 5, 3—4.

HAGEN, D. W. and GILBERTSON, L. G. (1972) 'Geographic variation and environmental selection in *Gasterosteus aculeatus* L. in the Pacific northwest, America', *Evolution, Lancaster, Pa.*, 26, 32—51.

HALBACH, U. (1971) 'Zum Adaptivwert der zyklomorphen Dornenbildung von *Brachionus calyciflorus* Pallas (Rotatoria). 1. Räuber-Beute-Beziehung in Kurzzeit-Versuchen', *Oecologia*, 6, 267—88.

HAMILTON, W. J., III and PETERMAN, R. M. (1971) 'Countershading in the colourful reef fish *Chaetodon lunula*: concealment, communication, or both?', *Anim. Behav.*, 19, 357—64.

HARDY, A. C. (1956) *The Open Sea. Its Natural History: Part 1, the World of Plankton*. London, Collins.

HARDY, A. C. (1967) *Great Waters*. London, Collins.

HAVERSCHMIDT, F. (1964) 'Potoo', in: *A New Dictionary of Birds*, ed. A. L. Thompson, 661—3, London and Edinburgh, Nelson.

HEATWOLE, H. (1965) 'Some aspects of the association of cattle egrets with cattle', *Anim. Behav.*, 13, 79—83.

HECHT, M. K. and MARIEN, D. (1956) 'The coral snake mimic problem: a reinterpretation', *J. Morph.*, 98, 335—56.

HERMANN, H. R. (1971) 'Sting autotomy, a defensive mechanism in certain social Hymenoptera', *Insectes soc.*, 18, 111—20.

HERREBOUT, W. M., KUYTEN, P. J. and RUITER, L. de (1963) 'Observations on colour patterns and behaviour of caterpillars feeding on scots pine', *Archs néerl. Zool.*, 15, 315—57.

HERRING, P. J. and CLARKE, M. R. (1971) *Deep Oceans*. London, Arthur Barker.

HINDE, R. A. (1952) 'The behaviour of the great tit (*Parus major*) and some other related species', *Behaviour, supplement* 2, 1—201.

HINDE, R. A. (1954) 'Factors governing the changes in strength of a partially inborn response, as shown by the mobbing behaviour of the chaffinch (*Fringilla coelebs*). 1. The nature of the response, and an examination of its course', *Proc. R. Soc., B*, 142, 306—31.

HINDSBRO, O. (1972) 'Effects of *Polymorphus* (Acanthocephala) on colour and behaviour of *Gammarus lacustris*', *Nature, Lond.*, 238, 333.

HINGSTON, R. W. G. (1927a) 'Protective devices in spiders' snares, with a description of seven new species of orb-weaving spiders', *Proc. zool. Soc. Lond.* (1927) (1), 259—93.

HINGSTON, R. W. G. (1927b) 'Field observations on spider mimics', *Proc. zool. Soc. Lond.* (1927) (2), 841—58.

HINTON, H. E. (1946) 'The "gin-traps" of some beetle pupae; a protective device which appears to be unknown', *Trans. R. ent. Soc. Lond.*, 97, 473—96.

HINTON, H. E. (1948) 'Sound production in lepidopterous pupae', *Entomologist*, 81, 254—69.

HINTON, H. E. (1951) 'On a little-known protective device of some chrysomelid pupae (Coleoptera)', *Proc. R. ent. Soc. Lond.*, A, 26, 67—73.

HINTON, S. (1962) 'Unusual defense movements in *Scorpaena plumieri mystes*', *Copeia* (1962), 842.

HOCKING, B. (1964) 'Fire melanism in some African grasshoppers', *Evolution, Lancaster, Pa.*, 18, 332—5.

HÖLLDOBLER, B. (1971) 'Communications between ants and their guests', *Scient. Am.*, 224(3), 86—93.

HOLLING, C. S. (1965) 'The functional response of predators to prey density and its role in mimicry and population regulation', *Mem. ent. Soc. Can.*, 45, 1—60.

HOLMES, W. (1940) 'The colour changes and colour patterns of *Sepia officinalis* L.', *Proc. zool. Soc. Lond.*, 110, 17—35.

HOOGLAND, R., MORRIS, D. and TINBERGEN, N. (1957) 'The spines of sticklebacks (*Gasterosteus* and *Pygosteus*) as means of defence against predators (*Perca* and *Esox*)', *Behaviour*, 10, 205—36.

HORN, M. H. (1970) 'The swimbladder as a juvenile organ in stromateoid fishes', *Breviora*, No. 359, 1—9.

HOSKING, E. (1970) *An Eye for a Bird*. London, Hutchinson.

HOWARD, R. R. and BRODIE, E. D. (1971) 'Experimental study of mimicry in salamanders involving *Notophthalmus viridescens viridescens* and *Pseudotriton ruber schencki*', *Nature, Lond.*, 233, 277.

HUGHES, R. N. and HUGHES, H. P. I. (1971) 'A study of the gastropod *Cassis tuberosa* (L.) preying upon sea urchins', *J. exp. mar. Biol. Ecol.*, 7, 305—14.

HUHEEY, J. E. (1960) 'Mimicry in the color pattern of certain Appalachian salamanders', *J. Elisha Mitchell scient. Soc.*, 76, 246—51.

HUHEEY, J. E. (1961) 'Studies in warning coloration and mimicry. III. Evolution of müllerian mimicry', *Evolution, Lancaster, Pa.*, 15, 567—8.

HUMPHRIES, D. A. and DRIVER, P. M. (1967) 'Erratic display as a device against predators', *Science, N.Y.*, 156, 1767—8.

HUMPHRIES, D. A. and DRIVER, P. M. (1971) 'Protean defence by prey animals', *Oecologia*, 5, 285—302.

IKIN, M. and TURNER, J. R. G. (1972) 'Experiments on mimicry: *Gestalt* perception and the evolution of genetic linkage', *Nature, Lond.*, 239, 525—7.

ISELY, F. B. (1938) 'Survival value of acridian protective coloration', *Ecology*, 19, 370–89.

JANZEN, D. H. (1966) 'Coevolution of mutualism between ants and acacias in Central America', *Evolution, Lancaster, Pa.*, 20, 249–75.

JANZEN, D. H. (1972) 'Escape in space by *Sterculia apetala* seeds from the bug *Dysdercus fasciatus* in a Costa Rican deciduous forest', *Ecology*, 53, 350–61.

JEANNE, R. L. (1970) 'Chemical defence of brood by a social wasp', *Science, N.Y.*, 168, 1465–6.

JEANNE, R. L. (1972) 'Social biology of the neotropical wasp *Mischocyttarus drewseni*', *Bull. Mus. comp. Zool. Harv.*, 144, 63–150.

JENSEN, M. (1966) 'The response of two sea-urchins to the sea-star *Marthasterias glacialis* (L.) and other stimuli', *Ophelia* (1966), 209–19.

JONES, D. A., PARSONS, J. and ROTHSCHILD, M. (1962) 'Release of hydrocyanic acid from crushed tissues of all stages in the life-cycle of species of the Zygaeninae (Lepidoptera)', *Nature, Lond.*, 193, 52–3.

JOVANCIC, L. (1960) 'Genèse des pigments tégumentaires et leur rôle physiologique chez la mante religieuse et chez d'autres espèces animales', *Mus. Hist. nat., Beograd*. 1–114.

KALMIJN, A. J. (1971) 'The electric sense of sharks and rays', *J. exp. Biol.*, 55, 371–83.

KETTLEWELL, H. B. D. (1955) 'Recognition of appropriate backgrounds by the pale and black phase of Lepidoptera', *Nature, Lond.*, 175, 934.

KETTLEWELL, H. B. D. (1956) 'Further selection experiments on industrial melanism in the Lepidoptera', *Heredity*, 10, 287–301.

KETTLEWELL, H. B. D. (1959) 'Brazilian insect adaptations', *Endeavour*, 18, 200–10.

KETTLEWELL, H. B. D., BERRY, R. J., CADBURY, C. J. and PHILLIPS, G. C. (1969) 'Differences in behaviour, dominance and survival within a cline. *Amathes glareosa* Esp. (Lep.) and its melanic *f. edda* Staud. in Shetland', *Heredity*, 24, 15–25.

KETTLEWELL, H. B. D., CADBURY, C. J. and LEES, D. R. (1971) 'Recessive melanism in the moth *Lasiocampa quercus* L. in industrial and non-industrial areas', in: *Ecological Genetics and Evolution*, ed. R. Creed, 175–201, Oxford and Edinburgh, Blackwell.

KISLOW, C. J. and EDWARDS, L. J. (1972) 'Repellent odour in aphids', *Nature, Lond.*, 235, 108–9.

KLOPFER, P. H. (1959) 'Social interactions in discrimination learning with special reference to feeding behaviour in birds', *Behaviour*, 14, 282–99.

KLOPFER, P. H. (1961) 'Observational learning in birds: the establishment of behavioural modes', *Behaviour*, 17, 71–80.

KOLENOSKY, G. B. (1972) 'Wolf predation on wintering deer in East-Central Ontario', *J. Wildl. Mgmt*, **36**, 357—69.

KRASNE, F. B. (1965) 'Escape from recurring tactile stimulation in *Branchiomma vesiculosum*', *J. exp. Biol.*, **42**, 307—22.

KREBS, C. J., GAINES, M. S., KELLER, B. L., MYERS, J. H. and TAMARIN, R. H. (1973) 'Population cycles in small rodents', *Science, N.Y.*, **179**, 35—41.

KRIEGER, R. I., FEENY, P. P. and WILKINSON, C. F. (1971) 'Detoxification enzymes in the guts of caterpillars: an evolutionary answer to plant defenses?', *Science, N.Y.*, **172**, 579—81.

KRUUK, H. (1964) 'Predators and anti-predator behaviour of the black-headed gull (*Larus ridibundus* L.)', *Behaviour*, supplement **11**, 1—129.

KRUUK, H. (1972) *The Spotted Hyaena*. Chicago and London, University of Chicago.

KWEI, E. A. (1969) *The Biology and Fisheries of the Horse Mackerel, Caranx hippos (Linné) in Ghanaian Waters*. MSc. thesis, University of Ghana.

LACK, D. (1971) *Ecological Isolation in Birds*. Oxford and Edinburgh, Blackwell.

LAGLER, K. F., BARDACH, J. E. and MILLER, R. R. (1962) *Ichthyology*. New York, John Wiley.

LAMBORN, W. A. (1913) 'On the relationship between certain West African insects, especially ants, Lycaenidae and Homoptera', *Trans. ent. Soc. Lond.* (1913), 436—98.

LAWICK-GOODALL, H. and LAWICK-GOODALL, J. van (1970) *Innocent Killers*. London and Glasgow, Collins.

LAWS, H. M. and LAWS, D. F. (1972) 'The escape response of *Donacilla angusta* Reeve (Mollusca: Bivalvia) in the presence of a naticed predator', *Veliger*, **14**, 289—90.

LESTON, D. (1972) *Insect Interrelations in Cocoa: a Contribution to Tropical Ecology*. PhD. thesis, University of Ghana.

LESTON, D. (1973) 'The flight behaviour of cocoa-capsids (Heteroptera: Miridae)', *Entomologia exp. appl.*, **16**, 91—100.

LEVIN, D. A. (1971) 'Plant phenolics: an ecological perspective', *Am. Nat.*, **105**, 157—81.

LEVIN, M. P. (1973) 'Preferential mating and the maintenance of the sex-limited dimorphism in *Papilio glaucus*: evidence from laboratory studies', *Evolution, Lancaster, Pa.*, **27**, 257—64.

LIMBAUGH, C. (1961) 'Cleaning symbiosis', *Scient. Am.*, **205**(2), 42—9.

LINDROTH, C. H. (1971) 'Disappearance as a protective factor. A supposed case of Batesian mimicry among beetles (Coleoptera: Carabidae and Chrysomelidae)', *Entomologia scand.*, **2**, 41—8.

LINSLEY, E. G., EISNER, T. and KLOTS, A. B. (1961) 'Mimetic

assemblages of sibling species of lycid beetles', *Evolution, Lancaster, Pa.,* 15, 15—29.

LISSMANN, H. W. (1958) 'On the function and evolution of electric organs in fish', *J. exp. Biol.,* 35, 156—91.

LISSMANN, H. W. (1963) 'Electric location by fishes', *Scient. Am.,* 208(3), 50—9.

LOOP, M. S. and SCOVILLE, S. A. (1972) 'Response of newborn *Eumeces inexpectatus* to prey-object extracts', *Herpetologica,* 28, 254—6.

LOPEZ, A. and QUESNEL, V. C. (1970) 'Defensive secretions of some papilionid caterpillars', *Caribb. J. Sci.,* 10, 5—7.

LOSEY, G. S. (1972) 'Predation protection in the poison-fang blenny, *Meiacanthus atrodorsalis,* and its mimics *Ecsenius bicolor* and *Runula laudandus* (Blenniidae)', *Pacif. Sci.,* 26, 129—39.

MACGINITIE, G. E. and MACGINITIE, N. (1949) *Natural History of Marine Animals.* New York, McGraw-Hill.

MACKINNON, J. (1970) 'Indications of territoriality in mantids', *Z. Tierpsychol.,* 27, 150—5.

McPHAIL, J. D. (1969) 'Predation and the evolution of a stickleback (*Gasterosteus*)', *J. Fish. Res. Bd Can.,* 26, 3183—208.

MAGNUS, B. E. (1963) 'Sex limited mimicry II — visual selection in the mate choice of butterflies', *Int. Congr. Zool.,* 16, 4, 179—83.

MAINARDI, D. and ROSSI, A. C. (1969) 'La distribuzione delle attinie *Calliactis parasitica* in rapporto allo stato sociale nel paguro *Dardanus arrosor*', *Pubbl. Staz. zool. Napoli,* 37, supplement, 200—2.

MALDONADO, H. (1970) 'The deimatic reaction in the praying mantis *Stagmatoptera biocellata*', *Z. vergl. Physiol.,* 68, 60—71.

MANSUETI, R. (1963) 'Symbiotic behaviour between small fishes and jellyfishes, with new data on that between the stromateid, *Peprilus alepidotus,* and the scyphomedusa *Chrysaora quinquecirrha*', *Copeia* (1963), 40—80.

MARGOLIN, A. S. (1964) 'The mantle response of *Diodora aspersa*', *Anim. Behav.,* 12, 187—94.

MARISCAL, R. N. (1970a) 'The nature of the symbiosis between Indo—Pacific anemone fishes and sea anemones', *Mar. Biol.,* 6, 58—65.

MARISCAL, R. N. (1970b) 'An experimental analysis of the protection of *Amphiprion xanthurus* Cuvier and Valenciennes and some other anemone fishes from sea anemones', *J. exp. mar. Biol. Ecol.,* 4, 134—49.

MARISCAL, R. N. (1972) 'Behaviour of symbiotic fishes and sea anemones', in: *Behaviour of Marine Animals,* Vol. 2, eds. H. E. Winn and B. L. Olla, 327—60, New York and London, Plenum Press.

MARLAR, P. (1956) 'Behaviour of the chaffinch *Fringilla coelebs*', *Behaviour,* supplement 5, 1—184.

MARLAR, P. (1957) 'Specific distinctiveness in the communication signals of birds', *Behaviour*, 11, 13—39.

MARPLES, B. J. (1969) 'Observations on decorated webs', *Bull. Br. arachnol. Soc.*, 1, 13—18.

MARSHALL, N. B. (1954) *Aspects of Deep Sea Biology*. London, Hutchinson.

MARSHALL, N. B. (1965) *The Life of Fishes*. London, Weidenfeld & Nicholson.

MARSON, J. E. (1947*a*) 'Some observations on the ecological variation and development of the cruciate zigzag camouflage device of *Argiope pulchella* (Thor.)', *Proc. zool. Soc. Lond.*, 117, 219—27.

MARSON, J. E. (1947*b*) 'The ant mimic *Myrmarachne plataleoides*', *Jl E. Africa nat. Hist. Soc.*, 19, 62—3.

MARTIN, R. and BRINCKMANN, A. (1963) 'Zum Brutparasitismus von *Phyllirrhoe bucephala* Per. & Les. (Gastropoda, Nudibranchia) auf der Meduse *Zanclea costata* Gegenb. (Hydrozoa, Anthomedusae)', *Pubbl. Staz. zool. Napoli*, 33, 206—23.

MATHEW, A. P. (1954) 'Observations on the habits of two spider mimics of the red ant, *Oecophylla smaragdina* (Fabr.)', *J. Bombay nat. Hist. Soc.*, 52, 249—63.

MECH, L. D. (1970) *The wolf: the Ecology and Behaviour of an Endangered Species*. New York, Natural History Press.

MELZACK, R. (1961) 'On the survival of mallard ducks after "habituation" to the hawk-shaped figure', *Behaviour*, 17, 9—16.

MERTENS, R. (1966) 'Das Problem der Mimikry bei Korallenschlangen', *Zool. Jb. Syst.*, 84, 541—76.

MILLER, L. A. (1971) 'Physiological responses of green lacewings (*Chrysopa*, Neuroptera) to ultrasound', *J. Insect Physiol.*, 17, 491—506.

MOMENT, G. B. (1962) 'Reflexive selection: a possible answer to an old puzzle', *Science, N.Y.*, 136, 262—3.

MOODIE, G. E. E. (1972) 'Predation, natural selection and adaptation in an unusual three spine stickleback', *Heredity*, 28, 155—67.

MORRELL, G. M. and TURNER, J. R. G. (1970) 'Experiments on mimicry: I. the response of wild birds to artificial prey', *Behaviour*, 36, 116—30.

MORRELL, R. (1969) 'Play snake for safety', *Animals*, 12, 154—5.

MORRIS, D. (1958) 'The reproductive behaviour of the ten-spined stickleback (*Pygosteus pungitius* L.)', *Behaviour*, supplement 6, 1—154.

MORRIS, D. (1967) *The Naked Ape*. London, Cape.

MORTON, E. S. (1971) 'Nest predation affecting the breeding season of the clay-colored robin, a tropical songbird', *Science, N.Y.*, 171, 920—1.

MOSS, A. M. (1920) 'Sphingidae of Para, Brazil', *Novit. zool.*, 27, 333—424.

MOSTLER, G. (1935) 'Beobachtungen zur Frage der Wespenmimikry', *Z. Morph. Ökol. Tiere*, 29, 381—454.

MOYNIHAN, M. (1968) 'Social mimicry; character convergence versus character displacement', *Evolution, Lancaster, Pa.*, 22, 315—31.

MUELLER, H. C. (1971) 'Oddity and specific searching image more important than conspicuousness in prey selection', *Nature, Lond.*, 233, 345—6.

MÜLLER-SCHWARZE, D. (1972) 'Responses of young black-tailed deer to predator odors', *J. Mammal.*, 53, 393—4.

MURRAY, J. (1972) *Genetic Diversity and Natural Selection.* Edinburgh, Oliver & Boyd.

NELLIS, C. H. and KEITH, L. B. (1968) 'Hunting activities and success of lynxes in Alberta', *J. Wildl. Mgmt*, 32, 718—22.

NELSON, C. E. and MILLER, G. A. (1971) 'A possible case of mimicry in frogs', *Herpetol. Rev.*, 3, 109.

NICOL, J. A. C. (1950) 'Responses of *Branchiomma vesiculosum* (Montagu) to photic stimulation', *J. mar. biol. Ass. U.K.*, 29, 303—20.

NICOL, J. A. C. (1960) *The Biology of Marine Animals.* London, Pitman.

NICOL, J. A. C. (1971) 'Physiological investigations of oceanic animals', in: *Deep Oceans*, eds. P. J. Herring and M. R. Clarke, 225—46, London, Barker.

NIKOLSKY, G. V. (1963) *The Ecology of Fishes.* London and New York, Academic Press.

O'DONALD, P. (1968) 'Natural selection by glow-worms in a population of *Cepaea nemoralis*', *Nature, Lond.*, 217, 194.

O'DONALD, P. and PILECKI, C. (1970) 'Polymorphic mimicry and natural selection', *Evolution, Lancaster, Pa.*, 24, 395—401.

ORR, L. P. (1967) 'Feeding experiments with a supposed mimetic complex in salamanders', *Am. Midl. Nat.*, 77, 147—55.

ORR, L. P. (1968) 'The relative abundance of mimics and models in a supposed mimetic complex of salamanders', *J. Elisha Mitchell scient. Soc.*, 84, 303—4.

OTTE, D. and WILLIAMS, K. (1972) 'Environmentally induced color dimorphisms in grasshoppers *Syrbula admirabilis*, *Dichromorpha viridis*, and *Chortophaga viridifasciata*', *Ann. ent. Soc. Am.*, 65, 1154—61.

OWEN, D. F. (1965) 'Density effects in polymorphic land snails', *Heredity*, 20, 312—15.

OWEN, D. F. (1969) 'Ecological aspects of polymorphism in an African land snail, *Limicolaria martensiana*', *J. Zool., Lond.*, 159, 79—96.

OWEN, D. F. (1970) 'Mimetic polymorphism and the palatability spectrum', *Oikos*, 21, 333—6.

OWEN, D. F. (1971) *Tropical Butterflies.* Oxford, Clarendon Press.

OWEN, D. F. and CHANTER, D. O. (1968) 'Population biology of tropical African butterflies. 2. Sex ratio and polymorphism in *Danaus chrysippus* L.', *Revue Zool. Bot. afr.*, **78**, 81—97.

OWEN, D. F. and CHANTER, D. O. (1969) 'Population biology of tropical African butterflies. Sex ratio and genetic variation in *Acraea encedon*', *J. Zool., Lond.*, **157**, 345—74.

OWEN, D. F. and CHANTER, D. O. (1971) 'Polymorphism in West African populations of the butterfly, *Acraea encedon*', *J. Zool., Lond.*, **163**, 481—8.

PACKARD, A. (1972) 'Cephalopods and fish: the limits of convergence', *Biol. Rev.*, **47**, 241—307.

PACKARD, A. and SANDERS, G. D. (1971) 'Body patterns of *Octopus vulgaris* and maturation of the response to disturbance', *Anim. Behav.*, **19**, 780—90.

PAGES, E. (1970) 'Sur l'écologie et les adaptations de l'oryctérope et des pangolins sympatriques du Gabon', *Biol. Gabon*, **6**, 27—92.

PAINE, R. T. (1963) 'Food recognition and predation on opisthobranchs by *Navanax inermis* (Gastropoda: Opisthobranchia)', *Veliger*, **6**, 1—9.

PAINE, R. T. (1965) 'Natural history, limiting factors and energetics of the opisthobranch *Navanax inermis*', *Ecology*, **46**, 603—19.

PAULSON, D. R. (1973) 'Predator polymorphism and apostatic selection', *Evolution, Lancaster, Pa.*, **27**, 269—77.

PEARSON, O. P. (1964) 'Carnivore—mouse predation: an example of its intensity and bioenergetics', *J. Mammal.*, **45**, 177—88.

PETERS, R. C. and BRETSCHNEIDER, F. (1972) 'Electric phenomena in the habitat of the catfish *Ictalurus nebulosus* LeS', *J. comp. Physiol.*, **81**, 345—62.

PETERSEN, B., TÖRNBLOM, O. and BODIN, N. O. (1952) 'Verhaltensstudien am Rapsweissling und Bergweissling (*Pieris napi* L. und *Pieris bryoniae* Ochs.)', *Behaviour*, **4**, 67—84.

PFEIFFER, W. (1961) 'Zur Biologie des Zitterrochens (*Torpedo marmorata* Risso)', *Pubbl. Staz. zool. Napoli*, **32**, 167—71.

PFEIFFER, W. (1962) 'The fright reaction of fish', *Biol. Rev.*, **37**, 495—511.

PFEIFFER, W. and LEMKE, J. (1973) 'Untersuchungen zur Isolierung und Identifizierung des Schreckstoffes aus der Haut der Elritze, *Phoxinus phoxinus* (L.) (Cyprinidae, Ostariophysi, Pisces)', *J. comp. Physiol.*, **82**, 407—10.

PICKENS, P. E. and McFARLAND, W. N. (1964) 'Electric discharge and associated behaviour in the stargazer', *Anim. Behav.*, **12**, 362—7.

PILECKI, C. and O'DONALD, P. (1971) 'The effects of predation on artificial mimetic polymorphisms with perfect and imperfect mimics at varying frequencies', *Evolution, Lancaster, Pa.*, **25**, 365—70.

PINHEY, E. (1960) *Hawk Moths of Central and Southern Africa*. Cape Town, Longman.

PLATT, A. P. and BROWER, L. P. (1968) 'Mimetic versus disruptive coloration in intergrading populations of *Limenitis arthemis* and *astyanax* butterflies', *Evolution, Lancaster, Pa.*, 22, 699–718.

PLATT, A. P., COPPINGER, R. P. and BROWER, L. P. (1971) 'Demonstration of the selective advantage of mimetic *Limenitis* butterflies presented to caged avian predators', *Evolution, Lancaster, Pa.*, 25, 692–701.

PLISKE, T. E. (1972) 'Sexual selection and dimorphism in female tiger swallow tails, *Papilio glaucus* L. (Lepidoptera: Papilionidae): a reappraisal', *Ann. ent. Soc. Am.*, 65, 1267–70.

POTTS, G. W. (1973) 'The ethology of *Labroides dimidiatus* (Cuv. & Val.) (Labridae, Pisces) on Aldabra', *Anim. Behav.*, 21, 250–91.

POUGH, F. H. (1971) 'Leech-repellent property of eastern red-spotted newts, *Notophthalmus viridescens*', *Science, N.Y.*, 174, 1144–6.

POUGH, F. H., BROWER, L. P., MECK, H. R. and KESSELL, S. R. (1973) 'Theoretical investigations of automimicry: multiple trial learning and the palatability spectrum', *Proc. natn. Acad. Sci. U.S.A.*, 70, 2261–5.

POULTON, E. B. (1888) 'Notes in 1887 upon lepidopterous larvae, etc., including a complete account of the life-history of the larvae of *Sphinx convolvuli* and *Aglia tan*', *Trans. ent. Soc. Lond.* (1888), 515–606.

POULTON, E. B. (1890) *The Colours of Animals*. London, Kegan Paul, Trench, Trübner.

PROUHO, H. (1890) 'Du rôle des pédicellaires gemmiformes des oursins', *C. r. hebd. Séanc. Acad. Sci., Paris*, 111, 62–4.

PURCHON, R. D. (1968) *The Biology of the Mollusca*. Oxford, Pergamon.

RAND, A. S. (1967) 'Predator–prey interactions and the evolution of aspect diversity', *Atas Simp. Biota Amazonica*, 5 (Zoologica), 73–83.

RASKE, A. G. (1967) 'Morphological and behavioural mimicry among beetles of the genus *Moneilema*', *Pan–Pacif. Ent.*, 43, 239–44.

REES, W. J. (1966) '*Cyanea lamarcki* Péron & Lesueur (Scyphozoa) and its association with young *Gadus merlangus* L. (Pisces)', *Ann. Mag. nat. Hist.*, (13) 9, 285–7.

REESE, E. S. (1963) 'The behavioural mechanisms underlying shell selection by hermit crabs', *Behaviour*, 21, 78–126.

REGNIER, F. E. and WILSON, E. O. (1971) 'Chemical communication and "propaganda" in slave-maker ants', *Science, N.Y.*, 172, 267–9.

REICHSTEIN, T., EUW, J. V., PARSONS, J. A. and ROTHSCHILD, M. (1968) 'Heart poisons in the monarch butterfly', *Science, N.Y.*, 161, 861–6.

REMINGTON, C. L. (1963) 'Historical backgrounds of mimicry', *Int. Congr. Zool.*, 16, 4, 145–9.

REMOLD, H. (1963) 'Scent-glands of land-bugs, their physiology and biological function', *Nature, Lond.*, 198, 764—8.

RETTENMEYER, C. W. (1970) 'Insect mimicry', *A. Rev. Ent.*, 15, 43—74.

REUTTER, K. and PFEIFFER, W. (1973) 'Fluoreszenzmikroskopischer Nachweis des Schreckstoffes in den Schreckstoffzellen der Elritze, *Phoxinus phoxinus* (L.) (Cyprinidae, Ostariophysi, Pisces)', *J. comp. Physiol.*, 82, 411—18.

RICHARDS, O. W. (1947) 'Observations on *Trypoxylon placidum* Cam. (Hym., Sphecoidea)', *Entomologist's mon. Mag.*, 83, 53.

RILLING, S., MITTELSTAEDT, H. and ROEDER, K. D. (1959) 'Prey recognition in the praying mantis', *Behaviour*, 14, 164—84. .

ROBILLIARD, G. A. (1972) 'A new species of *Dendronotus* from the northeastern Pacific with notes on *Dendronotus nanus* and *Dendronotus robustus* (Mollusca: Opisthobranchia)', *Can. J. Zool.*, 50, 421—32.

ROBINSON, M. H. (1965) 'The Javanese stick insect, *Orxines macklotti* De Haan (Phasmatodea, Phasmidae)', *Entomologist's mon. Mag.*, 100, 253—9.

ROBINSON, M. H. (1968a) 'The defensive behaviour of the stick insect *Oncotophasma martini* (Griffini) (Orthoptera: Phasmatidae)', *Proc. R. ent. Soc. Lond.*, 43, 183—7.

ROBINSON, M. H. (1968b) 'The defensive behaviour of the Javanese stick insect, *Orxines macklotti* De Haan, with a note on the startle display of *Metriotes diocles* Westw. (Phasmatodea, Phasmidae)', *Entomologist's mon. Mag.*, 104, 46—54.

ROBINSON, M. H. (1968c) 'The defensive behavior of *Pterinoxylus spinulosus* Redtenbacher, a winged stick insect from Panama (Phasmatodea)', *Psyche, Camb.*, 75, 195—207.

ROBINSON, M. H. (1969a) 'The· defensive behaviour of some orthopteroid insects from Panama', *Trans. R. ent. Soc. Lond.*, 121, 281—303.

ROBINSON, M. H. (1969b) 'Defenses against visually hunting predators', in: *Evolutionary Biology 3*, eds. T. Dobzhansky, M. K. Hecht and W. C. Steere, 225—59, New York, Meredith Corporation.

ROBINSON, M. H. (1973) 'Insect anti-predator adaptations and the behavior of predatory primates', *Actas Congr. Latinamericano Zool.*, 4, 2, 811—36.

ROBINSON, M. H., ABELE, L. G. and ROBINSON, B. (1970) 'Attack autotomy: a defense against predators', *Science, N.Y.*, 169, 300—1.

ROBINSON, M. H. and ROBINSON, B. (1970) 'The stabilimentum of the orb-web spider, *Argiope argentata*: an improbable defence against predators', *Can. Ent.*, 102, 641—55.

ROBSON, E. A. (1961) 'The swimming response and its pacemaker system in the anemone *Stomphia coccinea*', *J. exp. Biol.*, 38, 685—94.

ROBSON, E. A. (1966) 'Swimming in Actiniaria', *Symp. zool. Soc. Lond.*, **10**, 333—60.

ROBSON, E. A. (1971) 'The behaviour and neuromuscular system of *Gonactinia prolifera*, a swimming sea anemone', *J. exp. Biol.*, **55**, 611—40.

ROEDER, K. D. (1962) 'The behaviour of free-flying moths in the presence of artificial ultrasonic pulses', *Anim. Behav.*, **10**, 300—4.

ROEDER, K. D. (1965) 'Moths and ultrasound', *Scient. Am.*, **212**(4), 94—102.

ROEDER, K. D., TREAT, A. E. and VANDEBERG, J. S. (1968) 'Auditory sense in certain sphingid moths', *Science, N.Y.*, **159**, 331—3.

ROEDER, K. D., TREAT, A. E. and VANDEBERG, J. S. (1970) 'Distal lobe of the pilifer: an ultrasonic receptor in choerocampine hawk-moths', *Science, N.Y.*, **170**. 1098—9.

ROSIN, R. (1969) 'Escape response of the sea-anemone *Anthopleura nigrescens* (Verrill) to its predatory eolid nudibranch *Herviella baba* spec. nov.', *Veliger*, **12**, 74—7.

ROSS, D. M. (1967) 'Behavioural and ecological relationships between sea anemones and other invertebrates', *Oceanogr. Mar. Biol. Ann. Rev.*, **5**, 291—316.

ROSS, D. M. (1970) 'The commensal association of *Calliactis polypus* and the hermit crab *Dardanus gemmatus* in Hawaii', *Can. J. Zool.*, **48**, 351—7.

ROSS, D. M. (1971) 'Protection of hermit crabs (*Dardanus* spp) from octopus by commensal sea anemones (*Calliactis* spp)', *Nature Lond.*, **230**, 401—2.

ROSS, D. M. and SUTTON, L. (1961a) 'The response of the sea anemone *Calliactis parasitica* to shells of the hermit crab *Pagurus bernhardus*', *Proc. R. Soc.*, *B*, **155**, 266—81.

ROSS, D. M. and SUTTON, L. (1961b) 'The association between the hermit crab *Dardanus arrosor* (Herbst) and the sea anemone *Calliactis parasitica* (Couch)', *Proc. R. Soc.*, *B*, **155**, 282—91.

ROSS, D. M. and SUTTON, L. (1967) 'Swimming sea anemones of Puget Sound: swimming of *Actinostola* new species in response to *Stomphia coccinea*', *Science, N.Y.*, **155**, 1419—21.

ROTH, L. M. and EISNER, T. (1962) 'Chemical defences of arthropods', *A. Rev. Ent.*, **7**, 107—36.

ROTHSCHILD, M. (1962) 'Defensive odours and müllerian mimicry among insects', *Int. Congr. Ent.*, *11*, 3, 257.

ROTHSCHILD, M. (1963) 'Is the buff ermine (*Spilosoma lutea* (Huf.)) a mimic of the white ermine (*Spilosoma lubricipeda* (L.))?', *Proc. R. ent. Soc. Lond.*, *A*, **38**, 159—64.

ROTHSCHILD, M. (1964) 'An extension of Dr Lincoln Brower's theory on bird predation and food specificity, together with some observations on bird memory in relation to aposematic colour patterns', *Entomologist*, 97, 73—8.

ROTHSCHILD, M. (1971) 'Speculations about mimicry', in: *Ecological Genetics and Evolution*, ed. R. Creed, 202—23, Oxford and Edinburgh, Blackwell.

ROTHSCHILD, M. (1972) 'Some observations on the relationship between plants, toxic insects and birds', in: *Phytochemical Ecology*, ed. J. B. Harborne, 1—12, London and New York, Academic Press.

ROTHSCHILD, M., EUW, J. V. and REICHSTEIN, T. (1970) 'Cardiac glycosides in the oleander aphid, *Aphis nerii*', *J. Insect Physiol.*, 16, 1141—5.

ROTHSCHILD, M. and KELLETT, D. N. (1972) 'Reactions of various predators to insects storing heart poisons (cardiac glycosides) in their tissues', *J. Ent.*, A, 46, 103—10.

ROWELL, C. H. F. (1967) 'Experiments on aggregations of *Phymateus purpurascens* (Orthoptera, Acrididae, Pyrgomorphinae)', *J. Zool., Lond.*, 152, 179—93.

ROWELL, C. H. F. (1971) 'The variable coloration of acridid grasshoppers', *Adv. Insect Physiol.*, 8, 146—98.

ROZIN, P. and KALAT, J. W. (1971) 'Specific hungers and poison avoidance as adaptive specializations of learning', *Psychol. Rev.*, 78, 459—86.

RUBINOFF, I. and KROPACH, C. (1970) 'Differential reactions of Atlantic and Pacific predators to sea snakes', *Nature, Lond.*, 228, 1288—90.

RUITER, L. D. (1952) 'Some experiments on the camouflage of stick caterpillars', *Behaviour*, 4, 222—32.

RUITER, L. D. (1959) 'Some remarks on problems of the ecology and evolution of mimicry', *Archs néerl. Zool.*, 13, supplement, 351—68.

RUSSEL, R. J. (1972) 'Defensive responses of the aphid *Drepanosiphum platanoides* in encounters with the bug *Anthocoris nemorum*', *Oikos*, 23, 264—7.

SARGENT, T. D. (1968) 'Cryptic moths: effects on background selections of painting the circumocular scales', *Science, N.Y.*, 159, 100—1.

SARGENT, T. D. (1969*a*) 'Behavioural adaptations of cryptic moths. II. Experimental studies on bark-like species', *Jl N.Y. ent. Soc.*, 77, 75—9.

SARGENT, T. D. (1969*b*) 'Behavioural adaptations of cryptic moths. V. Preliminary studies on an anthophilous species, *Schinia florida* (Noctuidae)', *Jl N.Y. ent. Soc.*, 77, 123—8.

SARGENT, T. D. (1969*c*) 'Behavioural adaptations of cryptic moths. III. Resting attitudes of two bark-like species, *Melanolophia canadaria* and *Catocala ultronia*', *Anim. Behav.*, 17, 670—2.

SCHALLER, G. B. (1967) *The Deer and the Tiger*. Chicago and London, University of Chicago.

SCHALLER, G. B. (1972) *The Serengeti Lion.* Chicago and London, University of Chicago.

SCHILDKNECHT, H. (1971) 'Evolutionary peaks in the defensive chemistry of insects', *Endeavour*, 30, 136—41.

SCHLICHTER, D. (1968) 'Das Zusammenleben von Riffanemonen und Anemonenfischen', Z. *Tierpsychol.*, 25, 933—54.

SCHLICHTER, D. (1970) *'Thalassoma amblycephalus* ein neuer Anemonenfisch-Typ. Allgemeine Aspekte zur Beurteilung der Vergesellschaftung von Riffanemonen und ihren Partnern', *Mar. Biol.*, 7, 269—72.

SCHLICHTER, D. (1972) 'Chemische Tarnung. Die stoffliche Grundlage der Anpassung von Anemonenfischen an Riffanemonen', *Mar. Biol.*, 12, 137—50.

SCHMIDT, R. S. (1958) 'Behavioural evidence on the evolution of Batesian mimicry', *Anim. Behav.*, 6, 129—38.

SCHMIDT, R. S. (1960) 'Predator behaviour and the perfection of incipient mimetic resemblances', *Behaviour*, 16, 149—58.

SEMLER, D. E. (1971) 'Some aspects of adaptation in a polymorphism for breeding colours in the threespine stickleback *(Gasterosteus aculeatus)'*, *J. Zool., Lond.*, 165, 291—302.

SEXTON, O. J. (1960) 'Experimental studies of artificial batesian mimics', *Behaviour*, 15, 244—52.

SEXTON, O. J., HOGER, C. and ORTLEB, E. (1966) *'Anolis carolinensis*: effects of feeding on reactions to aposematic prey', *Science, N.Y.*, 153, 1140.

SHEPPARD, P. M. (1951) 'Fluctuations in the selective value of certain phenotypes in the polymorphic land snail, *Cepaea nemoralis* (L.)', *Heredity*, 5, 125—34.

SHEPPARD, P. M. (1962) 'Some aspects of the geography, genetics, and taxonomy of a butterfly', in: *Taxonomy and Geography*, ed. D. Nichols, 135—52, London, The Systematics Association.

SHETTLEWORTH, S. J. (1972) 'The role of novelty in learned avoidance of nonpalatable "prey" by domestic chicks *(Gallus gallus)'*, *Anim. Behav.*, 20, 29—35.

SHOUP, J. B. (1968) 'Shell opening by crabs of the genus *Calappa'*, *Science, N.Y.*, 160, 887—8.

SILBERGLIED, R. E. and EISNER, T. (1969) 'Mimicry of Hymenoptera by beetles with unconventional flight', *Science, N.Y.*, 163, 486—8.

SIMMONS, J. A., WEVER, E. G. and PYLKA, T. M. (1971) 'Periodical cicada: sound production and hearing', *Science, N.Y.*, 171, 212—13.

SIMMONS, K. E. L. (1952) 'The nature of the predator-reactions of breeding birds', *Behaviour*, 4, 161—72.

SIMMONS, K. E. L. (1955) 'The nature of the predator-reactions of waders towards humans; with special reference to the role of the aggressive-, escape- and brooding-drives', *Behaviour*, 8, 130—73.

SINGER, M. C., EHRLICH, P. R. and GILBERT, L. E. (1971) 'Butter-fly feeding on lycopsid', *Science N.Y.*, **172**, 1341—2.

SKINNER, D. M. and GRAHAM, D. E. (1970) 'Molting in land crabs: stimulation by leg removal', *Science, N.Y.*, **169**, 383—5.

SMITH, C. C. (1970) 'The coevolution of pine squirrels (*Tamiasciurus*) and conifers', *Ecol. Monogr.*, **40**, 349—71.

SMITH, D. A. S. (1973*a*) 'Batesian mimicry between *Danaus chrysippus* and *Hypolimnas misippus* (Lepidoptera) in Tanzania', *Nature, Lond.*, **242**, 129—31.

SMITH, D. A. S. (1973*b*) 'Negative non-random mating in the poly-morphic butterfly *Danaus chrysippus* in Tanzania', *Nature, Lond.*, **242**, 131—2.

SMITH, J. N. M. and DAWKINS, R. (1971) 'The hunting behaviour of individual great tits in relation to spatial variations in their food density', *Anim. Behav.*, **19**, 695—706.

SMITH, L. S. (1961) 'Clam-digging behaviour in the starfish *Pisaster brevispinus* (Stimpson 1857)', *Behaviour*, **18**, 148—51.

SMITH, N. G. (1968) 'The advantage of being parasitized', *Nature, Lond.*, **219**, 690—4.

SMYTH, J. D. (1962) *Introduction to Animal Parasitology*. London, English Universities Press.

SMYTHE, N. (1970) 'On the existence of "pursuit invitation" signals in mammals', *Am. Nat.*, **104**, 491—4.

SNYDER, N. and SNYDER, H. (1970) 'Alarm response of *Diadema antillarum*', *Science, N.Y.*, **168**, 276—8.

SOANE, I. D. and CLARKE, B. (1973) 'Evidence for apostatic selec-tion by predators using olfactory cues', *Nature, Lond.*, **241**, 62—4.

SOUTHERN, H. N. (1964) *The Handbook of British Mammals*. Oxford, Blackwell.

SPARROWE, R. D. (1972) 'Prey-catching behaviour in the sparrow hawk', *J. Wildl. Mgmt*, **36**, 297—308.

SPRINGER, V. G. and SMITH-VANIZ, W. F. (1972) 'Mimetic relation-ships involving fishes of the family Blenniidae', *Smithson. Contr. Zool.*, **112**, 1—36.

STAMM, R. A. (1968) 'Zur Abwehr von Raubfeinden durch *Lobiger serradifalci* (Calcara), 1840, und *Oxynoe olivacea* Rafinesque, 1819 (Gastropoda, Opisthobranchia)', *Revue suisse Zool.*, **75**, 661—5.

STASEK, C. R. (1967) 'Autotomy in the Mollusca', *Occ. Pap. Calif. Acad. Sci.*, No. 61, 1—44.

STAUBER, L. A. (1945) '*Pinnotheres ostreum*, parasitic on the American oyster, *Ostraea* (*Gryphaea*) *virginica*', *Biol. Bull. mar. biol. Lab., Woods Hole*, **88**, 269—91.

STRIDE, G. O. (1956*a*) 'On the mimetic association between certain species of *Phonoctonus* (Hemiptera, Reduviidae) and the Pyrrho-coridae', *J. ent. Soc. sth. Afr.*, **19**, 12—28.

STRIDE, G. O. (1956*b*) 'On the courtship behaviour of *Hypolimnas* *misippus* L. (Lepidoptera, Nymphalidae), with notes on the mimetic association with *Danaus chrysippus* L. (Lepidoptera, Danaidae)', *Br. J. Anim. Behav.*, 4, 52–68.

STRIDE, G. O. (1957) 'Investigations into the courtship behaviour of the male of *Hypolimnas misippus* L. (Lepidoptera, Nymphalidae), with special reference to the role of visual stimuli', *Br. J. Anim. Behav.*, 5, 153–67.

STRIDE, G. O. (1958) 'Further studies on the courtship behaviour of African mimetic butterflies', *Anim. Behav.*, 6, 224–30.

SUDD, J. H. (1967) *An Introduction to the Behaviour of Ants.* London, Arnold.

SUMNER, F. B. (1934) 'Does "protective coloration" protect? Results of some experiments with fishes and birds', *Proc. natn. Acad. Sci. U.S.A.*, 20, 559–64.

SUMNER, F. B. (1935) 'Studies of protective color change. III. Experiments with fishes both as predators and prey', *Proc. natn. Acad. Sci. U.S.A.*, 21, 345–53.

SWYNNERTON, C. F. M. (1915*a*) 'A brief preliminary statement of a few of the results of five years' special testing of the theories of mimicry', *Trans. ent. Soc. Lond.* (1915), xxxii–xliii.

SWYNNERTON, C. F. M. (1915*b*) 'Birds in relation to their prey: experiments on wood-hoopoes, small hornbills and a babbler', *Jl S. Afr. Orn. Un.*, 11, 32–108.

SWYNNERTON, C. F. M. (1919) 'Experiments and observations bearing on the explanation of form and colouring, 1908–1913', *J. Linn. Soc. (Zool.)*, 33, 203–385.

SZAL, R. (1971) ' "New" sense organ of primitive gastropods', *Nature, Lond.*, 229, 490–2.

SZOLLOSI, D. (1969) 'Unique envelope of a jellyfish ovum: the armed egg', *Science, N.Y.*, 163, 586–7.

TAYLOR, P. B. and CHEN, L. (1969) 'The predator–prey relationship between the octopus (*Octopus bimaculatus*) and the California scorpionfish (*Scorpaena guttata*)', *Pacif. Sci.*, 23, 311–16.

THOMAS, G. E. and GRUFFYDD, Ll. D. (1971) 'The types of escape reactions elicited in the scallop *Pecten maximus* by selected sea-star species', *Mar. Biol.*, 10, 87–93.

THOMPSON, T. E. (1960*a*) 'Defensive acid-secretion in marine gastropods', *J. mar. biol. Ass. U.K.*, 39, 115–22.

THOMPSON, T. E. (1960*b*) 'Defensive adaptations in opisthobranchs', *J. mar. biol. Ass. U.K.*, 39, 123–34.

THOMPSON, T. E. (1964) 'Grazing and the life cycles of British nudibranchs', in: *Grazing in Terrestrial and Marine Environments*, ed. D. J. Crisp, 275–97, Oxford, Blackwell.

THOMPSON, T. E. (1967) 'Adaptive significance of gastropod torsion', *Malacologia*, 5, 423–30.

THOMPSON, T. E. (1969) 'Acid secretion in Pacific Ocean gastropods', *Aust. J. Zool.*, 17, 755—64.

THOMPSON, T. E. and BENNETT, I. (1970) 'Observations on Australian Glaucidae (Mollusca: Opisthobranchia)', *Zool. J. Linn. Soc.*, 49, 187—97.

THOMPSON, V. (1973) 'Spittlebug polymorphic for warning coloration', *Nature, Lond.*, 242, 126—8.

TINBERGEN, N. (1951) *The Study of Instinct.* Oxford, Clarendon Press.

TINBERGEN, N. (1958) *Curious Naturalists.* London, Country Life.

TINBERGEN, N., IMPEKOVEN, M. and FRANCK, D. (1967) 'An experiment on spacing-out as a defence against predation', *Behaviour*, 28, 307—21.

TRIPLETT, N. B. (1901) 'The educability of the perch', *Am. J. Psychol.*, 12, 354—60.

TSCHINKEL, W. R. (1972) '6-alkyl-1,4-naphthoquinones from the defensive secretion of the tenebrionid beetle, *Argopsis alutacea*', *J. Insect Physiol.*, 18, 711—22.

TURNER, J. R. G. (1971a) 'Experiments on the demography of tropical butterflies. II. Longevity and home-range behaviour in *Heliconius erato*', *Biotropica*, 3, 21—31.

TURNER, J. R. G. (1971b) 'Studies of Müllerian mimicry and its evolution in burnet moths and heliconid butterflies', in: *Ecological Genetics and Evolution*, ed. R. Creed, 224—60, Oxford and Edinburgh, Blackwell.

TURNER, J. R. G. (1973) 'Passion flower butterflies', *Animals*, 15, 15—21.

TWEEDIE, M. W. F. (1960) 'The malayan gliding reptiles', *Proc. S. Lond. ent. nat. Hist. Soc.* (1959), 97—103.

TYSHCHENKO, U. P. (1961) 'Ob otnoshenii nekotorykh paukov semeistva Thomisidae k mimikriruyushchim nasekomym i ikh modelyam', *Vest. lening. Gosud. Univ.* (*Ser. biol.*), 3, 133—9. (Abstract seen.)

VERHEIJEN, F. J. and REUTER, J. H. (1969) 'The effect of alarm substance on predation among cyprinids', *Anim. Behav.*, 17, 551—4.

VERTS, B. J. (1967) *The Biology of the Striped Skunk.* Urbana, Chicago and London, University of Illinois Press.

WALDBAUER, G. P. and SHELDON, J. K. (1971) 'Phenological relationships of some aculeate Hymenoptera, their dipteran mimics, and insectivorous birds', *Evolution, Lancaster, Pa.*, 25, 371—82.

WALKER, E. P. (1964) *Mammals of the World.* Vols. 1 and 2. Baltimore, Johns Hopkins.

WALLACE, J. B. and BLUM, M. S. (1971) 'Reflex bleeding: a highly refined defence mechanism in *Diabrotica* larvae (Coleoptera: Chrysomelidae)', *Ann. ent. Soc. Am.*, 64, 1021—4.

WALTHER, F. R. (1969) 'Flight behaviour and avoidance of predators in Thomson's gazelle (*Gazella Thomsoni* Guenther 1884)', *Behaviour*, 34, 184—221.

WARMKE, G. L. and ALMODOVAR, L. R. (1972) 'Observations on the life cycle and regeneration in *Oxynoe antillarum* Mörch, an ascoglossan opisthobranch from the Caribbean, *Bull. mar. Sci.*, 22, 67—74.

WATT, W. B. (1968) 'Adaptive significance of pigment polymorphisms in *Colias* butterflies. 1. Variation of melanin pigment in relation to thermoregulation', *Evolution, Lancaster, Pa.*, 22, 437—58.

WAY, M. J. (1963) 'Mutualism between ants and honey-dew producing Homoptera', *A. Rev. Ent.*, 8, 307—44.

WEATHERSTON, J. and PERCY, J. E. (1970) 'Arthropod defensive secretions', in: *Chemicals Controlling Insect Behaviour*, ed. M. Beroza, 95—144, New York and London, Academic Press.

WELLS, M. J. (1962) 'Early learning in *Sepia*', *Symp. zool. Soc. Lond.*, 8, 149—69.

WELLS, M. J. (1968) *Lower Animals*. London, Weidenfeld & Nicholson.

WELLS, M. J. and BUCKLEY, S. K. L. (1972) 'Snails and trails', *Anim. Behav.*, 20, 345—55.

WELTY, J. C. (1934) 'Experiments in group behaviour of fishes', *Physiol. Zoöl.*, 7, 85—128.

WEST, D. A., SNELLINGS, W. M. and HERBEK, T. A. (1972) 'Pupal color dimorphism and its environmental control in *Papilio polyxenes* Stoll (Lepidoptera: Papilionidae)', *Jl N.Y. ent. Soc.*, 80, 205—11.

WHITTAKER, R. H. and FEENY, P. P. (1971) 'Allelochemics: chemical interactions between species', *Science, N.Y.*, 171, 757—70.

WICKLER, W. (1968) *Mimicry in Plants and Animals*. London, Weidenfeld & Nicholson.

WIEHLE, H. (1927) 'Beiträge zur Kenntnis des Radnetzbaues der Epeiriden, Tetragnathiden und Uloboriden', *Z. Morph. Ökol. Tiere*, 8, 468—537.

WILLIS, I. (1972) 'Adapting to a way of life', *Birds*, 4, 11—15.

WILLOWS, A. O. D. (1971) 'Giant brain cells in mollusks', *Scient. Am.*, 224(2), 68—75.

WILSON, E. O. and REGNIER, F. E. (1971) 'The evolution of the alarm-defense system in formicine ants', *Am. Nat.*, 105, 279—89.

WILSON, M. C. L. (1971) 'The morphology and mechanism of the pupal gin-traps of *Tenebrio molitor* L. (Col., Tenebrionidae)', *Jl Stored—Prod. Res.*, 7, 21—30.

WINDECKER, W. (1939) '*Euchelia* (*Hypocrita*) *jacobaeae* L. und das Schutztrachtenproblem', *Z. Morph. Ökol. Tiere*, 35, 84—138.

WOBBER, D. R. (1970) 'A report on the feeding of *Dendronotus iris* on the anthozoan *Cerianthus* sp. from Monterey Bay, California', *Veliger*, 12, 383—7.

WOLDA, H. (1963) 'Natural populations of the polymorphic land snail

Cepaea nemoralis (L.). Factors affecting their size and their genetic constitution', *Archs néerl. Zool.,* 15, 381–471.

WOLDA, H. (1965) 'Some preliminary observations on the distribution of the various morphs within natural populations of the polymorphic landsnail *Cepaea nemoralis* (L.)', *Archs néerl. Zool.,* 16, 280–92.

WOLDA, H. (1967) 'The effect of temperature on reproduction in some morphs of the landsnail *Cepaea nemoralis* (L.)', *Evolution, Lancaster, Pa.,* 21, 117–29.

WRIGHT, H. O. and MATTHEWS, G. A. (1973) 'Effect of commensal hydroids on hermit crab competition in the littoral zone of Texas', *Nature, Lond.,* 241, 139–40.

WRIGHT, T. S. (1858) 'On the cnidae or thread-cells of the Eolidae', *Proc. R. phys. Soc. Edinb.,* 2, 38–40.

YOUNG, A. M. (1971) 'Wing coloration and reflectance in *Morpho* butterflies as related to reproductive behaviour and escape from avian predators', *Oecologia,* 7, 209–22.

YOUNG, J. Z. (1950) *The Life of Vertebrates.* Oxford, Clarendon Press.

YOUNG, J. Z. (1959) 'Observations on *Argonauta* and especially its method of feeding', *Proc. zool. Soc. Lond.,* 133, 471–9.

ZARET, T. M. (1972) 'Predator–prey interactions in a tropical lacustrine ecosystem', *Ecology,* 53, 248–57.

ZENTALL, T. R. and LEVINE, J. M. (1972) 'Observational learning and social facilitation in the rat', *Science, N.Y.,* 178, 1220–1.

ZIEGLER, A. P. (1971) 'The strange case of the look-alike birds', *Animals,* 13, 736–7.

ZINNER, H. (1971) 'On the ecology and the significance of semantic coloration in the nocturnal desert-elapid *Walterinnesia aegyptia* Lataste (Reptiles, Ophidia)', *Oecologia,* 7, 267–75.

Author index
Index of animals
Subject index

Author index

Index of animals

Each animal has been classified to phylum or class, and family, with some intermediate taxa such as order or suborder in most cases. For mammals and birds the full classification and reference to pages are found under the English name. For most other groups the page numbers are given after the scientific name.

Subject index

Principal references to a particular subject are indicated by italic figures